Polaris ATV Owners Workshop Manual

Disclaimer

There are risks associated with automotive repairs. The ability to make repairs depends on the individual's skill, experience and proper tools. Individuals should act with due care and acknowledge and assume the risk of performing automotive repairs.

The purpose of this manual is to provide comprehensive, useful and accessible automotive repair information, to help you get the best value from your vehicle. However, this manual is not a substitute for a professional certified technician or mechanic.

This repair manual is produced by a third party and is not associated with an individual vehicle manufacturer. If there is any doubt or discrepancy between this manual and the owner's manual or the factory service manual, please refer to the factory service manual or seek assistance from a professional certified technician or mechanic.

Even though we have prepared this manual with extreme care and every attempt is made to ensure that the information in this manual is correct, neither the publisher nor the author can accept responsibility for loss, damage or injury caused by any errors in, or omissions from, the information given.

The manufacturer's authorised representative in the EU for product safety is:

HaynesPro BV
Stationsstraat 79 F, 3811MH Amersfoort
The Netherlands
gpsr@haynes.co.uk

by Alan Ahlstrand and John H Haynes
Member of the Guild of Motoring Writers

Models covered:
400 6X6, Big Boss, Cyclone, Magnum, Scrambler, Sport, Sportsman, Trail Blazer, Trail Boss, Xplorer, Xpress
2-stroke and 4-stroke
2-wheel, 4-wheel and 6-wheel drive
Includes information on 3-wheel models

(2302 - 8U4)

Haynes Group Limited
Haynes North America, Inc.

www.haynes.com

Acknowledgments

Our thanks to Bike World, Sunnyvale, CA, for providing the facilities used for these photographs; to Carlos Amaral, service manager, for arranging the facilities and fitting the mechanical work into his shop's busy schedule; and to Joe Ortiz, service technician, for doing the mechanical work and providing valuable technical information.

© **Haynes North America, Inc. 1999**
With permission from Haynes Group Limited

A book in the Haynes Owners Workshop Manual Series

All rights reserved. No part of this book may be reproduced or transmitted in any form or by any means, electronic or mechanical, including photocopying, recording or by any information storage or retrieval system, without permission in writing from the copyright holder.

ISBN-10: 1-56392-302-5
ISBN-13: 978-1-56392-302-9

Library of Congress Catalog Card Number 99-63080

We take great pride in the accuracy of information given in this manual, but motorcycle manufacturers make alterations and design changes during the production run of a particular motorcycle of which they do not inform us. No liability can be accepted by the authors or publishers for loss, damage or injury caused by any errors in, or omissions from, the information given.

Contents

Introductory pages

About this manual	0-5
Introduction to the Polaris ATV	0-5
Identification numbers	0-6
Buying parts	0-8
General specifications	0-8
Maintenance techniques, tools and working facilities	0-22
Safety first!	0-28
ATV chemicals and lubricants	0-30
Troubleshooting	0-31

Chapter 1
Tune-up and routine maintenance — 1-1

Chapter 2 Part A
2-stroke engines — 2A-1

Chapter 2 Part B
4-stroke engines — 2B-1

Chapter 2 Part C
Polaris Variable Transmission — 2C-1

Chapter 2 Part D
Transmissions — 2D-1

Chapter 3
Cooling system — 3-1

Chapter 4
Fuel and exhaust systems — 4-1

Chapter 5
Electrical and ignition systems — 5-1

Chapter 6
Steering, suspension and final drive — 6-1

Chapter 7
Brakes, wheels and tires — 7-1

Chapter 8
Frame and bodywork — 8-1

Chapter 9
Wiring diagrams — 9-1

Conversion factors — 9-11

Index — IND-1

Polaris ATV

About this manual

Its purpose

The purpose of this manual is to help you get the best value from your motorcycle. It can do so in several ways. It can help you decide what work must be done, even if you choose to have it done by a dealer service department or a repair shop; it provides information and procedures for routine maintenance and servicing; and it offers diagnostic and repair procedures to follow when trouble occurs.

We hope you use the manual to tackle the work yourself. For many simpler jobs, doing it yourself may be quicker than arranging an appointment to get the vehicle into a shop and making the trips to leave it and pick it up. More importantly, a lot of money can be saved by avoiding the expense the shop must pass on to you to cover its labor and overhead costs. An added benefit is the sense of satisfaction and accomplishment that you feel after doing the job yourself.

Using the manual

The manual is divided into Chapters. Each Chapter is divided into numbered Sections, which are headed in bold type between horizontal lines. Each Section consists of consecutively numbered paragraphs or steps.

At the beginning of each numbered Section you will be referred to any illustrations which apply to the procedures in that Section. The reference numbers used in illustration captions pinpoint the pertinent Section and the Step within that Section. That is, illustration 3.2 means the illustration refers to Section 3 and Step (or paragraph) 2 within that Section.

Procedures, once described in the text, are not normally repeated. When it's necessary to refer to another Chapter, the reference will be given as Chapter and Section number. Cross references given without use of the word 'Chapter' apply to Sections and/or paragraphs in the same Chapter. For example, 'see Section 8' means in the same Chapter.

References to the left or right side of the vehicle assume you are sitting on the seat, facing forward.

Motorcycle manufacturers continually make changes to specifications and recommendations, and these, when notified, are incorporated into our manuals at the earliest opportunity.

Even though we have prepared this manual with extreme care, neither the publisher nor the authors can accept responsibility for any errors in, or omissions from, the information given.

NOTE

A **Note** provides information necessary to properly complete a procedure or information which will make the procedure easier to understand.

CAUTION

A **Caution** provides a special procedure or special steps which must be taken while completing the procedure where the Caution is found. Not heeding a Caution can result in damage to the assembly being worked on.

WARNING

A **Warning** provides a special procedure or special steps which must be taken while completing the procedure where the Warning is found. Not heeding a Warning can result in personal injury.

Introduction to the Polaris ATV

The Polaris ATV range includes all types of all-terrain vehicles, from high-performance sport models to 6wd utility designs.

The engine on all models is either a single-cylinder two-stroke or a single-cylinder four-stroke with an overhead camshaft. Air or liquid cooling is used, depending on model.

All models use the Polaris Variable Transmission (PVT). This separate unit uses a V-belt and two variable-diameter pulleys to transmit power from the engine to the transmission. Several different transmissions have been used. They include high and reverse ranges and on some models, low.

Fuel is delivered to the cylinder by a single constant velocity carburetor.

The front suspension on all models uses a coil spring-shock absorber unit and a single control arm on each side of the vehicle.

The rear suspension on all models uses a single shock absorber and coil spring. 6wd models use two swing arms, one for each of the two rear axles. The center (forward) swingarm is supported by the shock/spring unit, while the rear swingarm is supported by a scissors-type stabilizer.

Early models use a drum brake at each front wheel; the rear uses a single mechanically-operated disc brake, mounted on the axle shaft inboard of the wheels. Later models use hydraulic disc brakes, one at each front wheel and one on the transmission output shaft.

Shaft final drive is used on the Sportsman 500 and Xplorer 500 models. Chain final drive is used on all other models covered in this manual.

Identification numbers

The model number and frame serial number are stamped into the frame in various locations, depending on model. The engine number is stamped into the crankcase. All of these numbers should be recorded and kept in a safe place so they can be furnished to law enforcement officials in the event of a theft.

The frame serial number, engine serial number and carburetor identification number should also be kept in a handy place (such as with your driver's license) so they are always available when purchasing or ordering parts for your machine.

The engine number is on the crankcase of all models.

The vehicle model number location varies according to model and year:

1985 and 1986 - on frame rail near exhaust pipe.

1987 and 1988 except R/ES - model number on upper center of frame, serial number on left side of frame near cylinder.

1988 R/ES - model and serial numbers on left side of frame near cylinder.

1989 through 1996 - model number on left side of frame near oil tank, serial number on frame near exhaust pipe, on frame forward of shift selector box, or on rear shock upper crossmember (under seat)

1997 - On right side of frame near air box.

Decoding 1985 through 1996 model numbers

The model number on consists of a letter followed by six digits, which are actually three two-digit codes, such as W968527. The initial letter in most cases is W; an initial letter X identifies a limited-production pilot build machine.

The first two numbers indicate the model year; 85 for 1985, 96 for 1996, and so on.

The second two numbers indicate the basic chassis designation, as listed in the following table.

The frame serial number is located at various points on the frame, depending on model

Typical engine serial number location (two-stroke)

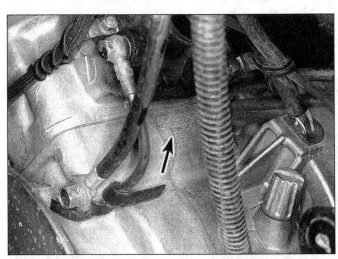

Typical engine serial number location (four-stroke)

Identification numbers

Basic chassis designation	Model
70	Scrambler three-wheeler
72	Trail Blazer
75	1985 through 1987, Trail Boss; later years, Trail Boss from that year's first production; Magnum 2x4
76	Trail Boss from that year's second production
78	1987 Cyclone; Trail Blazer; Scrambler
80	Trail Boss 4x4 from that year's first production; Sportsman 4x4
81	Trail Boss 4x4 from that year's second production; Magnum 4x4
83	Trail Boss 4x4 from that year's third production
85	1988, Trail Boss R/ES; later years, through 1995, Trail Boss from that year's first production; 1996, Trail Boss/Sport
86	Trail Boss 4x6, 250
87	Big Boss 6x6, 250, 350L; 400 6x6; Magnum 6x6
91	Xplorer 4x4; Xplorer 300; Xplorer 400
92	Sportsman 500
95	Xpress 300; Xpress 400

The third pair of numbers indicates the engine, as listed in the following table.

Engine code	Engine
21, 27, 28	244 cc 2-stroke
30	283 cc 2-stroke
39	283 cc air-cooled or 352 cc liquid-cooled 2-stroke
40	379 cc 2-stroke
44	425 cc or 498 cc 4-stroke (models through 1996)
42	425 cc 4-stroke (1997)
50	498 cc 4-stroke (1997)

Decoding 1997 and later model numbers

The 1997 numbering system is similar to the earlier one. The W is still used as the first digit, and the first pair of numbers indicates the model year (in other words, 1997 model numbers start with W97). However, the following digits indicate the body style, powertrain, engine and emissions package. For example, model number W97AC38C is decoded as follows:

W - initial letter of normal-production machines.
97 - model year.
A - Body style. This letter indicates one of the three basic body styles. A = Gen II body style; B = Gen III; C = Gen IV. The body styles included in each of these three categories are listed in Chapter 8.
B - powertrain. A = 2wd models, C = 4wd chain drive models, D = 4wd shaft drive models and E = 6wd chain drive models.
38 - These two numbers indicate the engine.
Refer to the table above for engine codes.
C - This letter indicates the emissions package. A = approved by the California Air Resources Board (CARB); C = not CARB approved; D indicates Norwegian and E indicates Swedish.

Decoding 1997-on vehicle identification numbers (VIN)

A 17-digit model number is used on 1997 and later models (for example, 4XAAB38C0V2000000). It decodes as follows:

4XA - these first three digits, which are always the same, are the world manufacturer identification assigned to Polaris.
AB38C - these are the same as the last five digits of the model number.
0 - check digit.
V - model year.
2 - plant number.
000000 - the serial number of the individual machine.

Once you have found all the identification numbers, record them for reference when buying parts. Since the manufacturers change specifications, parts and vendors (companies that manufacture various components on the machine), providing the ID numbers is the only way to be reasonably sure that you are buying the correct parts.

Whenever possible, take the worn part to the dealer so direct comparison with the new component can be made. Along the trail from the manufacturer to the parts shelf, there are numerous places that the part can end up with the wrong number or be listed incorrectly.

The two places to purchase new parts for your vehicle - the accessory store and the franchised dealer - differ in the type of parts they carry. While dealers can obtain virtually every part for your vehicle, the accessory dealer is usually limited to normal high wear items such as shock absorbers, tune-up parts, various engine gaskets, cables, chains, brake parts, etc. Rarely will an accessory outlet have major suspension components, cylinders, transmission gears, or cases.

Used parts can be obtained for roughly half the price of new ones, but you can't always be sure of what you're getting. Once again, take your worn part to the wrecking yard for direct comparison.

Whether buying new, used or rebuilt parts, the best course is to deal directly with someone who specializes in parts for your particular make.

Buying parts

Once you have found all the identification numbers, record them for reference when buying parts. Since the manufacturers change specifications, parts and vendors (companies that manufacture various components on the machine), providing the ID numbers is the only way to be reasonably sure that you are buying the correct parts.

Whenever possible, take the worn part to the dealer so direct comparison with the new component can be made. Along the trail from the manufacturer to the parts shelf, there are numerous places that the part can end up with the wrong number or be listed incorrectly.

The two places to purchase new parts for your motorcycle - the accessory store and the franchised dealer - differ in the type of parts they carry. While dealers can obtain virtually every part for your motorcycle, the accessory dealer is usually limited to normal high wear items such as shock absorbers, tune-up parts, various engine gaskets, cables, chains, brake parts, etc. Rarely will an accessory outlet have major suspension components, cylinders, transmission gears, or cases.

Used parts can be obtained for roughly half the price of new ones, but you can't always be sure of what you're getting. Once again, take your worn part to the wrecking yard (breaker) for direct comparison.

Whether buying new, used or rebuilt parts, the best course is to deal directly with someone who specializes in parts for your particular make.

General specifications

1985
    Scrambler three-wheeler
        Model number W857027
        Wheelbase .... 48 inches
        Overall length .... 74 inches
        Overall width .... 43 inches
        Overall height .... 42 inches
        Seat height .... 33.5 inches
        Footpad height .... 9.5 inches
        Ground clearance .... 5.75 inches
        Weight .... 380 lbs
    Trail Boss
        Model number W857527
        Wheelbase .... 48 inches
        Overall length .... 74 inches
        Overall width .... 43 inches
        Overall height .... 42 inches
        Seat height .... 33.5 inches
        Footpad height .... 9.5 inches
        Ground clearance .... 5.75 inches
        Weight .... 440 lbs
1986
    Scrambler three-wheeler
        Model number W867027
        Wheelbase .... 46 inches
        Overall length .... 74 inches
        Overall width .... 43 inches
        Overall height .... 42 inches
        Seat height .... 33.5 inches
        Footpad height .... 9.5 inches
        Ground clearance .... 5.75 inches
        Weight .... 380 lbs
    Trail Boss
        Model number W867527, W867627
        Wheelbase .... 46 inches
        Overall length .... 74 inches
        Overall width .... 43 inches
        Overall height .... 42 inches
        Seat height .... 33.5 inches
        Footpad height .... 9.5 inches
        Ground clearance .... 5.75 inches
        Weight .... 440 lbs

General specifications

1987
- Trail Boss 250
  - Model number W877527
  - Wheelbase .... 45.5 inches
  - Overall length .... 70 inches
  - Overall width .... 43.5 inches
  - Overall height .... 43 inches
  - Seat height .... 33 inches
  - Footpad height .... 9.5 inches
  - Ground clearance .... 6 inches
  - Weight .... 440 lbs
- Trail Boss 2x4
  - Model number W877828
  - Wheelbase .... 50 inches
  - Overall length .... 78 inches
  - Overall width .... 43.5 inches
  - Overall height .... 43 inches
  - Seat height .... 33 inches
  - Footpad height .... 9.5 inches
  - Ground clearance .... 6 inches
  - Weight .... 400 lbs
- Trail Boss 4x4
  - Model number W878027, W878127
  - Wheelbase .... 47.5 inches
  - Overall length .... 70 inches
  - Overall width .... 44.5 inches
  - Overall height .... 46 inches
  - Seat height .... 35.5 inches
  - Footpad height .... 11 inches
  - Ground clearance .... 6.75 inches
  - Weight .... 490 lbs

1988
- Trail Boss 2x4
  - Model number W887527
  - Wheelbase .... 49.5 inches
  - Overall length .... 70 inches
  - Overall width .... 43.5 inches
  - Overall height .... 43 inches
  - Seat height .... 33 inches
  - Footpad height .... 9.5 inches
  - Ground clearance .... 6 inches
  - Weight .... 440 lbs
- Trail Boss 250 R/ES
  - Model number W888528
  - Wheelbase .... 49.5 inches
  - Overall length .... 73.2 inches
  - Overall width .... 43.5 inches
  - Overall height .... 44 inches
  - Seat height .... 33 inches
  - Footpad height .... 11.5 inches
  - Ground clearance .... 5.5 inches
  - Weight .... 400 lbs
- Trail Boss 4x4
  - Model number W888127
  - Wheelbase .... 49.5 inches
  - Overall length .... 70 inches
  - Overall width .... 44.5 inches
  - Overall height .... 46 inches
  - Seat height .... 35.5 inches
  - Footpad height .... 11 inches
  - Ground clearance .... 6.75 inches
  - Weight .... 490 lbs

1989
- Trail Boss 250
  - Model number W898527
  - Wheelbase .... 49.5 inches
  - Overall length .... 73.2 inches
  - Overall width .... 44 inches
  - Overall height .... 44 inches
  - Seat height .... 33 inches

General specifications

1989 (continued)

- Trail Boss 250
  - Footpad height .... 11.5 inches
  - Ground clearance .... 5.5 inches
  - Weight .... 400 lbs
- Trail Boss 2x4
  - Model number W897527
  - Wheelbase .... 49.5 inches
  - Overall length .... 73.2 inches
  - Overall width .... 44 inches
  - Overall height .... 44 inches
  - Seat height .... 33 inches
  - Footpad height .... 11.5 inches
  - Ground clearance .... 5.5 inches
  - Weight .... 440 lbs
- Trail Boss 4x4
  - Model number W898127
  - Wheelbase .... 49.5 inches
  - Overall length .... 73.2 inches
  - Overall width .... 44.5 inches
  - Overall height .... 46 inches
  - Seat height .... 34 inches
  - Footpad height .... 11.5 inches
  - Ground clearance .... 6 inches
  - Weight .... 490 lbs
- Big Boss 4x6
  - Model number W898627
  - Wheelbase .... 75 inches
  - Overall length .... 97.5 inches
  - Overall width .... 44.5 inches
  - Overall height .... 44 inches
  - Seat height .... 33.5 inches
  - Footpad height .... 11.5 inches
  - Ground clearance .... 4.6 inches
  - Weight .... 650 lbs

1990

- Trail Blazer
  - Model number W907221
  - Wheelbase .... 49.5 inches
  - Overall length .... 73.2 inches
  - Overall width .... 44 inches
  - Overall height .... 44 inches
  - Seat height .... 33 inches
  - Footpad height .... 11.5 inches
  - Ground clearance .... 5.5 inches
  - Weight .... 390 lbs
- Trail Boss 250
  - Model number W908527
  - Wheelbase .... 49.5 inches
  - Overall length .... 73.2 inches
  - Overall width .... 44 inches
  - Overall height .... 44 inches
  - Seat height .... 33 inches
  - Footpad height .... 11.5 inches
  - Ground clearance .... 5.5 inches
  - Weight .... 425 lbs
- Trail Boss 2x4
  - Model number W907527
  - Wheelbase .... 49.5 inches
  - Overall length .... 73.2 inches
  - Overall width .... 44 inches
  - Overall height .... 44 inches
  - Seat height .... 33 inches
  - Footpad height .... 11.5 inches
  - Ground clearance .... 5.5 inches
  - Weight .... 440 lbs
- Trail Boss 4x4
  - Model number W908127
  - Wheelbase .... 49.75 inches
  - Overall length .... 73.2 inches

General specifications

- Overall width .... 44.5 inches
- Overall height .... 46 inches
- Seat height .... 34 inches
- Footpad height .... 11.75 inches
- Ground clearance .... 6 inches
- Weight .... 490 lbs

Big Boss 4x6
- Model number W908627
- Wheelbase .... 75 inches
- Overall length .... 97.5 inches
- Overall width .... 44.5 inches
- Overall height .... 44 inches
- Seat height .... 33.5 inches
- Footpad height .... 11.5 inches
- Ground clearance .... 4.6 inches
- Weight .... 650 lbs

Trail Boss 2x4 350L
- Model number W907539
- Wheelbase .... 49.75 inches
- Overall length .... 77 inches
- Overall width .... 44 inches
- Overall height .... 44 inches
- Seat height .... 33 inches
- Footpad height .... 11.5 inches
- Ground clearance .... 5.5 inches
- Weight .... 490 lbs

Trail Boss 4x4 350L
- Model number W908139
- Wheelbase .... 49.75 inches
- Overall length .... 77 inches
- Overall width .... 44.5 inches
- Overall height .... 46 inches
- Seat height .... 34 inches
- Footpad height .... 11.5 inches
- Ground clearance .... 6 inches
- Weight .... 560 lbs

1991

Trail Blazer
- Model number W917221
- Wheelbase .... 49.5 inches
- Overall length .... 73.2 inches
- Overall width .... 44 inches
- Overall height .... 44 inches
- Seat height .... 33 inches
- Footpad height .... 11.5 inches
- Ground clearance .... 5.5 inches
- Weight .... 390 lbs

Trail Boss 250
- Model number W918527
- Wheelbase .... 49.5 inches
- Overall length .... 73.2 inches
- Overall width .... 44 inches
- Overall height .... 44 inches
- Seat height .... 33 inches
- Footpad height .... 11.5 inches
- Ground clearance .... 5.5 inches
- Weight .... 425 lbs

Trail Boss 2x4
- Model number W917527
- Wheelbase .... 49.5 inches
- Overall length .... 73.2 inches
- Overall width .... 44 inches
- Overall height .... 44 inches
- Seat height .... 33 inches
- Footpad height .... 11.5 inches
- Ground clearance .... 5.5 inches
- Weight .... 440 lbs

Trail Boss 4x4
- Model number W918127
- Wheelbase .... 49.75 inches

General specifications

1991 (continued)

- Trail Boss 4x4
  - Overall length .... 73.2 inches
  - Overall width .... 44.5 inches
  - Overall height .... 46 inches
  - Seat height .... 34 inches
  - Footpad height .... 11.75 inches
  - Ground clearance .... 6 inches
  - Weight .... 490 lbs
- Big Boss 4x6
  - Model number W918627
  - Wheelbase .... 75 inches
  - Overall length .... 97.5 inches
  - Overall width .... 44.4 inches
  - Overall height .... 44 inches
  - Seat height .... 33.5 inches
  - Footpad height .... 11.5 inches
  - Ground clearance .... 4.6 inches
  - Weight .... 650 lbs
- Big Boss 6x6
  - Model number W918727
  - Wheelbase .... 75 inches
  - Overall length .... 97.5 inches
  - Overall width .... 45.7 inches
  - Overall height .... 46 inches
  - Seat height .... 34 inches
  - Footpad height .... 11.5 inches
  - Ground clearance .... 5 inches
  - Weight .... 750 lbs
- Trail Boss 2x4 350L
  - Model number W917539
  - Wheelbase .... 49.75 inches
  - Overall length .... 77 inches
  - Overall width .... 44 inches
  - Overall height .... 44 inches
  - Seat height .... 33 inches
  - Footpad height .... 11.5 inches
  - Ground clearance .... 5.5 inches
  - Weight .... 490 lbs
- Trail Boss 4x4 350L
  - Model number W918139
  - Wheelbase .... 49.75 inches
  - Overall length .... 77 inches
  - Overall width .... 44.5 inches
  - Overall height .... 46 inches
  - Seat height .... 34 inches
  - Footpad height .... 11.75 inches
  - Ground clearance .... 6 inches
  - Weight .... 560 lbs

1992

- Trail Blazer
  - Model number W927221
  - Wheelbase .... 49.5 inches
  - Overall length .... 73.2 inches
  - Overall width .... 44 inches
  - Overall height .... 44 inches
  - Seat height .... 33 inches
  - Footpad height .... 11.5 inches
  - Ground clearance .... 5.5 inches
  - Weight .... 390 lbs
- Trail Boss 250
  - Model number W928527
  - Wheelbase .... 49.5 inches
  - Overall length .... 73.2 inches
  - Overall width .... 44 inches
  - Overall height .... 44 inches
  - Seat height .... 33 inches
  - Footpad height .... 11.5 inches
  - Ground clearance .... 5.5 inches
  - Weight .... 425 lbs

General specifications

Trail Boss 2x4
- Model number: W927527
- Wheelbase: 49.5 inches
- Overall length: 73.2 inches
- Overall width: 44 inches
- Overall height: 44 inches
- Seat height: 33 inches
- Footpad height: 11.5 inches
- Ground clearance: 5.5 inches
- Weight: 440 lbs

Trail Boss 4x4
- Model number: W928127
- Wheelbase: 49.75 inches
- Overall length: 73.2 inches
- Overall width: 44.5 inches
- Overall height: 46 inches
- Seat height: 34 inches
- Footpad height: 11.75 inches
- Ground clearance: 6 inches
- Weight: 490 lbs

Big Boss 4x6
- Model number: W928627
- Wheelbase: 75 inches
- Overall length: 97.5 inches
- Overall width: 44.4 inches
- Overall height: 44 inches
- Seat height: 33.5 inches
- Footpad height: 11.5 inches
- Ground clearance: 4.6 inches
- Weight: 650 lbs

Big Boss 6x6
- Model number: W928727
- Wheelbase: 75 inches
- Overall length: 97.5 inches
- Overall width: 45.7 inches
- Overall height: 46 inches
- Seat height: 34 inches
- Footpad height: 11.5 inches
- Ground clearance: 5 inches
- Weight: 750 lbs

Trail Boss 2x4 350L
- Model number: W927539
- Wheelbase: 49.75 inches
- Overall length: 77 inches
- Overall width: 44 inches
- Overall height: 44 inches
- Seat height: 33 inches
- Footpad height: 11.5 inches
- Ground clearance: 5.5 inches
- Weight: 490 lbs

Trail Boss 4x4 350L
- Model number: W928139
- Wheelbase: 49.75 inches
- Overall length: 77 inches
- Overall width: 44.5 inches
- Overall height: 46 inches
- Seat height: 34 inches
- Footpad height: 11.75 inches
- Ground clearance: 6 inches
- Weight: 560 lbs

1993

Trail Blazer
- Model number: W937221
- Wheelbase: 49.5 inches
- Overall length: 73.2 inches
- Overall width: 44 inches
- Overall height: 44 inches
- Seat height: 33 inches
- Ground clearance: 5.5 inches
- Weight: 390 lbs

General specifications

1993 (continued)

Trail Boss 250
- Model number W938527
- Wheelbase .... 49.5 inches
- Overall length .... 73.2 inches
- Overall width .... 44 inches
- Overall height .... 44 inches
- Seat height .... 33 inches
- Ground clearance .... 5.5 inches
- Weight .... 425 lbs

2x4 250
- Model number W937527
- Wheelbase .... 49.5 inches
- Overall length .... 73.2 inches
- Overall width .... 44 inches
- Overall height .... 44 inches
- Seat height .... 33 inches
- Ground clearance .... 5.5 inches
- Weight .... 440 lbs

2x4 350L
- Model number W937539
- Wheelbase .... 49.5 inches
- Overall length .... 77 inches
- Overall width .... 44 inches
- Overall height .... 44 inches
- Seat height .... 33 inches
- Ground clearance .... 5.5 inches
- Weight .... 490 lbs

Trail Boss 4x4 250
- Model number W938127
- Wheelbase .... 49.75 inches
- Overall length .... 73.2 inches
- Overall width .... 44.5 inches
- Overall height .... 46 inches
- Seat height .... 34 inches
- Ground clearance .... 6 inches
- Weight .... 490 lbs

Trail Boss 4x4 350L/Sportsman
- Model number W938139
- Wheelbase .... 49.75 inches
- Overall length .... 77 inches
- Overall width .... 44.5 inches
- Overall height .... 46 inches
- Seat height .... 34 inches
- Ground clearance .... 6 inches
- Weight .... 560 lbs

Big Boss 6x6 250
- Model number W938727
- Wheelbase .... 75 inches
- Overall length .... 97.5 inches
- Overall width .... 45.7 inches
- Overall height .... 46 inches
- Seat height .... 34 inches
- Ground clearance .... 5 inches
- Weight .... 750 lbs

Big Boss 6x6 350L
- Model number W938739
- Wheelbase .... 75 inches
- Overall length .... 103 inches
- Overall width .... 45.7 inches
- Overall height .... 46 inches
- Seat height .... 37 inches
- Ground clearance .... 5.5 inches
- Weight .... 820 lbs

1994

Trail Blazer
- Model number W947221
- Wheelbase .... 49.5 inches
- Overall length .... 73.2 inches
- Overall width .... 44 inches

General specifications

Overall height	44 inches
Seat height	33 inches
Ground clearance	5.5 inches
Weight	390 lbs

Trail Boss 250
Model number	W948527
Wheelbase	49.5 inches
Overall length	73.2 inches
Overall width	44 inches
Overall height	44 inches
Seat height	33 inches
Ground clearance	5.5 inches
Weight	425 lbs

2x4 300
Model number	W947530
Wheelbase	49.5 inches
Overall length	73.2 inches
Overall width	44 inches
Overall height	44 inches
Seat height	33 inches
Ground clearance	5.5 inches
Weight	483 lbs

Sport 400L
Model number	W948540
Wheelbase	49.75 inches
Overall length	72 inches
Overall width	44 inches
Overall height	44 inches
Seat height	33 inches
Ground clearance	5.5 inches
Weight	479 lbs

2x4 400L
Model number	W947540
Wheelbase	49.75 inches
Overall length	77 inches
Overall width	44 inches
Overall height	44 inches
Seat height	33 inches
Ground clearance	5.5 inches
Weight	512 lbs

4x4 300
Model number	W948130
Wheelbase	49.75 inches
Overall length	73.2 inches
Overall width	44.5 inches
Overall height	46 inches
Seat height	34 inches
Ground clearance	6 inches
Weight	538 lbs

4x4 400L
Model number	W948140
Wheelbase	49.75 inches
Overall length	77 inches
Overall width	46 inches
Overall height	46 inches
Seat height	34 inches
Ground clearance	6 inches
Weight	572 lbs

Sportsman 4x4
Model number	W948040
Wheelbase	49.75 inches
Overall length	77 inches
Overall width	46 inches
Overall height	46 inches
Seat height	34 inches
Ground clearance	6 inches
Weight	585 lbs

Big Boss 6x6 300
Model number	W948730
Wheelbase	75 inches

General specifications

1994 (continued)

Big Boss 6x6 300
- Overall length .... 97.5 inches
- Overall width .... 45.7 inches
- Overall height .... 46 inches
- Seat height .... 34 inches
- Ground clearance .... 5 inches
- Weight .... 823 lbs

Big Boss 6x6 400L
- Model number W948740
- Wheelbase .... 75 inches
- Overall length .... 103 inches
- Overall width .... 45.7 inches
- Overall height .... 46 inches
- Seat height .... 37 inches
- Ground clearance .... 5.5 inches
- Weight .... 857 lbs

1995

Trail Blazer
- Model number W957221
- Wheelbase .... 49.5 inches
- Overall length .... 73.2 inches
- Overall width .... 44 inches
- Overall height .... 44 inches
- Seat height .... 33 inches
- Ground clearance .... 5.5 inches
- Weight .... 390 lbs

Trail Boss 250
- Model number W958527
- Wheelbase .... 49.5 inches
- Overall length .... 73.2 inches
- Overall width .... 44 inches
- Overall height .... 44 inches
- Seat height .... 33 inches
- Ground clearance .... 5.5 inches
- Weight .... 425 lbs

2x4 300
- Model number W957530
- Wheelbase .... 49.5 inches
- Overall length .... 73.2 inches
- Overall width .... 44 inches
- Overall height .... 44 inches
- Seat height .... 33 inches
- Ground clearance .... 5.5 inches
- Weight .... 483 lbs

Sport 400L
- Model number W958540
- Wheelbase .... 49.75 inches
- Overall length .... 72 inches
- Overall width .... 44 inches
- Overall height .... 44 inches
- Seat height .... 33 inches
- Ground clearance .... 5.5 inches
- Weight .... 479 lbs

2x4 400L
- Model number W957540
- Wheelbase .... 49.75 inches
- Overall length .... 77 inches
- Overall width .... 44 inches
- Overall height .... 46 inches
- Seat height .... 33 inches
- Ground clearance .... 5.5 inches
- Weight .... 512 lbs

4x4 300
- Model number W958130
- Wheelbase .... 49.75 inches
- Overall length .... 73.2 inches
- Overall width .... 44.5 inches
- Overall height .... 44.5 inches
- Seat height .... 34 inches

General specifications

    Ground clearance .... 6 inches
    Weight .... 538 lbs
  Xplorer 4x4
    Model number W959140
    Wheelbase .... 49.75 inches
    Overall length .... 77 inches
    Overall width .... 46 inches
    Overall height .... 47 inches
    Seat height .... 34 inches
    Ground clearance .... 6 inches
    Weight .... 570 lbs
  Sportsman 4x4
    Model number W958040
    Wheelbase .... 49.75 inches
    Overall length .... 77 inches
    Overall width .... 46 inches
    Overall height .... 46 inches
    Seat height .... 34 inches
    Ground clearance .... 6 inches
    Weight .... 585 lbs
  Scrambler
    Model number W957840
    Wheelbase .... 48.5 inches
    Overall length .... 74.5 inches
    Overall width .... 45.5 inches
    Overall height .... 47 inches
    Seat height .... 33 inches
    Ground clearance .... 6 inches
    Weight .... 490 lbs
  6x6 400L
    Model number W958740
    Wheelbase .... 75 inches
    Overall length .... 103 inches
    Overall width .... 46 inches
    Overall height .... 47 inches
    Seat height .... 37 inches
    Ground clearance .... 5.6 inches
    Weight .... 857 lbs
  Magnum 2x4
    Model number W957544
    Wheelbase .... 49.75 inches
    Overall length .... 77 inches
    Overall width .... 46.5 inches
    Overall height .... 47 inches
    Seat height .... 33 inches
    Ground clearance .... 5.5 inches
    Weight .... 534 lbs
  Magnum 4x4
    Model number W958144
    Wheelbase .... 49.75 inches
    Overall length .... 77 inches
    Overall width .... 46 inches
    Overall height .... 46 inches
    Seat height .... 34 inches
    Ground clearance .... 6 inches
    Weight .... 595 lbs
1996
  Trail Boss
    Model number W968527
    Wheelbase .... 49.75 inches
    Overall length .... 73.2 inches
    Overall width .... 44 inches
    Overall height .... 44 inches
    Seat height .... 33 inches
    Ground clearance .... 5.5 inches
    Weight .... 425 lbs
  Trail Blazer
    Model number W967827
    Wheelbase .... 49.75 inches
    Overall length .... 74.5 inches

General specifications

1996 (continued)

Trail Blazer
- Overall width .... 46.5 inches
- Overall height .... 46 inches
- Seat height .... 34 inches
- Ground clearance .... 6 inches
- Weight .... 420 lbs

Xpress 300
- Model number W969530
- Wheelbase .... 49.75 inches
- Overall length .... 78.5 inches
- Overall width .... 46 inches
- Overall height .... 45.5 inches
- Seat height .... 33 inches
- Ground clearance .... 6.75 inches
- Weight .... 512 lbs

Xplorer 300
- Model number W969130
- Wheelbase .... 49.75 inches
- Overall length .... 81 inches
- Overall width .... 46 inches
- Overall height .... 45.5 inches
- Seat height .... 34 inches
- Ground clearance .... 6 inches
- Weight .... 567 lbs

Xpress 400
- Model number W969540
- Wheelbase .... 49.75 inches
- Overall length .... 78.5 inches
- Overall width .... 46 inches
- Overall height .... 47.5 inches
- Seat height .... 33 inches
- Ground clearance .... 6.75 inches
- Weight .... 543 lbs

Xplorer 400
- Model number W969140
- Wheelbase .... 49.75 inches
- Overall length .... 81 inches
- Overall width .... 46 inches
- Overall height .... 47.5 inches
- Seat height .... 34 inches
- Ground clearance .... 7.375 inches
- Weight .... 570 lbs

Sport 4x4
- Model number W968540
- Wheelbase .... 49.75 inches
- Overall length .... 74.5 inches
- Overall width .... 46.5 inches
- Overall height .... 47 inches
- Seat height .... 33 inches
- Ground clearance .... 6 inches
- Weight .... 479 lbs

Sportsman 4x4
- Model number W968040
- Wheelbase .... 49.75 inches
- Overall length .... 77 inches
- Overall width .... 46 inches
- Overall height .... 46 inches
- Seat height .... 34 inches
- Ground clearance .... 6 inches
- Weight .... 585 lbs

Scrambler
- Model number W967840
- Wheelbase .... 48.5 inches
- Overall length .... 74.5 inches
- Overall width .... 45.5 inches
- Overall height .... 47 inches
- Seat height .... 33 inches
- Ground clearance .... 6 inches
- Weight .... 490 lbs

General specifications

Magnum 2x4
- Model number W957544
- Wheelbase .... 49.75 inches
- Overall length .... 77 inches
- Overall width .... 46.5 inches
- Overall height .... 47 inches
- Seat height .... 33 inches
- Ground clearance .... 5.5 inches
- Weight .... 534 lbs

Magnum 4x4
- Model number W968144
- Wheelbase .... 49.75 inches
- Overall length .... 77 inches
- Overall width .... 46 inches
- Overall height .... 46 inches
- Seat height .... 34 inches
- Ground clearance .... 6 inches
- Weight .... 595 lbs

Sportsman 500
- Model number W969244
- Wheelbase .... 50.50 inches
- Overall length .... 77 inches
- Overall width .... 46 inches
- Overall height .... 47 inches
- Seat height .... 34 inches
- Ground clearance .... 10 inches
- Weight .... 649 lbs

Magnum 6x6
- Model number W968744
- Wheelbase .... 75 inches
- Overall length .... 103 inches
- Overall width .... 46 inches
- Overall height .... 47 inches
- Seat height .... 34 inches
- Ground clearance .... 5.5 inches
- Weight .... 852 lbs

6x6 400
- Model number W968740
- Wheelbase .... 75 inches
- Overall length .... 103 inches
- Overall width .... 46 inches
- Overall height .... 47 inches
- Seat height .... 34 inches
- Ground clearance .... 5.5 inches
- Weight .... 830 lbs

1997

Trail Boss
- Model number W97AA25C
- Wheelbase .... 49.75 inches
- Overall length .... 73.2 inches
- Overall width .... 44 inches
- Overall height .... 44 inches
- Seat height .... 33 inches
- Ground clearance .... 5.5 inches
- Weight .... 425 lbs

Trail Blazer
- Model number W97BA25C
- Wheelbase .... 49.75 inches
- Overall length .... 74.5 inches
- Overall width .... 46.5 inches
- Overall height .... 46 inches
- Seat height .... 34 inches
- Ground clearance .... 6 inches
- Weight .... 420 lbs

Xpress 300
- Model number W97VS28V
- Wheelbase .... 49.75 inches
- Overall length .... 78.5 inches
- Overall width .... 46 inches
- Overall height .... 45.5 inches

General specifications

1997 (continued)
- Xpress 300
  - Seat height .... 33 inches
  - Ground clearance .... 6.75 inches
  - Weight .... 512 lbs
- Xplorer 300
  - Model number W97CC28C
  - Wheelbase .... 49.75 inches
  - Overall length .... 81 inches
  - Overall width .... 46 inches
  - Overall height .... 45.5 inches
  - Seat height .... 34 inches
  - Ground clearance .... 6 inches
  - Weight .... 567 lbs
- Xpress 400
  - Model number W97CA38C
  - Wheelbase .... 49.75 inches
  - Overall length .... 78.5 inches
  - Overall width .... 46 inches
  - Overall height .... 47.5 inches
  - Seat height .... 33 inches
  - Ground clearance .... 6.75 inches
  - Weight .... 543 lbs
- Xplorer 400
  - Model number W972238C
  - Wheelbase .... 49.75 inches
  - Overall length .... 81 inches
  - Overall width .... 46 inches
  - Overall height .... 47.5 inches
  - Seat height .... 34 inches
  - Ground clearance .... 7.375 inches
  - Weight .... 570 lbs
- Sport 400
  - Model number W97BA38C
  - Wheelbase .... 49.75 inches
  - Overall length .... 74.5 inches
  - Overall width .... 46.5 inches
  - Overall height .... 47 inches
  - Seat height .... 33 inches
  - Ground clearance .... 5.5 inches
  - Weight .... 479 lbs
- Sportsman 4x4
  - Model number W97AC38C
  - Wheelbase .... 49.75 inches
  - Overall length .... 77 inches
  - Overall width .... 46 inches
  - Overall height .... 46 inches
  - Seat height .... 34 inches
  - Ground clearance .... 6 inches
  - Weight .... 585 lbs
- Scrambler 4x4
  - Model number W97BC38C
  - Wheelbase .... 48.5 inches
  - Overall length .... 74.5 inches
  - Overall width .... 45.5 inches
  - Overall height .... 47 inches
  - Seat height .... 33 inches
  - Ground clearance .... 6 inches
  - Weight .... 490 lbs
- Magnum 2x4
  - Model number W97AA42A
  - Wheelbase .... 49.75 inches
  - Overall length .... 77 inches
  - Overall width .... 46.5 inches
  - Overall height .... 47 inches
  - Seat height .... 33 inches
  - Ground clearance .... 5.5 inches
  - Weight .... 534 lbs
- Magnum 4x4
  - Model number W97AC42A

General specifications

Wheelbase	49.75 inches
Overall length	77 inches
Overall width	46 inches
Overall height	46 inches
Seat height	34 inches
Ground clearance	6 inches
Weight	595 lbs

Sportsman 500

Model number	W97CH50A
Wheelbase	50.5 inches
Overall length	77 inches
Overall width	46 inches
Overall height	47 inches
Seat height	34 inches
Ground clearance	10 inches
Weight	649 lbs

Scrambler 500

Model number	W97BC50A
Wheelbase	48.5 inches
Overall length	74.5 inches
Overall width	45.5 inches
Overall height	47 inches
Seat height	33 inches
Ground clearance	6 inches
Weight	547 lbs

Xplorer 500

Model number	W97CD50A
Wheelbase	50.5 inches
Overall length	77 inches
Overall width	46 inches
Overall height	47 inches
Seat height	34 inches
Ground clearance	10 inches
Weight	649 lbs

Magnum 6x6

Model number	W97AE42A
Wheelbase	75 inches
Overall length	103 inches
Overall width	46 inches
Overall height	47 inches
Seat height	34 inches
Ground clearance	5.5 inches
Weight	852 lbs

400 6x6

Model number	W97AE38A
Wheelbase	75 inches
Overall length	103 inches
Overall width	46 inches
Overall height	47 inches
Seat height	34 inches
Ground clearance	5.5 inches
Weight	830 lbs

Maintenance techniques, tools and working facilities

Basic maintenance techniques

There are a number of techniques involved in maintenance and repair that will be referred to throughout this manual. Application of these techniques will enable the amateur mechanic to be more efficient, better organized and capable of performing the various tasks properly, which will ensure that the repair job is thorough and complete.

Fastening systems

Fasteners, basically, are nuts, bolts and screws used to hold two or more parts together. There are a few things to keep in mind when working with fasteners. Almost all of them use a locking device of some type (either a lock washer, locknut, locking tab or thread adhesive). All threaded fasteners should be clean, straight, have undamaged threads and undamaged corners on the hex head where the wrench fits. Develop the habit of replacing all damaged nuts and bolts with new ones.

Rusted nuts and bolts should be treated with a penetrating oil to ease removal and prevent breakage. Some mechanics use turpentine in a spout type oil can, which works quite well. After applying the rust penetrant, let it -work for a few minutes before trying to loosen the nut or bolt. Badly rusted fasteners may have to be chiseled off or removed with a special nut breaker, available at tool stores.

If a bolt or stud breaks off in an assembly, it can be drilled out and removed with a special tool called an E-Z out (or screw extractor). Most dealer service departments and motorcycle repair shops can perform this task, as well as others (such as the repair of threaded holes that have been stripped out).

Flat washers and lock washers, when removed from an assembly, should always be replaced exactly as removed. Replace any damaged washers with new ones. Always use a flat washer between a lock washer and any soft metal surface (such as aluminum), thin sheet metal or plastic. Special locknuts can only be used once or twice before they lose their locking ability and must be replaced.

Tightening sequences and procedures

When threaded fasteners are tightened, they are often tightened to a specific torque value (torque is basically a twisting force). Over-tightening the fastener can weaken it and cause it to break, while under-tightening can cause it to eventually come loose. Each bolt, depending on the material it's made of, the diameter of its shank and the material it is threaded into, has a specific torque value, which is noted in the Specifications. Be sure to follow the torque recommendations closely.

Fasteners laid out in a pattern (i.e. cylinder head bolts, engine case bolts, etc.) must be loosened or tightened in a sequence to avoid warping the component. Initially, the bolts/nuts should go on finger tight only. Next, they should be tightened one full turn each, in a criss-cross or diagonal pattern. After each one has been tightened one full turn, return to the first one tightened and tighten them all one half turn, following the same pattern. Finally, tighten each of them one quarter turn at a time until each fastener has been tightened to the proper torque. To loosen and remove the fasteners the procedure would be reversed.

Disassembly sequence

Component disassembly should be done with care and purpose to help ensure that the parts go back together properly during reassembly. Always keep track of the sequence in which parts are removed. Take note of special characteristics or marks on parts that can be installed more than one way (such as a grooved thrust washer on a shaft). It's a good idea to lay the disassembled parts out on a clean surface in the order that they were removed. It may also be helpful to make sketches or take instant photos of components before removal.

When removing fasteners from a component, keep track of their locations. Sometimes threading a bolt back in a part, or putting the washers and nut back on a stud, can prevent mix-ups later. If nuts and bolts can't be returned to their original locations, they should be kept in a compartmented box or a series of small boxes. A cupcake or muffin tin is ideal for this purpose, since each cavity can hold the bolts and nuts from a particular area (i.e. engine case bolts, valve cover bolts, engine mount bolts, etc.). A pan of this type is especially helpful when working on assemblies with very small parts (such as the carburetors and the valve train). The cavities can be marked with paint or tape to identify the contents.

Whenever wiring looms, harnesses or connectors are separated, it's a good idea to identify the two halves with numbered pieces of masking tape so they can be easily reconnected.

Gasket sealing surfaces

Throughout any motorcycle, gaskets are used to seal the mating surfaces between components and keep lubricants, fluids, vacuum or pressure contained in an assembly.

Many times these gaskets are coated with a liquid or paste type gasket sealing compound before assembly. Age, heat and pressure can sometimes cause the two parts to stick together so tightly that they are very difficult to separate. In most cases, the part can be loosened by striking it with a soft-faced hammer near the mating surfaces. A regular hammer can be used if a block of wood is placed between the hammer and the part. Do not hammer on cast parts or parts that could be easily damaged. With any particularly stubborn part, always recheck to make sure that every fastener has been removed.

Avoid using a screwdriver or bar to pry apart components, as they can easily mar the gasket sealing surfaces of the parts (which must remain smooth). If prying is absolutely necessary, use a piece of wood, but keep in mind that extra clean-up will be necessary if the wood splinters.

After the parts are separated, the old gasket must be carefully scraped off and the gasket surfaces cleaned. Stubborn gasket material can be soaked with a gasket remover (available in aerosol cans) to soften it so it can be easily scraped off. A scraper can be fashioned from a piece of copper tubing by flattening and sharpening one end. Copper is recommended because it is usually softer than the surfaces to be scraped, which reduces the chance of gouging the part. Some gaskets can be removed with a wire brush, but regardless of the method used, the mating surfaces must be left clean and smooth. If for some reason the gasket surface is gouged, then a gasket sealer thick enough to fill scratches will have to be used during reassembly of the components. For most applications, a non-drying (or semi-drying) gasket sealer is best.

Hose removal tips

Hose removal precautions closely parallel gasket removal precautions. Avoid scratching or gouging the surface that the hose mates against or the connection may leak. Because of various chemical reactions, the rubber in hoses can bond itself to the metal spigot that the hose fits over. To remove a hose, first loosen the hose clamps that secure it to the spigot. Then, with slip joint pliers, grab the hose at the clamp and rotate it around the spigot. Work it back and forth until it is completely free, then pull it off (silicone or other lubricants will ease removal if they can be applied between the hose and the outside of the spigot). Apply the same lubricant to the inside of the hose and the outside of the spigot to simplify installation.

If a hose clamp is broken or damaged, do not reuse it. Also, do not reuse hoses that are cracked, split or torn.

Maintenance techniques, tools and working facilities

Spark plug gap adjusting tool

Feeler gauge set

Control cable pressure luber

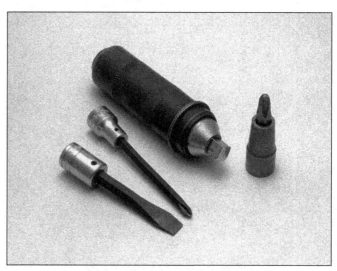

Hand impact screwdriver and bits

Tools

A selection of good tools is a basic requirement for anyone who plans to maintain and repair a motorcycle. For the owner who has few tools, if any, the initial investment might seem high, but when compared to the spiraling costs of routine maintenance and repair, it is a wise one.

To help the owner decide which tools are needed to perform the tasks detailed in this manual, the following tool lists are offered: *Maintenance and minor repair*, *Repair and overhaul* and *Special*. The newcomer to practical mechanics should start off with the *Maintenance and minor repair* tool kit, which is adequate for the simpler jobs. Then, as confidence and experience grow, the owner can tackle more difficult tasks, buying additional tools as they are needed. Eventually the basic kit will be built into the *Repair and overhaul* tool set. Over a period of time, the experienced do-it-yourselfer will assemble a tool set complete enough for most repair and overhaul procedures and will add tools from the *Special* category when it is felt that the expense is justified by the frequency of use.

Maintenance and minor repair tool kit

The tools in this list should be considered the minimum required for performance of routine maintenance, servicing and minor repair work. We recommend the purchase of combination wrenches (box end and open end combined in one wrench); while more expensive than

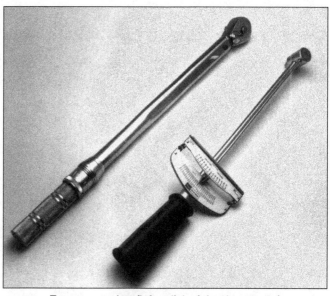

Torque wrenches (left - click; right - beam type)

0-24 Maintenance techniques, tools and working facilities

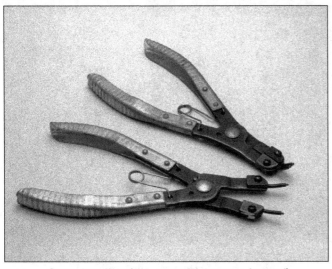

Snap-ring pliers (top - external; bottom - internal)

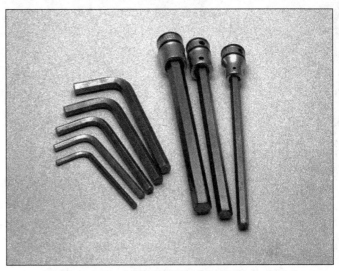

Allen wrenches (left), and Allen head sockets (right)

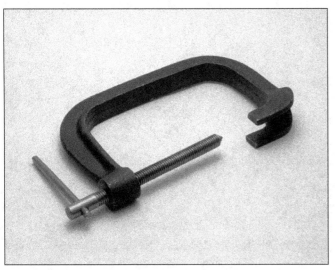

Valve spring compressor

Piston ring removal/installation tool

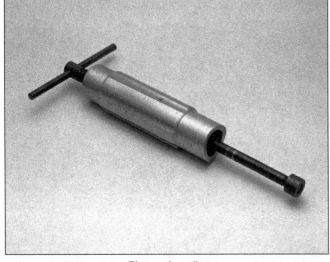

Piston pin puller

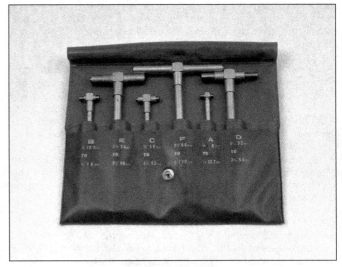

Telescoping gauges

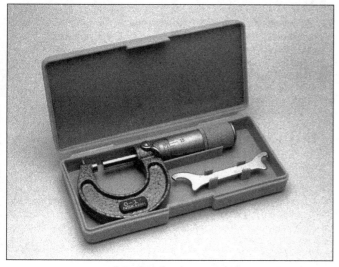
0-to-1 inch micrometer

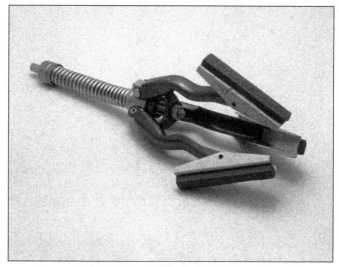
Cylinder surfacing hone

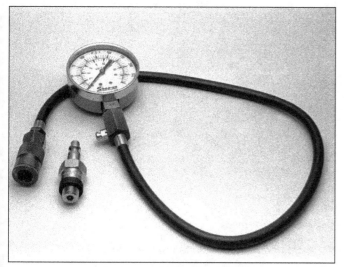
Cylinder compression gauge

Dial indicator set

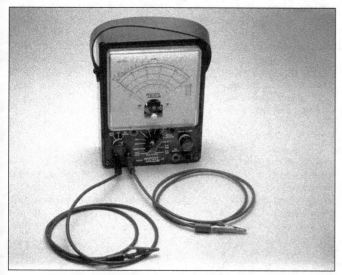
Multimeter (volt/ohm/ammeter)

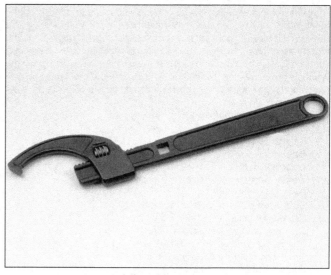
Adjustable spanner

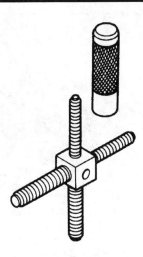

Alternator rotor puller

open-ended ones, they offer the advantages of both types of wrench.

    Combination wrench set (6 mm to 22 mm)
    Adjustable wrench - 8 in
    Spark plug socket (with rubber insert)
    Spark plug gap adjusting tool
    Feeler gauge set
    Standard screwdriver (5/16 in x 6 in)
    Phillips screwdriver (No. 2 x 6 in)
    Allen (hex) wrench set (4 mm to 12 mm)
    Combination (slip-joint) pliers - 6 in
    Hacksaw and assortment of blades
    Tire pressure gauge
    Control cable pressure luber
    Grease gun
    Oil can
    Fine emery cloth
    Wire brush
    Hand impact screwdriver and bits
    Funnel (medium size)
    Safety goggles
    Drain pan
    Work light with extension cord

Repair and overhaul tool set

These tools are essential for anyone who plans to perform major repairs and are intended to supplement those in the Maintenance and minor repair tool kit. Included is a comprehensive set of sockets which, though expensive, are invaluable because of their versatility (especially when various extensions and drives are available). We recommend the 3/8 inch drive over the 1/2 inch drive for general motorcycle maintenance and repair (ideally, the mechanic would have a 3/8 inch drive set and a 1/2 inch drive set).

    Alternator rotor removal tool
    Socket set(s)
    Reversible ratchet
    Extension - 6 in
    Universal joint
    Torque wrench (same size drive as sockets)
    Ball pein hammer - 8 oz
    Soft-faced hammer (plastic/rubber)
    Standard screwdriver (1/4 in x 6 in)
    Standard screwdriver (stubby - 5/16 in)
    Phillips screwdriver (No. 3 x 8 in)
    Phillips screwdriver (stubby - No. 2)
    Pliers - locking
    Pliers - lineman's
    Pliers - needle nose
    Pliers - snap-ring (internal and external)
    Cold chisel - 1/2 in
    Scriber
    Scraper (made from flattened copper tubing)
    Center punch
    Pin punches (1/16, 1/8, 3/16 in)
    Steel rule/straightedge - 12 in
    Pin-type spanner wrench
    A selection of files
    Wire brush (large)

Note: *Another tool which is often useful is an electric drill with a chuck capacity of 3/8 inch (and a set of good quality drill bits).*

Special tools

The tools in this list include those which are not used regularly, are expensive to buy, or which need to be used in accordance with their manufacturer's instructions. Unless these tools will be used frequently, it is not very economical to purchase many of them. A consideration would be to split the cost and use between yourself and a friend or friends (i.e. members of a motorcycle club).

This list primarily contains tools and instruments widely available to the public, as well as some special tools produced by the vehicle manufacturer for distribution to dealer service departments. As a result, references to the manufacturer's special tools are occasionally included in the text of this manual. Generally, an alternative method of doing the job without the special tool is offered. However, sometimes there is no alternative to their use. Where this is the case, and the tool can't be purchased or borrowed, the work should be turned over to the dealer service department or a motorcycle repair shop.

    Paddock stand (for models not fitted with a centerstand)
    Valve spring compressor
    Piston ring removal and installation tool
    Piston pin puller
    Telescoping gauges
    Micrometer(s) and/or dial/Vernier calipers
    Cylinder surfacing hone
    Cylinder compression gauge
    Dial indicator set
    Multimeter
    Adjustable spanner
    Manometer or vacuum gauge set
    Small air compressor with blow gun and tire chuck

Buying tools

For the do-it-yourselfer who is just starting to get involved in motorcycle maintenance and repair, there are a number of options available when purchasing tools. If maintenance and minor repair is the extent of the work to be done, the purchase of individual tools is satisfactory. If, on the other hand, extensive work is planned, it would be a good idea to purchase a modest tool set from one of the large retail chain stores. A set can usually be bought at a substantial savings over the individual tool prices (and they often come with a tool box). As additional tools are needed, add-on sets, individual tools and a larger tool box can be purchased to expand the tool selection. Building a tool set gradually allows the cost of the tools to be spread over a longer period of time and gives the mechanic the freedom to choose only those tools that will actually be used.

Tool stores and motorcycle dealers will often be the only source of some of the special tools that are needed, but regardless of where tools are bought, try to avoid cheap ones (especially when buying screwdrivers and sockets) because they won't last very long. There are plenty of tools around at reasonable prices, but always aim to purchase items which meet the relevant national safety standards. The expense involved in replacing cheap tools will eventually be greater than the initial cost of quality tools.

It is obviously not possible to cover the subject of tools fully here. For those who wish to learn more about tools and their use, there is a book entitled *Motorcycle Workshop Practice Manual* (Book no. 1454) available from the publishers of this manual. It also provides an introduction to basic workshop practice which will be of interest to a home mechanic working on any type of motorcycle.

Care and maintenance of tools

Good tools are expensive, so it makes sense to treat them with respect. Keep them clean and in usable condition and store them properly when not in use. Always wipe off any dirt, grease or metal chips before putting them away. Never leave tools lying around in the work area.

Some tools, such as screwdrivers, pliers, wrenches and sockets, can be hung on a panel mounted on the garage or workshop wall, while others should be kept in a tool box or tray. Measuring instruments, gauges, meters, etc. must be carefully stored where they can't be damaged by weather or impact from other tools.

When tools are used with care and stored properly, they will last a very long time. Even with the best of care, tools will wear out if used frequently. When a tool is damaged or worn out, replace it; subsequent jobs will be safer and more enjoyable if you do.

Working facilities

Not to be overlooked when discussing tools is the workshop. If anything more than routine maintenance is to be carried out, some sort of suitable work area is essential.

It is understood, and appreciated, that many home mechanics do not have a good workshop or garage available and end up removing an engine or doing major repairs outside (it is recommended, however, that the overhaul or repair be completed under the cover of a roof).

A clean, flat workbench or table of comfortable working height is an absolute necessity. The workbench should be equipped with a vise that has a jaw opening of at least four inches.

As mentioned previously, some clean, dry storage space is also required for tools, as well as the lubricants, fluids, cleaning solvents, etc. which soon become necessary.

Sometimes waste oil and fluids, drained from the engine or cooling system during normal maintenance or repairs, present a disposal problem. To avoid pouring them on the ground or into a sewage system, simply pour the used fluids into large containers, seal them with caps and take them to an authorized disposal site or service station. Plastic jugs (such as old antifreeze containers) are ideal for this purpose.

Always keep a supply of old newspapers and clean rags available. Old towels are excellent for mopping up spills. Many mechanics use rolls of paper towels for most work because they are readily available and disposable. To help keep the area under the motorcycle clean, a large cardboard box can be cut open and flattened to protect the garage or shop floor.

Whenever working over a painted surface (such as the fuel tank) cover it with an old blanket or bedspread to protect the finish.

Safety first!

Professional mechanics are trained in safe working procedures. However enthusiastic you may be about getting on with the job at hand, take the time to ensure that your safety is not put at risk. A moment's lack of attention can result in an accident, as can failure to observe simple precautions.

There will always be new ways of having accidents, and the following is not a comprehensive list of all dangers; it is intended rather to make you aware of the risks and to encourage a safe approach to all work you carry out on your bike.

Essential DOs and DON'Ts

DON'T start the engine without first ascertaining that the transmission is in neutral.
DON'T suddenly remove the pressure cap from a hot cooling system - cover it with a cloth and release the pressure gradually first, or you may get scalded by escaping coolant.
DON'T attempt to drain oil until you are sure it has cooled sufficiently to avoid scalding you.
DON'T grasp any part of the engine or exhaust system without first ascertaining that it is cool enough not to burn you.
DON'T allow brake fluid or antifreeze to contact the machine's paint work or plastic components.
DON'T siphon toxic liquids such as fuel, hydraulic fluid or antifreeze by mouth, or allow them to remain on your skin.
DON'T inhale dust - it may be injurious to health (see *Asbestos* heading).
DON'T allow any spilled oil or grease to remain on the floor - wipe it up right away, before someone slips on it.
DON'T use ill fitting wrenches or other tools which may slip and cause injury.
DON'T attempt to lift a heavy component which may be beyond your capability - get assistance.
DON'T rush to finish a job or take unverified short cuts.
DON'T allow children or animals in or around an unattended vehicle.
DON'T inflate a tire to a pressure above the recommended maximum. Apart from over stressing the carcase and wheel rim, in extreme cases the tire may blow off forcibly.
DO ensure that the machine is supported securely at all times. This is especially important when the machine is blocked up to aid wheel or fork removal.
DO take care when attempting to loosen a stubborn nut or bolt. It is generally better to pull on a wrench, rather than push, so that if you slip, you fall away from the machine rather than onto it.
DO wear eye protection when using power tools such as drill, sander, bench grinder etc.
DO use a barrier cream on your hands prior to undertaking dirty jobs - it will protect your skin from infection as well as making the dirt easier to remove afterwards; but make sure your hands aren't left slippery. Note that long-term contact with used engine oil can be a health hazard.
DO keep loose clothing (cuffs, ties etc. and long hair) well out of the way of moving mechanical parts.
DO remove rings, wristwatch etc., before working on the vehicle - especially the electrical system.
DO keep your work area tidy - it is only too easy to fall over articles left lying around.
DO exercise caution when compressing springs for removal or installation. Ensure that the tension is applied and released in a controlled manner, using suitable tools which preclude the possibility of the spring escaping violently.
DO ensure that any lifting tackle used has a safe working load rating adequate for the job.
DO get someone to check periodically that all is well, when working alone on the vehicle.
DO carry out work in a logical sequence and check that everything is correctly assembled and tightened afterwards.
DO remember that your vehicle's safety affects that of yourself and others. If in doubt on any point, get professional advice.
IF, in spite of following these precautions, you are unfortunate enough to injure yourself, seek medical attention as soon as possible.

Asbestos

Certain friction, insulating, sealing and other products - such as brake pads, clutch linings, gaskets, etc. - contain asbestos. *Extreme care must be taken to avoid inhalation of dust from such products since it is hazardous to health*. If in doubt, assume that they *do* contain asbestos.

Fire

Remember at all times that gasoline (petrol) is highly flammable. Never smoke or have any kind of naked flame around, when working on the vehicle. But the risk does not end there - a spark caused by an electrical short-circuit, by two metal surfaces contacting each other, by careless use of tools, or even by static electricity built up in your body under certain conditions, can ignite gasoline (petrol) vapor, which in a confined space is highly explosive. Never use gasoline (petrol) as a cleaning solvent. Use an approved safety solvent.

Always disconnect the battery ground (earth) terminal before working on any part of the fuel or electrical system, and never risk spilling fuel on to a hot engine or exhaust.

It is recommended that a fire extinguisher of a type suitable for fuel and electrical fires is kept handy in the garage or workplace at all times. Never try to extinguish a fuel or electrical fire with water.

Fumes

Certain fumes are highly toxic and can quickly cause unconsciousness and even death if inhaled to any extent. Gasoline (petrol) vapor comes into this category, as do the vapors from certain solvents such as trichloroethylene. Any draining or pouring of such volatile fluids should be done in a well ventilated area.

When using cleaning fluids and solvents, read the instructions carefully. Never use materials from unmarked containers - they may give off poisonous vapors.

Never run the engine of a motor vehicle in an enclosed space such as a garage. Exhaust fumes contain carbon monoxide which is extremely poisonous; if you need to run the engine, always do so in the open air or at least have the rear of the vehicle outside the workplace.

The battery

Never cause a spark, or allow a naked light near the vehicle's battery. It will normally be giving off a certain amount of hydrogen gas, which is highly explosive.

Always disconnect the battery ground (earth) terminal before working on the fuel or electrical systems (except where noted).

If possible, loosen the filler plugs or cover when charging the battery from an external source. Do not charge at an excessive rate or the battery may burst.

Take care when topping up, cleaning or carrying the battery. The acid electrolyte, even when diluted, is very corrosive and should not be allowed to contact the eyes or skin. Always wear rubber gloves and goggles or a face shield. If you ever need to prepare electrolyte yourself, always add the acid slowly to the water; never add the water to the acid.

Electricity

When using an electric power tool, inspection light etc., always ensure that the appliance is correctly connected to its plug and that, where necessary, it is properly grounded (earthed). Do not use such appliances in damp conditions and, again, beware of creating a spark or applying excessive heat in the vicinity of fuel or fuel vapor. Also ensure that the appliances meet national safety standards.

A severe electric shock can result from touching certain parts of the electrical system, such as the spark plug wires (HT leads), when the engine is running or being cranked, particularly if components are damp or the insulation is defective. Where an electronic ignition system is used, the secondary (HT) voltage is much higher and could prove fatal.

ATV chemicals and lubricants

A number of chemicals and lubricants are available for use in vehicle maintenance and repair. They include a wide variety of products ranging from cleaning solvents and degreasers to lubricants and protective sprays for rubber, plastic and vinyl.

Contact point/spark plug cleaner is a solvent used to clean oily film and dirt from points, grime from electrical connectors and oil deposits from spark plugs. It is oil free and leaves no residue. It can also be used to remove gum and varnish from carburetor jets and other orifices.

Carburetor cleaner is similar to contact point/spark plug cleaner but it usually has a stronger solvent and may leave a slight oily residue. It is not recommended for cleaning electrical components or connections.

Brake system cleaner is used to remove grease or brake fluid from brake system components (where clean surfaces are absolutely necessary and petroleum-based solvents cannot be used); it also leaves no residue.

Silicone-based lubricants are used to protect rubber parts such as hoses and grommets, and are used as lubricants for hinges and locks.

Multi-purpose grease is an all purpose lubricant used wherever grease is more practical than a liquid lubricant such as oil. Some multi-purpose grease is colored white and specially formulated to be more resistant to water than ordinary grease.

Gear oil (sometimes called gear lube) is a specially designed oil used in transmissions and final drive units, as well as other areas where high friction, high temperature lubrication is required. It is available in a number of viscosities (weights) for various applications. The transmission on Banshee models is lubricated by four-stroke motor oil.

Motor oil, of course, is the lubricant specially formulated for use in the engine. It normally contains a wide variety of additives to prevent corrosion and reduce foaming and wear. Motor oil comes in various weights (viscosity ratings) of from 5 to 80. The recommended weight of the oil depends on the seasonal temperature and the demands on the engine. Light oil is used in cold climates and under light load conditions; heavy oil is used in hot climates and where high loads are encountered. Multi-viscosity oils are designed to have characteristics of both light and heavy oils and are available in a number of weights from 5W-20 to 20W-50. On Banshees, the engine oil is premixed with the gasoline. On Warriors, the same oil supply is shared by the engine and transmission.

Gas additives perform several functions, depending on their chemical makeup. They usually contain solvents that help dissolve gum and varnish that build up on carburetor and intake parts. They also serve to break down carbon deposits that form on the inside surfaces of the combustion chambers. Some additives contain upper cylinder lubricants for valves and piston rings.

Brake fluid is a specially formulated hydraulic fluid that can withstand the heat and pressure encountered in brake systems. Care must be taken that this fluid does not come in contact with painted surfaces or plastics. An opened container should always be resealed to prevent contamination by water or dirt.

Chain lubricants are formulated especially for use on the final drive chains of vehicles so equipped. A good chain lube should adhere well and have good penetrating qualities to be effective as a lubricant inside the chain and on the side plates, pins and rollers. Most chain lubes are either the foaming type or quick drying type and are usually marketed as sprays.

Degreasers are heavy duty solvents used to remove grease and grime that may accumulate on engine and frame components. They can be sprayed or brushed on and, depending on the type, are rinsed with either water or solvent.

Solvents are used alone or in combination with degreasers to clean parts and assemblies during repair and overhaul. The home mechanic should use only solvents that are non-flammable and that do not produce irritating fumes.

Gasket sealing compounds may be used in conjunction with gaskets, to improve their sealing capabilities, or alone, to seal metal-to-metal joints. Many gasket sealers can withstand extreme heat, some are impervious to gasoline and lubricants, while others are capable of filling and sealing large cavities. Depending on the intended use, gasket sealers either dry hard or stay relatively soft and pliable. They are usually applied by hand, with a brush, or are sprayed on the gasket sealing surfaces.

Thread cement is an adhesive locking compound that prevents threaded fasteners from loosening because of vibration. It is available in a variety of types for different applications.

Moisture dispersants are usually sprays that can be used to dry out electrical components such as the fuse block and wiring connectors. Some types can also be used as treatment for rubber and as a lubricant for hinges, cables and locks.

Waxes and polishes are used to help protect painted and plated surfaces from the weather. Different types of paint may require the use of different types of wax polish. Some polishes utilize a chemical or abrasive cleaner to help remove the top layer of oxidized (dull) paint on older vehicles. In recent years, many non-wax polishes (that contain a wide variety of chemicals such as polymers and silicones) have been introduced. These non-wax polishes are usually easier to apply and last longer than conventional waxes and polishes.

Troubleshooting

Contents

Symptom	Section
Engine doesn't start or is difficult to start	
Starter motor doesn't rotate	1
Starter motor rotates but engine does not turn over	2
Starter works but engine won't turn over (seized)	3
No fuel flow	4
Engine flooded	5
No spark or weak spark	6
Compression low	7
Stalls after starting	8
Rough idle	9
Poor running at low speed	
Spark weak	10
Fuel/air mixture incorrect	11
Compression low	12
Poor acceleration	13
Poor running or no power at high speed	
Firing incorrect	14
Fuel/air mixture incorrect	15
Compression low	16
Knocking or pinging	17
Miscellaneous causes	18
Overheating	
Engine overheats	19
Firing incorrect	20
Fuel/air mixture incorrect	21
Compression too high	22
Engine load excessive	23
Lubrication inadequate	24
Miscellaneous causes	25
Polaris Variable Transmission (PVT) problems	
Drive belt slipping	26
Drive belt upside down in pulleys	27
Burn marks or thin spots on drive belt	28
Harsh engagement	29
Grabby or erratic engagement	30
Noisy operation	31
Melted PVT cover	32
Engine rpm too low when vehicle is driven	34
Engine rpm erratic when vehicle is driven	35

Symptom	Section
Gear shifting problems	
Doesn't go into gear	36
Abnormal engine noise	
Knocking or pinging	37
Piston slap or rattling	38
Valve noise (4-stroke)	39
Other noise	40
Abnormal driveline noise	
Chain noise	41
Polaris Variable Transmission (PVT) noise	42
Transmission noise	43
Front gearcase noise (shaft drive models)	44
Abnormal chassis noise	
Suspension noise	45
Driveaxle noise	46
Brake noise	47
Oil level light comes on (2-stroke)	
Engine lubrication system	48
Electrical system	49
Excessive exhaust smoke	
White smoke	50
Black smoke	51
Brown smoke	52
Poor handling or stability	
Handlebar hard to turn	53
Handlebar shakes or vibrates excessively	54
Handlebar pulls to one side	55
Poor shock absorbing qualities	56
Braking problems	
Brakes are spongy, don't hold (hydraulic brakes)	57
Brake lever pulsates	58
Brakes drag	59
Electrical problems	
Battery dead or weak	60
Battery overcharged	61

Troubleshooting

Engine doesn't start or is difficult to start

1 Starter motor does not rotate

1 Engine kill switch Off.
2 Circuit breaker tripped. Check circuit breaker (Chapter 5).
3 Battery voltage low. Check and recharge battery (Chapter 5).
4 Starter motor defective. Make sure the wiring to the starter is secure. Test starter solenoid (Chapter 8). If the solenoid is good, then the fault is in the wiring or motor.
5 Starter solenoid faulty. Check it according to the procedure in Chapter 5.
6 Starter switch not contacting. The contacts could be wet, corroded or dirty. Disassemble and clean the switch (Chapter 5).
7 Wiring open or shorted. Check all wiring connections and harnesses to make sure that they are dry, tight and not corroded. Also check for broken or frayed wires that can cause a short to ground (see wiring diagram, Chapter 5).
8 Ignition (main) switch defective. Check the switch according to the procedure in Chapter 5. Replace the switch with a new one if it is defective.
9 Engine kill switch defective. Check for wet, dirty or corroded contacts. Clean or replace the switch as necessary (Chapter 5).

2 Starter motor rotates but engine does not turn over

1 Starter motor drive defective. Inspect and repair or replace (Chapter 5).
2 Damaged starter gears. Inspect and replace the damaged parts (Chapter 2).

3 Starter works but engine won't turn over (seized)

Seized engine caused by one or more internally damaged components. Failure due to wear, abuse or lack of lubrication. Damage can include seized valves (4-stroke), valve lifters (4-stroke), camshaft (4-stroke), piston, or crankshaft or connecting rod bearings. Refer to Chapter 2 for engine disassembly.

4 No fuel flow

1 No fuel in tank.
2 Tank cap air vent obstructed. Usually caused by dirt or water. Remove it and clean the cap vent hole.
3 Clogged strainer in fuel tap or inside fuel tank. Remove and clean the strainer(s) (Chapter 1).
4 Fuel line clogged. Pull the fuel line loose and carefully blow through it.
5 Inlet needle valve clogged. A very bad batch of fuel with an unusual additive may have been used, or some other foreign material has entered the tank. Many times after a machine has been stored for many months without running, the fuel turns to a varnish-like liquid and forms deposits on the inlet needle valve and jets. The carburetor should be removed and overhauled if draining the float chamber doesn't solve the problem.

5 Engine flooded

1 Float level too high. Check as described in Chapter 4 and replace the float if necessary.
2 Inlet needle valve worn or stuck open. A piece of dirt, rust or other debris can cause the inlet needle to seat improperly, causing excess fuel to be admitted to the float bowl. In this case, the float chamber should be cleaned and the needle and seat inspected. If the needle and seat are worn, then the leaking will persist and the parts should be replaced with new ones (Chapter 4).
3 Starting technique incorrect. Under normal circumstances (i.e., if all the carburetor functions are sound) the machine should start with little or no throttle. When the engine is cold, the choke should be operated and the engine started without opening the throttle. When the engine is at operating temperature, only a very slight amount of throttle should be necessary. If the engine is flooded, turn the fuel tap off and hold the throttle open while cranking the engine. This will allow additional air to reach the cylinder. Remember to turn the fuel tap back on after the engine starts.

6 No spark or weak spark

1 Ignition switch Off.
2 Engine kill switch turned to the Off position.
3 Battery voltage low. Check and recharge battery as necessary (Chapter 4).
4 Spark plug dirty, defective or worn out. Locate reason for fouled plug using spark plug condition chart and follow the plug maintenance procedures in Chapter 1.
5 Spark plug cap or wire faulty. Check condition. Replace either or both components if cracks or deterioration are evident (Chapter 5).
6 Spark plug cap not making good contact. Make sure that the plug cap fits snugly over the plug end.
7 CDI box defective. Check the unit, referring to Chapter 5 for details.
8 Ignition pulse coil or combined exciter/pulse coil defective. Check the unit, referring to Chapter 5 for details.
9 Ignition coil defective. Check the coil, referring to Chapter 5.
10 Ignition or kill switch shorted. This is usually caused by water, corrosion, damage or excessive wear. The kill switch can be disassembled and cleaned with electrical contact cleaner. If cleaning does not help, replace the switches (Chapter 5).
11 Wiring shorted or broken between:
  a) Ignition switch and engine kill switch
  b) CDI box and engine kill switch
  c) CDI box and ignition coil
  d) Ignition coil and plug
  e) CDI box and pulse coil or pulse/exciter coil

Make sure that all wiring connections are clean, dry and tight. Look for chafed and broken wires (Chapters 5 and 9).

7 Compression low

1 Spark plug loose. Remove the plug and inspect the threads. Reinstall and tighten to the specified torque (Chapter 1).
2 Cylinder head not sufficiently tightened down. If the cylinder head is suspected of being loose, then there's a chance that the gasket or head is damaged if the problem has persisted for any length of time. The head nuts and bolts should be tightened to the proper torque in the correct sequence (Chapter 2).
3 Improper valve clearance (4-stroke). This means that the valve is not closing completely and compression pressure is leaking past the valve. Check and adjust the valve clearances (Chapter 1).
4 Cylinder and/or piston worn. Excessive wear will cause compression pressure to leak past the rings. This is usually accompanied by worn rings as well. A top end overhaul is necessary (Chapter 2).
5 Piston rings worn, weak, broken, or sticking. Broken or sticking piston rings usually indicate a lubrication or carburetion problem that causes excess carbon deposits or seizures to form on the pistons and rings. Top end overhaul is necessary (Chapter 2).
6 Piston ring-to-groove clearance excessive. This is caused by excessive wear of the piston ring lands. Piston replacement is necessary (Chapter 2).
7 Cylinder head gasket damaged. If the head is allowed to become

Troubleshooting 0-33

loose, or if excessive carbon build-up on a piston crown and combustion chamber causes extremely high compression, the head gasket may leak. Retorquing the head is not always sufficient to restore the seal, so gasket replacement is necessary (Chapter 2).
8 Cylinder head warped. This is caused by overheating or improperly tightened head nuts and bolts. Machine shop resurfacing or head replacement is necessary (Chapter 2).
9 Valve spring broken or weak (4-stroke). Caused by component failure or wear; the spring(s) must be replaced (Chapter 2).
10 Valve not seating properly (4-stroke). This is caused by a bent valve (from over-revving or improper valve adjustment), burned valve or seat (improper carburetion) or an accumulation of carbon deposits on the seat (from carburetion or lubrication problems). The valves must be cleaned and/or replaced and the seats serviced if possible (Chapter 2).

8 Stalls after starting

1 Improper choke action. Make sure the choke cable is getting a full stroke and staying in the out position.
2 Ignition malfunction. See Chapter 5.
3 Carburetor malfunction. See Chapter 4.
4 Fuel contaminated. The fuel can be contaminated with either dirt or water, or can change chemically if the machine is allowed to sit for several months or more. Drain the tank and float bowl and refill with fresh fuel (Chapter 4).
5 Intake air leak. Check for loose carburetor-to-intake joint connections or loose carburetor top (Chapter 4).
6 Engine idle speed incorrect. Turn throttle stop screw until the engine idles at the specified rpm (Chapter 1).

9 Rough idle

1 Ignition malfunction. See Chapter 5.
2 Idle speed incorrect. See Chapter 1.
3 Carburetor malfunction. See Chapter 4.
4 Idle fuel/air mixture incorrect. See Chapter 4.
5 Fuel contaminated. The fuel can be contaminated with either dirt or water, or can change chemically if the machine is allowed to sit for several months or more. Drain the tank and float bowls (Chapter 4).
6 Intake air leak. Check for loose carburetor-to-intake joint connections, loose or missing vacuum gauge access port cap or hose, or loose carburetor top (Chapter 4).
7 Air cleaner clogged. Service or replace air cleaner element (Chapter 1).

Poor running at low speed

10 Spark weak

1 Battery voltage low. Check and recharge battery (Chapter 5).
2 Spark plug fouled, defective or worn out. Refer to Chapter 1 for spark plug maintenance.
3 Spark plug cap or secondary wiring defective. Refer to Chapters 1 and 5 for details on the ignition system.
4 Spark plug cap not making contact.
5 Incorrect spark plug. Wrong type, heat range or cap configuration. Check and install correct plug listed in Chapter 1. A cold plug or one with a recessed firing electrode will not operate at low speeds without fouling.
6 CDI box defective. See Chapter 5.
7 Pulse coil or pulse/exciter coil defective. See Chapter 5.
8 Ignition coil defective. See Chapter 5.

11 Fuel/air mixture incorrect

1 Pilot screw out of adjustment (Chapter 4).
2 Pilot jet or air passage clogged. Remove and overhaul the carburetor (Chapter 4).
3 Air bleed holes clogged. Remove carburetor and blow out all passages (Chapter 4).
4 Air cleaner clogged, poorly sealed or missing.
5 Air cleaner-to-carburetor boot poorly sealed. Look for cracks, holes or loose clamps and replace or repair defective parts.
6 Float level too high or too low. Check and replace the float if necessary (Chapter 4).
7 Fuel tank air vent obstructed. Make sure that the air vent passage in the filler cap is open.
8 Carburetor intake joint loose. Check for cracks, breaks, tears or loose clamps or bolts. Repair or replace the rubber boot and its O-ring.

12 Compression low

1 Spark plug loose. Remove the plug and inspect the threads. Reinstall and tighten to the specified torque (Chapter 1).
2 Cylinder head not sufficiently tightened down. If the cylinder head is suspected of being loose, then there's a chance that the gasket and head are damaged if the problem has persisted for any length of time. The head nuts and bolts should be tightened to the proper torque in the correct sequence (Chapter 2).
3 Improper valve clearance (4-stroke). This means that the valve is not closing completely and compression pressure is leaking past the valve. Check and adjust the valve clearances (Chapter 1).
4 Cylinder and/or piston worn. Excessive wear will cause compression pressure to leak past the rings. This is usually accompanied by worn rings as well. A top end overhaul is necessary (Chapter 2).
5 Piston rings worn, weak, broken, or sticking. Broken or sticking piston rings usually indicate a lubrication or carburetion problem that causes excess carbon deposits or seizures to form on the pistons and rings. Top end overhaul is necessary (Chapter 2).
6 Piston ring-to-groove clearance excessive. This is caused by excessive wear of the piston ring lands. Piston replacement is necessary (Chapter 2).
7 Cylinder head gasket damaged. If the head is allowed to become loose, or if excessive carbon build-up on the piston crown and combustion chamber causes extremely high compression, the head gasket may leak. Retorquing the head is not always sufficient to restore the seal, so gasket replacement is necessary (Chapter 2).
8 Cylinder head warped. This is caused by overheating or improperly tightened head nuts and bolts. Machine shop resurfacing or head replacement is necessary (Chapter 2).
9 Valve spring broken or weak (4-stroke). Caused by component failure or wear; the spring(s) must be replaced (Chapter 2).
10 Valve not seating properly (4-stroke). This is caused by a bent valve (from over-revving or improper valve adjustment), burned valve or seat (improper carburetion) or an accumulation of carbon deposits on the seat (from carburetion, lubrication problems). The valves must be cleaned and/or replaced and the seats serviced if possible (Chapter 2).

13 Poor acceleration

1 Carburetor leaking or dirty. Overhaul the carburetor (Chapter 4).
2 Timing not advancing. The pulse coil or pulse/exciter coil or the CDI box may be defective. If so, they must be replaced with new ones, as they can't be repaired.
3 Engine oil viscosity too high. Using a heavier oil than that recommended in Chapter 1 can damage the oil pump or lubrication system and cause drag on the engine.
4 Brakes dragging. Usually caused by debris which has entered the

brake caliper sealing boots (hydraulic brakes), corroded calipers (hydraulic brakes), sticking brake cam (mechanical brakes) or from a warped drum, warped disc or bent axle. Repair as necessary (Chapter 6).

Poor running or no power at high speed

14 Firing incorrect

1 Air cleaner restricted. Clean or replace element (Chapter 1).
2 Spark plug fouled, defective or worn out. See Chapter 1 for spark plug maintenance.
3 Spark plug cap or wire defective. See Chapters 1 and 5 for details of the ignition system.
4 Spark plug cap not in good contact. See Chapter 5.
5 Incorrect spark plug. Wrong type, heat range or cap configuration. Check and install correct plugs listed in Chapter 1. A cold plug or one with a recessed firing electrode will not operate at low speeds without fouling.
6 CDI box defective. See Chapter 5.
7 Ignition coil defective. See Chapter 5.

15 Fuel/air mixture incorrect

1 Pilot screw out of adjustment. See Chapter 4 for adjustment procedures.
2 Main jet clogged. Dirt, water or other contaminants can clog the main jets. Clean the fuel tap strainer and in-tank strainer, the float bowl area, and the jets and carburetor orifices (Chapter 4).
3 Main jet wrong size. See Chapter 4 for jet specifications.
4 Throttle shaft-to-carburetor body clearance excessive. Refer to Chapter 4 for inspection and part replacement procedures.
5 Air bleed holes clogged. Remove and overhaul carburetor (Chapter 4).
6 Air cleaner clogged, poorly sealed, or missing.
7 Air cleaner-to-carburetor boot poorly sealed. Look for cracks, holes or loose clamps, and replace or repair defective parts.
8 Float level too high or too low. Check float level and replace the float if necessary (Chapter 4).
9 Fuel tank air vent obstructed. Make sure the air vent passage in the filler cap is open.
10 Carburetor intake joint loose. Check for cracks, breaks, tears or loose clamps or bolts. Repair or replace the rubber boots (Chapter 4).
11 Fuel tap clogged. Remove the tap and clean it (Chapter 1).
12 Fuel line clogged. Pull the fuel line loose and carefully blow through it.

16 Compression low

1 Spark plug loose. Remove the plug and inspect the threads. Reinstall and tighten to the specified torque (Chapter 1).
2 Cylinder head not sufficiently tightened down. If the cylinder head is suspected of being loose, then there's a chance that the gasket and head are damaged if the problem has persisted for any length of time. The head nuts and bolts should be tightened to the proper torque in the correct sequence (Chapter 2).
3 Improper valve clearance (4-stroke). This means that the valve is not closing completely and compression pressure is leaking past the valve. Check and adjust the valve clearances (Chapter 1).
4 Cylinder and/or piston worn. Excessive wear will cause compression pressure to leak past the rings. This is usually accompanied by worn rings as well. A top end overhaul is necessary (Chapter 2).
5 Piston rings worn, weak, broken, or sticking. Broken or sticking piston rings usually indicate a lubrication or carburetion problem that causes excess carbon deposits or seizures to form on the pistons and rings. Top end overhaul is necessary (Chapter 2).
6 Piston ring-to-groove clearance excessive. This is caused by excessive wear of the piston ring lands. Piston replacement is necessary (Chapter 2).
7 Cylinder head gasket damaged. If a head is allowed to become loose, or if excessive carbon build-up on the piston crown and combustion chamber causes extremely high compression, the head gasket may leak. Retorquing the head is not always sufficient to restore the seal, so gasket replacement is necessary (Chapter 2).
8 Cylinder head warped. This is caused by overheating or improperly tightened head nuts and bolts. Machine shop resurfacing or head replacement is necessary (Chapter 2).
9 Valve spring broken or weak (4-stroke). Caused by component failure or wear; the spring(s) must be replaced (Chapter 2).
10 Valve not seating properly (4-stroke). This is caused by a bent valve (from over-revving or improper valve adjustment), burned valve or seat (improper carburetion) or an accumulation of carbon deposits on the seat (from carburetion or lubrication problems). The valves must be cleaned and/or replaced and the seats serviced if possible (Chapter 2).

17 Knocking or pinging

1 Carbon build-up in combustion chamber. Use of a fuel additive that will dissolve the adhesive bonding the carbon particles to the crown and chamber is the easiest way to remove the build-up. Otherwise, the cylinder head will have to be removed and decarbonized (Chapter 2).
2 Incorrect or poor quality fuel. Old or improper grades of fuel can cause detonation. This causes the piston to rattle, thus the knocking or pinging sound. Drain old fuel and always use the recommended fuel grade.
3 Spark plug heat range incorrect. Uncontrolled detonation indicates the plug heat range is too hot. The plug in effect becomes a glow plug, raising cylinder temperatures. Install the proper heat range plug (Chapter 1).
4 Improper air/fuel mixture. This will cause the cylinder to run hot, which leads to detonation. Clogged jets or an air leak can cause this imbalance. See Chapter 4.

18 Miscellaneous causes

1 Throttle valve doesn't open fully. Adjust the cable slack (Chapter 1).
2 PVT belt slipping. May be caused by overheating or water on the belt. Check cooling air flow path for obstructions. Check PVT housing gaskets for leaks.
3 Timing not advancing.
4 Engine oil viscosity too high. Using a heavier oil than the one recommended in Chapter 1 can damage the oil pump or lubrication system and cause drag on the engine.
5 Brakes dragging. Usually caused by debris which has entered the brake piston sealing boot (hydraulic brakes), or from a warped disc or bent axle. Repair as necessary.

Overheating

19 Engine overheats

1 Coolant level low (liquid-cooled models). Check and add coolant (Chapter 1), then look for leaks (Chapter 3).
2 Crankcase oil level low (4-stroke models) or injector oil tank level low (2-stroke models). Check and add oil (Chapter 1).
3 Wrong type of oil (4-stroke models). If you're not sure what type

Troubleshooting 0-35

of oil is in the engine, drain it and fill with the correct type (Chapter 1).
4   Cooling fan not working. Check fan operation and repair as necessary (Chapter 4).
5   Air leak at carburetor intake joints. Check and tighten or replace as necessary (Chapter 3).
6   Worn oil pump or clogged oil passages. Replace pump or clean passages as necessary.
7   Carbon build-up in combustion chambers. Use of a fuel additive that will dissolve the adhesive bonding the carbon particles to the piston crown and chambers is the easiest way to remove the build-up. Otherwise, the cylinder head will have to be removed and decarbonized (Chapter 2).

20 Firing incorrect

1   Spark plug fouled, defective or worn out. See Chapter 1 for spark plug maintenance.
2   Incorrect spark plug (see Chapter 1).
3   Faulty ignition coil(s) (Chapter 4).

21 Fuel/air mixture incorrect

1   Pilot screw out of adjustment (Chapter 3).
2   Main jet clogged. Dirt, water and other contaminants can clog the main jets. Clean the fuel tap strainer and in-tank strainer, the float bowl area and the jets and carburetor orifices (Chapter 3).
3   Main jet wrong size. The standard jetting is for sea level atmospheric pressure and oxygen content. See Chapter 3 for high altitude settings.
4   Air cleaner poorly sealed or missing.
5   Air cleaner-to-carburetor boot poorly sealed. Look for cracks, holes or loose clamps and replace or repair.
6   Fuel level too low. Check float level and replace the float if necessary (Chapter 4).
7   Fuel tank air vent obstructed. Make sure that the air vent passage in the filler cap is open.
8   Carburetor intake joint loose. Check for cracks, breaks, tears or loose clamps or bolts. Repair or replace the rubber boot (Chapter 4).

22 Compression too high

1   Carbon build-up in combustion chamber. Use of a fuel additive that will dissolve the adhesive bonding the carbon particles to the piston crown and chamber is the easiest way to remove the build-up. Otherwise, the cylinder head will have to be removed and decarbonized (Chapter 2).
2   Improperly machined head surface or installation of incorrect gasket during engine assembly.

23 Engine load excessive

1   PVT belt slipping. Can be caused by damaged, loose or worn PVT components. Refer to Chapter 2 for overhaul procedures.
2   Engine oil level too high (4-stroke). The addition of too much oil will cause pressurization of the crankcase and inefficient engine operation. Check Specifications and drain to proper level (Chapter 1).
3   Engine oil viscosity too high (4-stroke). Using a heavier oil than the one recommended in Chapter 1 can damage the oil pump or lubrication system as well as cause drag on the engine.
4   Brakes dragging. Usually caused by debris which has entered the brake caliper sealing boots (hydraulic disc brakes), corroded calipers (hydraulic disc brakes), sticking brake cam (front drum or mechanical rear brakes) or from a warped disc, warped drum or bent axle. Repair as necessary (Chapter 6).

24 Lubrication inadequate

1   Engine oil level too low (4-stroke) or oil tank level too low (2-stroke). Friction caused by intermittent lack of lubrication or from oil that is overworked can cause overheating. The oil provides a definite cooling function in the engine. Check the oil level (Chapter 1).
2   Poor quality engine oil or incorrect viscosity or type. Oil is rated not only according to viscosity but also according to type. Some oils are not rated high enough for use in this engine. Check the Specifications section and change to the correct oil (Chapter 1).
3   Camshaft or journals worn (4-stroke). Excessive wear causing drop in oil pressure. Replace cam or cylinder head. Abnormal wear could be caused by oil starvation at high rpm from low oil level or improper viscosity or type of oil (Chapter 1).
4   Crankshaft and/or bearings worn. Same problems as paragraph 3. Check and replace crankshaft assembly if necessary (Chapter 2).

25 Miscellaneous causes

Modification to exhaust system. Most aftermarket exhaust systems cause the engine to run leaner, which makes it run hotter. When installing an accessory exhaust system, always rejet the carburetor.

Polaris Variable Transmission (PVT) problems

26 Drive belt slipping

1   Too much belt deflection. Adjust the belt (Chapter 1).
2   Worn belt. Replace the belt (Chapter 2).
3   Oil or grease on belt. Clean belt and check for leaking seals (Chapter 2).
4   Water (not engine coolant) on belt. Check PVT cover for proper sealing.

27 Drive belt upside down in pulleys

1   Wrong drive belt for that model ATV. Check part number on belt and replace with the correct belt if necessary.
2   Clutch out of alignment. Check the alignment and correct if necessary (Chapter 2).
3   Loose or broken engine mount. Check and tighten or replace as necessary (Chapter 2).

28 Burn marks or thin spots on drive belt

1   Excessive load on vehicle (weight on racks, heavy trailer, oversized accessory). Remove excessive load.
2   Brakes dragging. Usually caused by debris which has entered the brake caliper sealing boots (hydraulic disc brakes), corroded calipers (hydraulic disc brakes), sticking brake cam (front drum or mechanical rear brakes) or from a warped disc, warped drum or bent axle. Repair as necessary (Chapter 6).
3   Applying throttle and continuously raising engine speed when the machine is not moving.

29 Harsh engagement

1   Worn drive belt. Replace the belt (Chapter 2).
2   Excessive clearance between belt and sheave with new belt. Adjust the clearance by changing the number of shims (Chapter 2).

Troubleshooting

30 Grabby or erratic engagement

1   Thin spots or overall wear on drive belt. Inspect the belt and replace if necessary (Chapter 2). If there are thin spots, check possible causes described in Section 28 above.
2   Drive clutch bushings sticking. Inspect the bushings and replace if necessary (Chapter 2).

31 Noisy operation

1   Loose belt. Inspect the belt tension and adjust if necessary (Chapter 1).
2   Worn belt or separated belt plies. Inspect the belt and replace it if necessary (Chapter 2).
3   Thin spots on belt. Replace the belt and check for causes of thin spots listed in Section 28 above.

32 Melted PVT cover

1   Air intake or outlet clogged. Check the inlet and outlet for obstructions and clean as necessary.
2   Belt slipping due to contamination and rubbing on cover. Clean away contamination. Check the cover for proper sealing against outside water. Check the engine for sources of any oil or grease leaks.
3   Rotating mechanical components hitting cover. Check for damage and repair as necessary.

33 Engine rpm too low when vehicle is driven

1   Engine out of tune. Tune up engine (Chapter 1).
2   Belt slipping. Inspect belt and adjust or replace as necessary. Clean any excess grease from sheaves.
3   Driven clutch spring broken or installed incorrectly. Inspect the spring and reinstall or replace it.
4   Incorrect drive clutch shift weight for that model ATV. Verify part number and install correct shift weight if necessary.

34 Engine rpm too high when vehicle is driven

1   Incorrect drive clutch shift weight for that model ATV. Verify part number and install correct shift weight if necessary.
2   Incorrect drive clutch spring for that model ATV. Verify part number and install correct spring if necessary.
3   Binding drive clutch. Disassemble the clutch, clean away any dirt and inspect the buttons and shift weights. Reassemble the clutch without the spring and operate it through its full range by hand to check operation.
4   Binding driven clutch. Disassemble the clutch, noting whether the location of the helix spring is correct. Clean away any dirt and inspect the sheave bushing and ramp buttons.

35 Engine rpm erratic when vehicle is driven

1   Thin or burned spots on the drive belt. Replace the belt and check for causes of thin spots listed in Section 28 above.
2   Binding drive clutch. Disassemble the clutch, clean away any dirt and inspect the shift weights. Clean and polish the hub of the stationary shaft, then reassemble the clutch without the spring and operate it through its full range by hand to check operation.
3   Binding driven clutch. Disassemble the clutch, noting whether the location of the helix spring is correct. Clean away any dirt and inspect the sheave bushing and ramp buttons.
4   Wear groove in a sheave face. Replace the affected clutch.

Gear shifting problems

36 Doesn't go into gear

1   Insufficient or incorrect transmission oil. Check the oil and add or change it as necessary (Chapter 1).
2   Incorrect shift linkage adjustment. Check and adjust as necessary (Chapter 2).
3   Wear or damage to external linkage or internal transmission components. To isolate the problem, disconnect the linkage rods from the transmission. Shift into gear by hand and operate the vehicle. If it now operates correctly, the problem is in the linkage. If not, the problem is internal.
4   Linkage problem. Recheck and verify proper adjustment. Check for worn linkage rod ends, bent linkage rods or damaged bellcranks. Replace damaged components and readjust as necessary (Chapter 2).
5   Internal problem. Disassemble and inspect transmission. Replace worn or damaged components (Chapter 2).

Abnormal engine noise

37 Knocking or pinging

1   Carbon build-up in combustion chamber. Use of a fuel additive that will dissolve the adhesive bonding the carbon particles to the piston crown and chamber is the easiest way to remove the build-up. Otherwise, the cylinder head will have to be removed and decarbonized (Chapter 2).
2   Incorrect or poor quality fuel. Old or improper fuel can cause detonation. This causes the piston to rattle, thus the knocking or pinging sound. Drain the old fuel (Chapter 3) and always use the recommended grade fuel (Chapter 1).
3   Spark plug heat range incorrect. Uncontrolled detonation indicates that the plug heat range is too hot. The plug in effect becomes a glow plug, raising cylinder temperatures. Install the proper heat range plug (Chapter 1).
4   Improper air/fuel mixture. This will cause the cylinder to run hot and lead to detonation. Clogged jets or an air leak can cause this imbalance. See Chapter 4.

38 Piston slap or rattling

1   Cylinder-to-piston clearance excessive. Caused by improper assembly. Inspect and overhaul top end parts (Chapter 2).
2   Connecting rod bent. Caused by over-revving, trying to start a badly flooded engine or from ingesting a foreign object into the combustion chamber. Replace the damaged parts (Chapter 2).
3   Piston pin or piston pin bore worn or seized from wear or lack of lubrication. Replace damaged parts (Chapter 2).
4   Piston ring(s) worn, broken or sticking. Overhaul the top end (Chapter 2).
5   Piston seizure damage. Usually from lack of lubrication or overheating. Replace the pistons and bore the cylinder, as necessary (Chapter 2).
6   Connecting rod upper or lower end clearance excessive. Caused by excessive wear or lack of lubrication. Replace worn parts.

39 Valve noise (4-stroke)

1   Incorrect valve clearances. Adjust the clearances by referring to Chapter 1.
2   Valve spring broken or weak. Check and replace weak valve

springs (Chapter 2).
3 Camshaft or cylinder head worn or damaged. Lack of lubrication at high rpm is usually the cause of damage. Insufficient oil or failure to change the oil at the recommended intervals are the chief causes.

40 Other noise

1 Cylinder head gasket leaking.
2 Exhaust pipe leaking at cylinder head connection. Caused by improper fit of pipe, damaged gasket or loose exhaust flange. All exhaust fasteners should be tightened evenly and carefully. Failure to do this will lead to a leak.
3 Crankshaft runout excessive. Caused by a bent crankshaft (from over-revving) or damage from an upper cylinder component failure.
4 Engine mounting bolts or nuts loose. Tighten all engine mounting bolts and nuts to the specified torque (Chapter 2).
5 Crankshaft bearings worn (Chapter 2).
6 Camshaft chain tensioner defective (4-stroke). Replace according to the procedure in Chapter 2.
7 Camshaft chain, sprockets or guides worn (4-stroke) (Chapter 2).

Abnormal driveline noise

41 Chain noise

1 Dry or dirty chain. Inspect, clean and lubricate (Chapter 1).
2 Chain out of adjustment. Adjust chain slack (Chapter 1).
3 Chain and sprockets damaged or worn. Inspect the chain and sprockets and replace them as necessary (Chapter 6).
4 Sprockets loose (Chapter 6).

42 Polaris Variable Transmission (PVT) noise

1 Loose belt. Inspect the belt tension and adjust if necessary (Chapter 1).
2 Worn belt or separated belt plies. Inspect the belt and replace it if necessary (Chapter 2).
3 Thin spots on belt. Replace the belt and check for causes of thin spots listed in Section 28 above.

43 Transmission noise

1 Bearings worn. Also includes the possibility that the shafts are worn. Overhaul the transmission (Chapter 2).
2 Gears or chain worn or chipped (Chapter 2).
3 Metal chips jammed in gear teeth. Probably pieces from a broken gear or shift mechanism that were picked up by the gears. This will cause early bearing failure (Chapter 2).
4 Transmission oil level too low. Causes a howl from transmission. Also affects engine power and PVT operation (Chapter 1).

44 Front gearcase noise (shaft drive models)

1 Bearings worn. Also includes the possibility that the shafts are worn. Overhaul the gearcase (Chapter 6).
2 Gears worn or chipped (Chapter 6).
3 Metal chips jammed in gear teeth. This will cause early bearing failure (Chapter 6).
4 Gearcase oil level too low. Causes a howl from gearcase. Also affects engine power and PVT operation (Chapter 1).

Abnormal chassis noise

45 Suspension noise

1 Spring weak or broken. Makes a clicking or scraping sound.
2 Steering shaft bearings worn or damaged. Clicks when braking. Check and replace as necessary (Chapter 5).
3 Shock absorber fluid level incorrect. Indicates a leak caused by defective seal. Shock will be covered with oil. Replace shock (Chapter 5).
4 Defective shock absorber with internal damage (four-wheel models). This is in the body of the shock and can't be remedied. The shock must be replaced with a new one (Chapter 5).
5 Bent or damaged shock body (four-wheel models). Replace the shock with a new one (Chapter 5).
6 Low fork oil or internal fork damage (three-wheel models). Inspect the fork and add oil or repair as necessary (Chapter 6).

46 Driveaxle noise

1 Worn or damaged outer joint. Makes clicking noise in turns. Check for cut or damaged seals and repair as necessary (see Chapter 6).
2 Worn or damaged inner joint. Makes knock or clunk when accelerating after coasting. Check for cut or damaged seals and repair as necessary (see Chapter 6).

47 Brake noise

1 Squeal caused by disc brake pad shim not installed or installed incorrectly (Chapter 7).
2 Squeal caused by dust on disc brake pads. Usually found in combination with glazed pads. Clean using brake cleaning solvent (Chapter 7). If the pads are glazed, replace them.
3 Contamination of disc brake pads. Oil, brake fluid or dirt causing pads to chatter or squeal. Clean or replace pads (see Chapter 7).
4 Disc brake pads glazed. Caused by excessive heat from prolonged use or from contamination. Do not use sandpaper, emery cloth or carborundum cloth or any other abrasives to roughen pad surface as abrasives will stay in the pad material and damage the disc, A very fine flat file can be used, but pad replacement is suggested as a cure (see Chapter 7).
5 Disc warped. Can cause a chattering, clicking or intermittent squeal. Usually accompanied by a pulsating lever and uneven braking. Replace the disc (see Chapter 7).
8 Drum brake linings worn or contaminated. Can cause scraping or squealing. Replace the shoes (Chapter 7).
9 Drum brake linings warped or worn unevenly. Can cause chattering. Replace the linings (Chapter 7).
10 Brake drum out of round. Can cause chattering. Replace brake drum (Chapter 7).
11 Loose or worn knuckle or rear axle bearings. Check and replace as needed (Chapter 6).

Oil level light comes on (2-stroke)

48 Engine lubrication system

Low oil level in tank. Check and add the specified oil (see Chapter 1).

49 Electrical system

1 Oil level sensor defective. Check the sensor according to the procedure in Chapter 8. Replace it if it's defective.
2 Oil level indicator light circuit defective. Check for pinched, shorted, disconnected or damaged wiring (Chapter 8).

Excessive exhaust smoke

50 White smoke

1 Piston oil ring worn (4-stroke). The ring may be broken or damaged, causing oil from the crankcase to be pulled past the piston into the combustion chamber. Replace the rings with new ones (Chapter 2).
2 Cylinders worn, cracked, or scored. Caused by overheating or oil starvation. If worn or scored, the cylinders will have to be rebored and new pistons installed. If cracked, the cylinder block will have to be replaced (see Chapter 2).
3 Valve oil seal damaged or worn (4-stroke). Replace oil seals with new ones (Chapter 2).
4 Valve guide worn (4-stroke). Perform a complete valve job (Chapter 2).
5 Four-stroke engine oil level too high, which causes the oil to be forced past the rings. Drain oil to the proper level (Chapter 1).
6 Head gasket broken between oil return and cylinder (4-stroke). Causes oil to be pulled into the combustion chamber. Replace the head gasket and check the head for warpage (Chapter 2).
7 Abnormal crankcase pressurization, which forces oil past the rings (4-stroke). Clogged breather or hoses usually the cause (Chapter 2).
8 Oil/fuel mixture too rich (2-stroke). Check oil injection pump adjustment (Chapter 2).

51 Black smoke

1 Air cleaner clogged. Clean or replace the element (Chapter 1).
2 Main jet too large or loose. Compare the jet size to the Specifications (Chapter 4).
3 Choke stuck, causing fuel to be pulled through choke circuit (Chapter 4).
4 Fuel level too high. Check the float level and replace the float if necessary (Chapter 4).
5 Inlet needle held off needle seat. Clean the float chamber and fuel line and replace the needle and seat if necessary (Chapter 4).

52 Brown smoke

1 Main jet too small or clogged. Lean condition caused by wrong size main jet or by a restricted orifice. Clean float chamber and jets and compare jet size to Specifications (Chapter 4).
2 Fuel flow insufficient. Fuel inlet needle valve stuck closed due to chemical reaction with old fuel. Float level incorrect; check and replace float if necessary. Restricted fuel line. Clean line and float chamber.
3 Carburetor intake tube loose (Chapter 4).
4 Air cleaner poorly sealed or not installed (Chapter 1).

Poor handling or stability

53 Handlebar hard to turn

1 Steering shaft nut too tight (Chapter 6).
2 Steering shaft bushing(s) damaged. Roughness can be felt as the bars are turned from side-to-side. Replace bushing(s) (Chapter 6).
3 Steering shaft bent. Caused by a collision, hitting a pothole or by rolling the machine. Replace damaged part. Don't try to straighten the steering shaft (Chapter 5).
4 Front tire air pressure too low (Chapter 1).

54 Handlebar shakes or vibrates excessively

1 Tires worn or out of balance (Chapter 1 or 6).
2 Swingarm bearings worn. Replace worn bearings by referring to Chapter 6.
3 Wheel rim(s) warped or damaged. Inspect wheels (Chapter 6).
4 Wheel bearings worn. Worn front or rear wheel bearings can cause poor tracking. Worn front bearings will cause wobble (Chapter 6).
5 Wheel hubs installed incorrectly (Chapter 6).
6 Handlebar clamp bolts or bracket nuts loose (Chapter 6).
7 Steering shaft nut or bolts loose. Tighten them to the specified torque (Chapter 6).
8 Motor mount bolts loose. Will cause excessive vibration with increased engine rpm (Chapter 2).

55 Handlebar pulls to one side

1 Uneven tire pressures (Chapter 1).
2 Frame bent. Definitely suspect this if the machine has been rolled. May or may not be accompanied by cracking near the bend. Replace the frame (Chapter 8).
3 Wheel out of alignment. Caused by incorrect toe-in adjustment (Chapter 1) or bent tie-rod (Chapter 6).
4 Swingarm bent or twisted. Caused by age (metal fatigue) or impact damage. Replace the swingarm (Chapter 6).
5 Steering shaft bent. Caused by impact damage or by rolling the vehicle. Replace the steering stem (Chapter 6).

56 Poor shock absorbing qualities

1 Too hard:
   a) *Shock internal damage.*
   b) *Tire pressure too high (Chapters 1 and 7).*
2 Too soft:
   a) *Shock oil insufficient and/or leaking (Chapter 6).*
   b) *Fork springs weak or broken (Chapter 6).*

Braking problems

57 Front brakes are spongy, don't hold (hydraulic brakes)

1 Air in brake line. Caused by inattention to master cylinder fluid level or by leakage. Locate problem and bleed brakes (Chapter 6).
2 Linings worn (Chapters 1 and 6).
3 Brake fluid leak. See paragraph 1.
4 Contaminated linings. Caused by contamination with oil, grease, brake fluid, etc. Clean or replace linings. Clean drum thoroughly with brake cleaner (Chapter 6).
5 Brake fluid deteriorated. Fluid is old or contaminated. Drain system, replenish with new fluid and bleed the system (Chapter 6).
6 Master cylinder internal parts worn or damaged causing fluid to bypass (Chapter 6).
7 Master cylinder bore scratched by foreign material or broken spring. Repair or replace master cylinder (Chapter 6).
8 Disc warped. Replace drum (Chapter 6).

58 Brake lever pulsates

1. Axle bent. Replace axle (Chapter 6).
2. Wheel warped or otherwise damaged (Chapter 7).
3. Hub or axle bearings damaged or worn (Chapter 7).
4. Brake disc warped or drum out of round. Replace brake disc or drum (Chapter 6).

59 Brakes drag

1. Master cylinder piston seized (hydraulic brakes). Caused by wear or damage to piston or cylinder bore (Chapter 7).
2. Lever balky or stuck. Check pivot and lubricate (Chapter 7).
3. Caliper piston seized in bore (hydraulic brakes). Caused by wear or ingestion of dirt past deteriorated seal (Chapter 7).
4. Brake shoes damaged (drum brakes). Lining material separated from shoes. Usually caused by faulty manufacturing process or from contact with chemicals. Replace shoes (Chapter 7).
5. Shoes improperly installed (Chapter 7).
6. Rear brake pedal or lever free play insufficient (Chapter 1).
7. Rear brake springs weak. Replace brake springs (Chapter 7).
8. Rear brake adjuster problem (mechanical rear brake). Disassemble and clean adjuster; replace damaged parts as necessary (see Chapter 7).

Electrical problems

60 Battery dead or weak

1. Battery faulty. Caused by sulfated plates which are shorted through sedimentation or low electrolyte level. Also, broken battery terminal making only occasional contact (Chapter 5).
2. Battery cables making poor contact (Chapter 5).
3. Load excessive. Caused by addition of high wattage lights or other electrical accessories.
4. Ignition switch defective. Switch either grounds internally or fails to shut off system. Replace the switch (Chapter 5).
5. Regulator/rectifier defective (Chapter 5).
6. Stator coil open or shorted (Chapter 5).
7. Wiring faulty. Wiring grounded or connections loose in ignition, charging or lighting circuits (Chapter 5).

61 Battery overcharged

1. Regulator/rectifier defective. Overcharging is noticed when battery gets excessively warm or boils over (Chapter 5).
2. Battery defective. Replace battery with a new one (Chapter 5).
3. Battery amperage too low, wrong type or size. Install manufacturer's specified amp-hour battery to handle charging load (Chapter 5).

Notes

Chapter 1
Tune-up and routine maintenance

Contents

	Section		Section
Air cleaner drain tube - cleaning	5	Front gearcase oil (shaft drive models) - change	16
Air cleaner - pre-filter check and main element replacement	4	Front hub fluid - change	29
Battery - electrolyte level/specific gravity - check	7	Fuel system - check and filter cleaning	17
Brake fluid - change	33	Idle speed and throttle operation - check and adjustment	12
Brake lever and pedal freeplay - check and adjustment	9	Ignition timing - check	19
Brake system - general check	8	Introduction to tune-up and routine maintenance	2
Choke - check and freeplay adjustment	13	Lubrication - general	14
Cooling system - draining, flushing and refilling	31	PVT - check and clean	32
Cooling system - inspection	30	Recoil housing - draining	6
Counterbalancer fluid - change	28	Routine maintenance intervals	1
Cylinder compression - check	21	Spark plug - replacement	20
Drive chain and sprockets - check, adjustment and lubrication	26	Steering system - inspection and toe-in adjustment	25
Driveaxle boots - inspection	11	Suspension - check	24
Engine oil/filter - change	15	Tires/wheels - general check	10
Exhaust system - inspection	18	Transmission oil - change	27
Fasteners - check	23	Valve clearances - check and adjustment	22
Fluid levels - check	3		

Specifications

Engine

Ignition timing check speed
    All except 425 and 500 .... 3000 rpm
    425 and 500 .... 3500 rpm

Chapter 1 Tune-up and routine maintenance

Spark plugs
1985 through 1988
    Type .... NGK BPR8ES, Champion RN4YC
    Gap .... 0.020 inch
1988
    Type .... Champion RN4YC
    Gap .... 0.0-25 inch
1989
    Type .... NGK BR8ES, Champion RN4YC
    Gap .... 0.028 inch
1990 through 1995 (2-stroke)
    Type .... NGK BR8ES
    Gap .... 0.028 inch
1990 through 1995 (4-stroke)
    Type .... NGK BR6ES
    Gap .... 0.025 inch
1996-on (2-stroke)
    Type .... NGK BR8ES
    Gap .... 0.028 inch
1996-on (4-stroke)
    Type .... NGK BR5ES
    Gap .... 0.025 inch
Engine idle speed
    1985 through 1989 models .... 800 rpm
    1990-on
        2-stroke.. 700 rpm
        4-stroke.. 1200 rpm
Compression (2-stroke) .... 115 psi minimum
Cylinder leakage (4-stroke) .... 10 percent maximum
Valve clearance (COLD engine) .... 0.006 inch

Miscellaneous
Drum brake shoe lining thickness .... Not specified
Mechanical rear caliper pad thickness .... Replace when worn to indicator grooves
Hydraulic caliper pad thickness limit
    1985 through 1995 .... 0.075 inch
    1996-on .... 3/64 -inch
Front brake lever freeplay .... 3/32 to 1/8 inch
Rear brake pedal freeplay
    Mechanical rear caliper ....
    Auxiliary brake .... 1/2 to 3/4 inch
Rear brake lever freeplay .... 1/4 inch
Throttle lever freeplay .... 1/16 inch
Choke freeplay .... 1/8 to 1/4 inch
Minimum tire tread depth .... 1/8 inch

Tire pressures (psi, cold)
Trail Boss 250, Trail Boss 2x4, Trail Blazer, 2x4 300 .... 3 (all wheels)
All other models with four wheels except Sportsman 500
    Front 4
    Rear 3
Sportsman 500 .... 5 (all wheels)
Six wheel models (4x6 and 6x6) .... 5 (all wheels)
Drive chain slack
    Rear chain (four-wheel models) .... 1/4 inch (1)
    Center front and front chains .... 1/4 to 1/2 inch
    Center rear chain (six-wheel models) .... 1-1/4 to 1-1/2 inch
    Rear chain (six-wheel models) .... 1/4 to 1/2 inch
Battery specific gravity Minimum) 1.270
Front wheel toe-out .... 1/8 to 1/4-inch

Chapter 1 Tune-up and routine maintenance

Torque specifications

Ft-lbs (unless otherwise indicated)

Oil tank drain plug	14 ft-lbs
Crankcase oil drain plug (4-stroke)	14 ft-lbs
Valve adjusting screw locknuts	5.8 to 7.2 ft-lbs (2)
Spark plugs	
New	11 ft-lbs
Used	18 ft-lbs
Front gear case filler and drain plug	Not specified
Transmission drain plug	14 ft-lbs
Tie rod locknuts	12 to 14 ft-lbs
Rear chain eccentric lockbolts and nuts	60 ft-lbs
Center and front chain eccentric lockbolts and nuts	50 ft-lbs
Master cylinder cover screws (metal reservoir)	45 inch-lbs

Recommended lubricants and fluids

Engine	
2-stroke models	
Type	Polaris 2-stroke oil or equivalent
Capacity	2 qt
4-stroke models	
Type	Polaris synthetic 0W-40 or API grade SH 5W-30
Capacity	2 qt
Transmission oil	
Type	Motor oil, API grade SG or SH
Viscosity	10W-30
Capacity	
Type III with high, low and reverse	20 fl oz
Type III with high and reverse	16 fl oz
Type IV	32 fl oz
Counterbalancer fluid	SAE 10W-30 engine oil
Capacity	3-1/4 fl oz
Front gearcase oil	
Type	Non-hypoid gear oil
Viscosity	SAE 80-90, API GL-5
Capacity	3-1/4 fl oz
Shift selector box	
Type	Polaris 0W-40 engine oil or 10W non-detergent engine oil
Capacity	
Air cleaner (foam element)	2-stroke oil or foam filter oil
Front hubs	Polaris Demand Drive Hub Fluid or Type F automatic transmission fluid
Coolant	Ethylene glycol-based antifreeze and water, 60/40 mixture
Brake fluid	DOT 3 or DOT 4
Level (metal reservoir)	Within 1/4 inch of the top

Miscellaneous

Drive chain	O-ring chain lube
Greasing points	Medium weight, lithium-based multi-purpose grease (NLGI no. 2)
Cables and lever pivots	Chain and cable lubricant or 10W30 motor oil

1. *The rear suspension must be compressed so the countershaft, swingarm pivot and rear axle are in a straight line.*
2. *Using special Polaris torque wrench adapter.*

Polaris ATV Routine maintenance intervals

Routine maintenance intervals

Note: *The pre-ride inspection outlined in the owner's manual covers checks and maintenance that should be carried out on a daily basis. It's condensed and included here to remind you of its importance. Always perform the pre-ride inspection at every maintenance interval (in addition to the procedures listed). The intervals listed below are the shortest intervals recommended by the manufacturer for each particular operation during the model years covered in this manual. Your owner's manual may have different intervals for your model.*

Daily or before riding

Clean the air cleaner pre-filter and the air box sediment tube
Check the coolant level
Check the operation of the headlight, taillight and brake light
Check the fuel level and oil tank level
Check the operation of both brakes - check the brake fluid level and look
for leakage; check the brake lever and pedal for correct freeplay
Check the tires for damage, the presence of foreign objects and correct air pressure
Check the throttle for smooth operation and correct freeplay
Make sure the steering operates smoothly
Check for proper operation of the headlight, taillight, brake light, indicator lights and speedometer (if equipped)
Make sure the engine kill switch works properly
Check the driveaxle boots for damage or deterioration
Make sure any cargo is properly loaded and securely fastened
Check all fasteners, including wheel nuts and axle nuts, for tightness; make sure axle nut cotter pins are in place
Check the underbody for mud or debris that could start a fire or interfere with vehicle operation

Weekly

Check the main air cleaner element and replace it if necessary
Drain the recoil starter housing*
Every 10 operating hours, 100 miles or monthly
Check the brake pads for wear
Every 20 operating hours or monthly
Inspect the battery; top up the electrolyte and clean the terminals if necessary
Every 25 operating hours
Check the fuel system for leaks
Every 25 operating hours or monthly
Check the transmission oil level and top up if necessary; change the transmission oil once a year

Monthly

Lubricate the output shaft bearing

Every 50 operating hours or monthly

Check fluid level in the front hubs

Every 50 operating hours or 3 months

Lubricate all fittings, pivot points and control cables

Every 1000 miles or 6 months

Check the front gearcase oil (shaft drive models) and top up if necessary; change the front gearcase oil once a year

Every 50 operating hours or 6 months

Adjust and lubricate the oil pump cable (2-stroke models) (see Chapter 2 Part A)
Lubricate the throttle cable, check its adjustment and check operation of the electronic throttle control system
Lubricate the choke cable and check its adjustment
Drain the carburetor float chamber

Every 100 operating hours or 6 months

Change the engine oil and filter (4-stroke models)
Measure coolant specific gravity and pressure test the cooling system

Every 100 operating hours or 12 months

Change the oil filter (2-stroke models)
Check the valve clearance (4-stroke models)
Clean and gap the spark plug; replace it if necessary
Check the ignition timing
Change the counterbalancer fluid
Replace the fuel filter
Clean the outside of the radiator and check it for damage; inspect the cooling system hoses
Inspect the engine mounts and tighten or replace them as necessary (see Chapter 2)
Change the front hub fluid
Inspect and clean the PVT clutches

Every 200 operating hours or 24 months

Check the cleanliness of the fuel system and inspect the fuel lines
Change the transmission oil
Change the brake fluid

*More often if the machine is operated in water.

Chapter 1 Tune-up and routine maintenance

2.1a Decals on the vehicle include maintenance information . . .

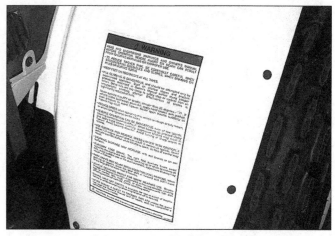

2.1b . . . and safety information

2 Introduction to tune-up and routine maintenance

Refer to illustrations 2.1a and 2.1b

This Chapter covers in detail the checks and procedures necessary for the tune-up and routine maintenance of your vehicle. Section 1 includes the routine maintenance schedule, which is designed to keep the machine in proper running condition and prevent possible problems. The remaining Sections contain detailed procedures for carrying out the items listed on the maintenance schedule, as well as additional maintenance information designed to increase reliability. Maintenance information is also printed on decals, which are mounted in various locations on the vehicle **(see illustrations)**. Where information on the decals differs from that presented in this Chapter, use the decal information.

Since routine maintenance plays such an important role in the safe and efficient operation of your vehicle, it is presented here as a comprehensive check list. For the rider who does all his own maintenance, these lists outline the procedures and checks that should be done on a routine basis.

Deciding where to start or plug into the routine maintenance schedule depends on several factors. If you have a vehicle whose warranty has recently expired, and if it has been maintained according to the warranty standards, you may want to pick up routine maintenance as it coincides with the next mileage or calendar interval. If you have owned the machine for some time but have never performed any maintenance on it, then you may want to start at the nearest interval and include some additional procedures to ensure that nothing important is overlooked. If you have just had a major engine overhaul, then you may want to start the maintenance routine from the beginning. If you have a used machine and have no knowledge of its history or maintenance record, you may desire to combine all the checks into one large service initially and then settle into the maintenance schedule prescribed.

The Sections which actually outline the inspection and maintenance procedures are written as step-by-step comprehensive guides to the actual performance of the work. They explain in detail each of the routine inspections and maintenance procedures on the check list. References to additional information in applicable Chapters is also included and should not be overlooked.

Before beginning any actual maintenance or repair, the machine should be cleaned thoroughly, especially around the oil tank, spark plug, engine covers, carburetor, etc. Cleaning will help ensure that dirt does not contaminate the engine and will allow you to detect wear and damage that could otherwise easily go unnoticed.

3 Fluid levels - check

Engine oil

4-stroke models

Refer to illustrations 3.3a and 3.3b

1 Park the vehicle in a level position, then start the engine and run it for 20 to 30 seconds to achieve the correct level in the oil tank. **Caution:** *Do not run the engine in an enclosed space such as a garage or shop.*
2 Stop the engine.
3 With the engine off, unscrew the dipstick from the oil tank on the left side of the vehicle **(see illustration)**. Pull it out, wipe it off with a

3.3a The coolant reservoir and 4-stroke engine oil dipstick (arrow) are on the left side of the machine

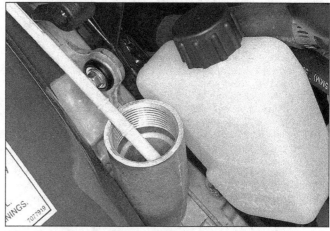

3.3b The engine oil level on 4-strokes must be between the upper and lower marks on the dipstick

1-6 Chapter 1 Tune-up and routine maintenance

3.6 Fill the 2-stroke oil tank through the filler cap

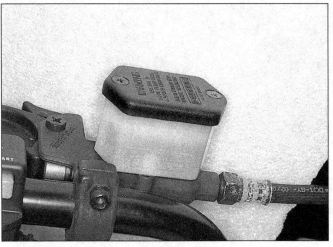

3.11 The brake fluid level on plastic reservoirs must be between the marks on the reservoir; remove the cover to add fluid

clean rag, and screw it all the way back in (don't just let the dipstick rest on the threads). Unscrew the dipstick, pull it out and check the oil level on the dipstick scale. The oil level should be within the normal range on the scale **(see illustration)**.

4   If the level is below the Add 8 oz mark, add oil through the dipstick hole. Add enough oil of the recommended grade and type to bring the level up to the Full mark. Do not overfill. **Note:** *If you're operating the vehicle in cold weather, write down the oil level. If it's higher at the next level check - and you haven't added oil - water from condensation may be collecting in the oil tank.*

2-stroke models

Refer to illustration 3.6

5   The 2-stroke engines used on these models are equipped with a variable-ratio oil injection system. The oil is not pre-mixed with the gasoline; instead, it is stored in a separate tank and injected by a pump into the intake stream.

6   Unscrew the oil tank cap and look into the hole to check the level **(see illustration)**. Fill the tank with the 2-stroke oil listed in this Chapter's Specifications.

Engine coolant

7   Start the engine and warm it to normal operating temperature.

8   Check the coolant level in the reservoir **(see illustration 3.3a)**. It should be between the upper and lower lines. If it's too low, remove the cap from the reservoir. Fill to the upper line with a 50/50 mixture of water and the antifreeze listed in this Chapter's Specifications.

Brake fluid

Refer to illustration 3.11

9   In order to ensure proper operation of the hydraulic brakes on models so equipped, the fluid level in the master cylinder reservoir must be properly maintained.

10 With the vehicle in a level position, turn the handlebars until the top of the brake master cylinder is as level as possible. Before removing the master cylinder cover, place rags beneath the reservoir (to protect the paint from brake fluid spills) and remove all dust and dirt from the area around the cap.

11 If the vehicle has a metal brake fluid reservoir, remove the cover and look at the fluid level inside. It should be the distance below the top of the reservoir listed in this Chapter's Specifications. If the vehicle has a translucent plastic master cylinder reservoir, look through it to check the fluid level. Make sure it's between the Min and Max marks on the reservoir **(see illustration)**.

12 If the level is low, the fluid must be replenished.

13 Remove the cover screws, then lift off the cover, rubber diaphragm and float (if equipped). **Note:** *Don't operate the brake lever with the cover removed.*

14 Add new, clean brake fluid of the recommended type until the level is 1/4-inch below the top (aluminum reservoir) or up to the Max line (plastic reservoir). Don't mix different brands of brake fluid in the reservoir, as they may not be compatible. Also, don't mix different specifications (DOT 3 or 4 with DOT 5).

15 Reinstall the float (if equipped), rubber diaphragm and cover. Tighten the cover screws to the torque listed in this Chapter's Specifications.

16 Wipe any spilled fluid off the reservoir body.

17 If the brake fluid level was low, inspect the brake system for leaks.

Counterbalancer oil (350 and 400 models)

Refer to illustrations 3.19, 3.20a and 3.20b

18 Park the vehicle on a level surface. If it's warmed up, let it cool to room temperature.

19 If you're working on a 350, unscrew the counterbalancer fill plug **(see illustration)**. The oil level should be up to the bottom of the fill plug threads.

20 If you're working on a 400, unscrew the dipstick with a long screwdriver **(see illustration)**. Pull it out, wipe it clean and reinsert it (screw it in, don't just let it rest on the threads). Pull the dipstick back out and check oil level **(see illustration)**. It should be in the knurled area of the dipstick. Add oil if necessary of the type recommended in

3.19 On 350 models, remove the counterbalancer filler plug to check oil level

Chapter 1 Tune-up and routine maintenance

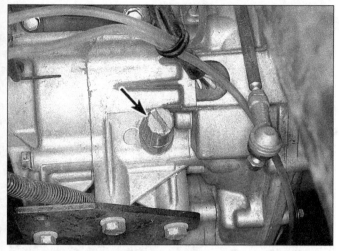

3.20a On 400 models, unscrew the counterbalancer dipstick (arrow) . . .

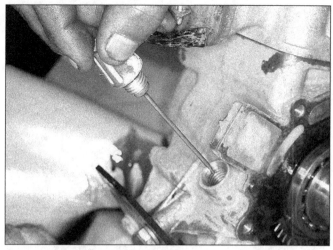

3.20b . . . and check the oil level on the dipstick

this Chapter's Specifications.
21 Reinstall the dipstick and tighten securely.

Front hub fluid (4wd models)

Refer to illustration 3.22

22 Park the vehicle on a level surface and roll it until the hub filler plug is in the 4 o'clock or 8 o'clock position **(see illustration)**.
23 Remove the filler plug with an Allen wrench. Oil should flow out of the hole.
24 If no oil flows, add the fluid listed in this Chapter's Specifications until it just trickles out of the hole. **Caution:** *Don't force oil in with a pressurized filler or the hub seals may be damaged.*
25 After the oil stops flowing, install the filler plug and tighten it securely, but don't overtighten.

Front gearcase oil (shaft drive models)

Refer to illustration 3.27

26 Park the machine on a level surface.
27 Remove the filler plug **(see illustration)**. Oil should flow out of the hole.
28 If no oil flows, add the oil recommended in this Chapter's Specifications until it just trickles out of the hole. **Caution:** *Don't force oil in with a pressurized filler or the hub seals may be damaged.*
29 After the oil stops flowing, install the filler plug and tighten it securely, but don't overtighten.

Transmission oil

Refer to illustration 3.31

30 Park the vehicle in a level position.
31 With the engine off, unscrew the dipstick from the transmission and pull it out **(see illustration)**. If the end of the dipstick is equipped

3.22 Place the hub filler screw (arrow) in the 4 o'clock or 8 o'clock position, then remove it to check fluid level

3.27 Remove the front gearcase filler plug to check oil level and add oil

3.31 Pull out the transmission dipstick to check oil level

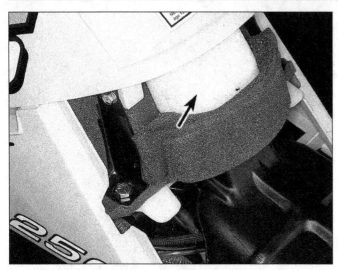

4.1 On some models, the pre-filter (arrow) fits over the end of the intake duct

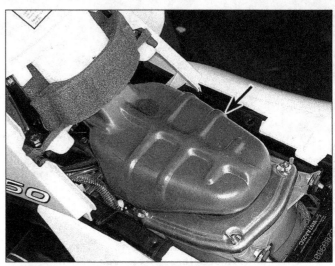

4.9a Separate the intake duct from the air cleaner cover...

4.9b ... remove the wing nuts ...

4.9c ... and lift the cover off ...

with a magnet, clean off any metal filings. Wipe the dipstick off with a clean rag and put it all the way back in (don't thread it in, though). Pull the dipstick out and check the oil level on the dipstick scale. The oil level should be within the normal range on the scale.

32 If the level is below the lower mark, add oil through the dipstick hole. Add enough oil of the recommended grade and type to bring the level up to the Full mark. Do not overfill.

4 Air cleaner - pre-filter check and main element replacement

Pre-filter check

Refer to illustration 4.1

1 An air pre-filter is used on all 1987 and later models. Two designs are used; one fits over the end of the intake duct and the other is slipped over the main element **(see illustration)**.
2 Remove the seat (see Chapter 7).
3 If the pre-filter is located over the upper end of the air cleaner duct, carefully pull it off.
4 If the pre-filter is not on the intake duct, it's in the main air box

body. You'll need to remove the cover for access. Pull out the air filter and slip the foam pre-filter off the outside.
5 Clean the pre-filter in a high flash point solvent, squeeze the solvent out of the foam and let it dry completely.
6 Soak the foam element in the amount and type of oil listed in this Chapter's Specifications, then squeeze it firmly to remove the excess oil.
7 Place the element on the main element or on the end of the intake duct. On models with the element on the end of the intake duct, make sure the wire clip is positioned correctly. Also make sure the air dam doesn't block the duct.
8 The remainder of installation is the reverse of the removal steps.

Main element replacement

Refer to illustrations 4.9a through 4.9g

9 Remove the cover wing nuts and lift off the cover **(see illustrations)**. Some models have two covers, one for access and one to retain the element. On 500 models, the element is secured by a clamp.
10 Lift the element out. If there's a foam pre-filter on the outside of the element, remove and clean it as described above. Tap the element on a hard surface to shake out dirt. If available, blow compressed air from the inside of the filter to the outside. Use a dispersion nozzle and

Chapter 1 Tune-up and routine maintenance

4.9d ... lift out the element retainer and the element

4.9e Remove the access cover wing nuts ...

4.9f ... and the element cover wing nut ...

4.9g ... then lift off the cover and remove the element

don't exceed 40 psi to prevent damage to the element.
11 If the element has been soaked in water, oil or gasoline, replace it with a new one.
12 Check the cover gasket for cracking or deterioration and replace it if necessary.
13 Install the pre-filter (if equipped) on the main filter.
14 Lightly coat the sealing rubber on top and bottom of the filter with multi-purpose grease.
15 Installation is the reverse of the removal steps.

5 Air cleaner drain tube - cleaning

Refer to illustration 5.2

1 A closed drain tube at the bottom of the air cleaner housing collects water and crankcase vapors that accumulate in the air cleaner housing.
2 Check the drain tube for accumulated water and oil **(see illustration)**. If oil or water has built up in the tube, remove its plug and let it drain. Reinstall the plug.

5.2 Remove the plug from the bottom of the air cleaner drain tube to let out oil and water

Chapter 1 Tune-up and routine maintenance

6.2 Two-stroke drain plugs

- A Recoil housing drain plug
- B Counterbalancer drain plug location (350 and 400 models)
- C Crankcase drain plug

6 Recoil housing - draining

Refer to illustration 6.2

1 The recoil starter housing should be drained of water at the specified interval.
2 To drain the housing, remove its plug **(see illustration)**.
3 Reinstall the plug and tighten securely.

7 Battery - electrolyte level/specific gravity - check

Warning: *Be extremely careful when handling or working around the battery. The electrolyte is very caustic and an explosive gas (hydrogen) is given off when the battery is charging.*

Refer to illustrations 7.2, 7.3 and 7.7

1 This procedure applies to batteries that have removable filler caps, which can be removed to add water to the battery. If the original equipment battery has been replaced by a sealed maintenance-free battery, specific gravity can't be checked and water can't be added. The state of charge on a sealed battery can be checked using an open-circuit voltage test, which is described in Chapter 8.
2 Locate the battery under the rear fender. Detach the retaining

7.3 Always undo the negative cable first and reconnect it last

- A Negative cable
- B Positive cable
- C Vent hose

7.2 Unhook the rubber retainer strap and lift the cover off the battery

strap and remove the cover **(see illustration)**.
3 Remove the screws securing the battery cables to the battery terminals (remove the negative cable first, positive cable last). Detach the vent hose, then remove the battery **(see illustration)**. **Warning:** *Always disconnect the negative cable first and reconnect it last to prevent sparks which could cause a battery explosion.*
4 The electrolyte level is visible through the translucent battery case - it should be between the Upper and Lower marks **(see illustration 7.2)**.
5 If the electrolyte level is low, remove the cell caps and fill each cell to the upper level mark with distilled water (don't add more electrolyte). Do not use tap water (except in an emergency) and do not overfill. The cell holes are quite small, so it may help to use a plastic squeeze bottle with a small spout to add the water. If the level is within the marks on the case, additional water is not necessary.
6 Next, check the specific gravity of the electrolyte in each cell with a small hydrometer made for motorcycle and ATV batteries. These are available from most dealer parts departments or ATV accessory stores.
7 Remove the caps, draw some electrolyte from the first cell into the hydrometer **(see illustration)**, then note the specific gravity. Compare the reading to the value listed in this Chapter's Specifications. **Note:** *Add 0.004 points to the reading for every 10-degrees F above 68-degrees F - subtract 0.004 points from the reading for every 10-degrees below 68-degrees F.*
8 Return the electrolyte to the appropriate cell and repeat the check for the remaining cells. When the check is complete, rinse the hydrometer thoroughly with clean water.
9 If the specific gravity of the electrolyte in each cell is as specified, the battery is in good condition and is apparently being charged by the

7.7 Check the battery specific gravity with a hydrometer

Chapter 1 Tune-up and routine maintenance

8.8 Check the brake pad friction material (arrow) for wear

machine's charging system.
10 If the specific gravity is low, the battery is not fully charged. This may be due to corroded battery terminals, a dirty battery case, a malfunctioning charging system, or loose or corroded wiring connections. On the other hand, it may be that the battery is worn out, especially if the machine is old, or that infrequent use of the machine prevents normal charging from taking place.
11 Be sure to correct any problems and charge the battery if necessary. Refer to Chapter 8 for additional battery maintenance and charging procedures.
12 Install the battery cell caps, tightening them securely. Reconnect the cables to the battery, attaching the positive cable first and the negative cable last. Make sure to install the insulating boot(s) over the terminal(s).
13 Install all components removed for access and route the battery vent tube correctly. Be very careful not to pinch or otherwise restrict the tube, as the battery may build up enough pressure during normal charging system operation to explode.
14 If the vehicle will be stored for an extended time, fully charge the battery, then disconnect the negative cable before storage.

8 Brake system - general check

1 A routine general check of the brakes will ensure that any problems are discovered and remedied before the rider's safety is jeopardized.
2 Check the brake lever and pedal (if equipped) for loose connections, excessive play, bends, and other damage. Replace any damaged parts with new ones (see Chapter 7).
3 Make sure all brake fasteners are tight. Check the brakes for wear as described below and make sure the fluid level in the reservoir (if equipped) is correct (see Section 3). Look for leaks at the hose connections and check for cracks in the hoses.
4 Make sure the brake light operates when the brake lever is depressed. The brake light switch is not adjustable. If it fails to operate properly, replace it with a new one (see Chapter 8).
5 Operate the brake lever(s). If the machine has mechanical brakes, check for rough or sticky operation. If a problem is found, refer to Section 14 and lubricate the cable(s). If the brakes are hydraulic, check for sponginess or a lever that travels too close to the handlebar. If a problem is found, refer to Chapter 7 and bleed the brakes.
6 Operate the auxiliary brake pedal (if equipped). If operation is rough or sticky, check the brake rod for damage or obstructions (mud, branches, etc.). Lubricate the brake rod pivots.

Front drum brakes

7 Remove the brake drums (see Chapter 7). Inspect the thickness of the lining material on the brake shoes. Polaris doesn't specify a wear limit, but if it's worn to near the metal backing, refer to Chapter 7 and replace the brake shoes.

Disc brakes

Refer to illustration 8.8

8 Look at the edges of the brake pads **(see illustration)**. On some models, you may need to remove the wheels or the caliper so you can see the pads. If they're worn to the wear line or to the limit listed in this Chapter's Specifications, refer to Chapter 7 and replace the pads. Also check the discs for scoring or other damage.

9 Brake lever and pedal freeplay - check and adjustment

Front brake lever (drum brakes)

1 Squeeze the front brake lever and note how far the lever travels. If it exceeds the limit listed in this Chapter's Specifications, adjust the front brakes as described below.
2 Securely block the rear wheels so the vehicle can't roll. Jack up the front end and support it securely on jackstands.
3 Loosen the locknut on the lever adjuster. Turn the adjuster until lever freeplay is within the range listed in this Chapter's Specifications, then tighten the locknut.
4 Tighten the adjuster at each front wheel until the brakes drag slightly when the tire is turned by hand, then back it off slightly. Spin the tire by hand to make sure the brake lining isn't dragging on the drum; if it is, back off the adjuster just enough so the dragging stops. The adjustment should be even at each wheel; you can confirm this by checking the position of the brake cable joint. It should be horizontal. If it's tilted, repeat the adjustments until it becomes horizontal.
5 Remove the jackstands and lower the vehicle.

Mechanical rear brake

6 Check the rear brake lever play at the left handlebar in the same way as for front brake lever play. If it exceeds the limit listed in this Chapter's Specifications, adjust it as described below.
7 Pull back the rubber cover from the handlebar adjuster. Loosen the lockwheel and turn the adjuster until there's a fairly large amount of freeplay at the lever, then operate the lever 15 to 20 times.
8 Roll the vehicle slightly and squeeze the brake lever to make sure the brake is operating.
9 Turn the adjuster at the brake lever to set freeplay to the value listed in this Chapter's Specifications, then tighten the lockwheel.

Auxiliary rear brake

Refer to illustrations 9.11a, 9.11b, 9.11c and 9.11d

10 Models equipped with hydraulic brakes also have an auxiliary mechanical linkage that will operate the rear brake in the event of a hydraulic system failure.
11 Check the play of the auxiliary brake pedal **(see illustration)**. If it exceeds the limit listed in this Chapter's Specifications, place the

9.11a Measure brake pedal height above the floorboard

9.11b The auxiliary brake pedal is adjusted with a bolt and locknut . . .

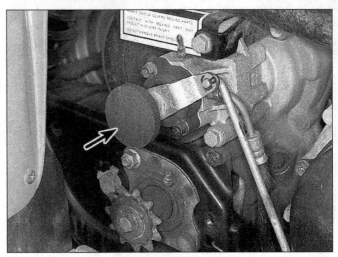

9.11c . . . a knob (arrow) and locknut . . .

9.11d . . . or an Allen bolt (arrow)

10.4 Check tire pressure with a gauge that will read accurately at the low pressures used in ATV tires

transmission in neutral. Loosen the adjuster bolt locknut (if equipped). Turn the adjuster bolt until the rear brake drags when the vehicle is rolled, then back off the adjuster bolt 1/4 turn **(see illustrations)**.
12 Recheck pedal play and repeat the adjustment if necessary.

Hydraulic disc brakes

13 The brakes adjust themselves automatically; there is no means of mechanical adjustment. If there's too much freeplay or the lever comes too close to the handlebar, check fluid level in the master cylinder (Section 3) and bleed the brakes (Chapter 7).

10 Tires/wheels - general check

Refer to illustration 10.4

1 Routine tire and wheel checks should be made with the realization that your safety depends to a great extent on their condition.
2 Check the tires carefully for cuts, tears, embedded nails or other sharp objects and excessive wear. Operation of the vehicle with excessively worn tires is extremely hazardous, as traction and handling are directly affected. Measure the tread depth at the center of the tire and replace worn tires with new ones when the tread depth is less than that listed in this Chapter's Specifications.
3 Repair or replace punctured tires as soon as damage is noted. Do not try to patch a torn tire, as wheel balance and tire reliability may be impaired.
4 Check the tire pressures when the tires are cold and keep them properly inflated **(see illustration)**. Proper air pressure will increase tire life and provide maximum stability and ride comfort. Keep in mind that low tire pressures may cause the tire to slip on the rim or come off, while high tire pressures will cause abnormal tread wear and unsafe handling.
5 The steel wheels used on this machine are virtually maintenance free, but they should be kept clean and checked periodically for cracks, bending and rust. Never attempt to repair damaged wheels; they must be replaced with new ones.
6 Check the valve stem locknuts to make sure they're tight. Also, make sure the valve stem cap is in place and tight. If it is missing, install a new one made of metal or hard plastic.

11 Driveaxle boots - inspection

Refer to illustration 11.1

1 If the vehicle is equipped with driveaxles, clean the rubber boots so they can be inspected **(see illustration)**.
2 Check the boots for cracks, deterioration and loose clamps.

Chapter 1 Tune-up and routine maintenance

11.1 Check the driveaxle boots for cuts, cracks or deteriorated rubber

12.3a Two-stroke pilot screw (left arrow) and throttle stop screw (right arrow)

12.3b Four-stroke throttle stop screw (left arrow) and float chamber drain screw (right arrow)

12.8 Measure throttle freeplay at the lever tip

    A   Locknut   B   Throttle cable adjuster

Make sure the boots haven't been running on any part of the vehicle.
3   Tighten loose clamps. If the boots have any of the other problems, or if the lubricant has leaked out, refer to Chapter 6 for replacement procedures.

12 Idle speed and throttle operation - check and adjustment

Idle speed

Refer to illustrations 12.3a and 12.3b

1   Start the engine and warm it to normal operating temperature. Ten to 15 minutes of stop-and-go riding is usually sufficient.
2   Connect a tune-up tachometer to the engine, following the tachometer manufacturer's instructions.
3   Check idle speed on the tachometer. Adjust if necessary by turning the idle speed screw **(see illustrations)**.
4   After adjusting the idle speed, always check throttle operation and freeplay as described below. If you're working on a 2-stroke, always adjust the oil pump cable as well (Chapter 2 Part A).

Throttle check

5   Make sure the throttle lever moves easily from fully closed to fully open with the front wheel turned at various angles. The lever should return automatically from fully open to fully closed when released. If the throttle sticks, check the throttle cable for cracks or kinks in the housings. Also, make sure the inner cable is clean and well-lubricated.
6   With the engine idling, turn the handlebars all the way in one direction, then all the way in the other direction. Idle speed should not increase. If it does, check the cable for incorrect routing. If the cable is routed correctly, insufficient freeplay may be the problem. **Warning:** *This condition can cause you to lose control of the vehicle. Don't operate it until the problem has been found and corrected.*

Throttle adjustment

Refer to illustrations 12.8, 12.9, 12.13a and 12.13b

7   Engine idle speed must be set correctly before making this adjustment.
8   Minor freeplay adjustments can be made at the throttle lever end of the accelerator cable. Pull back the adjuster boot and loosen the locknut on the cable **(see illustration)**.

1-14   Chapter 1   Tune-up and routine maintenance

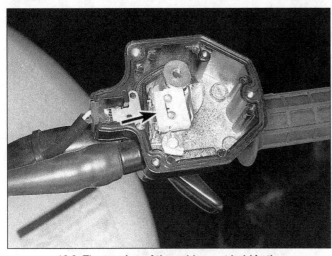

12.9 The tension of the cable must hold in the switch plunger (arrow)

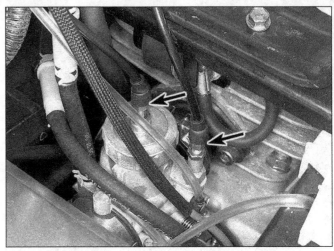

12.13a Two-stroke throttle cable adjuster (left arrow) and choke cable adjuster (right arrow)

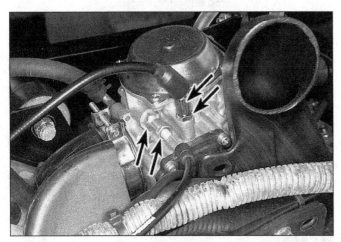

12.13b Four-stroke choke cable adjuster and locknut (left arrows) and throttle cable adjuster and locknut (right arrows)

12.15 Loosen the screw (arrow) to adjust the throttle limiter

9   On later models with electronic throttle control, remove the cover from the throttle housing **(see illustration)**.
10   Operate the throttle lever and check for a small amount of freeplay at the lever (this is the amount the lever moves before engine speed starts to increase). Compare the freeplay to the value listed in this Chapter's Specifications **(see illustration 12.8)**. If it's incorrect, loosen the locknut. Back out the adjuster until engine speed starts to increase, then turn it in until freeplay is as specified. If the vehicle has electronic throttle control, the switch plunger must be held in by the tension of the throttle cable **(see illustration 12.9)**.
11   Once freeplay is set correctly, tighten the locknut. Install the throttle housing cover if it was removed.
12   If the freeplay can't be adjusted at the grip end, adjust the cable at the carburetor end. To do this, first remove the fuel tank (see Chapter 4).
13   Remove the boot from the cable adjuster and loosen the locknut on the throttle cable **(see illustrations)**. Turn the adjusting nut to set freeplay, then tighten the locknut securely.

Throttle lever stop adjustment

Refer to illustration 12.15

14   The throttle lever stop can be used to restrict throttle lever movement.
15   To adjust the stop position, loosen the screw (accessible from beneath the throttle housing) **(see illustration)**. Slide the stop to the desired position and tighten the screw.

13   Choke - check and freeplay adjustment

Check

Refer to illustrations 13.1a, 13.1b, 13.1c and 13.3
1   Operate the choke lever while you feel for smooth operation **(see illustrations)**. If the lever doesn't move smoothly, refer to Section 14

13.1a The choke lever should stay in the full on position . . .

Chapter 1 Tune-up and routine maintenance

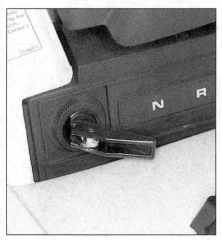

13.1b ... the half on position ...

13.1c ... and the off position

13.3 With the choke in the off position, check the freeplay between the choke lever and the cable retaining nut

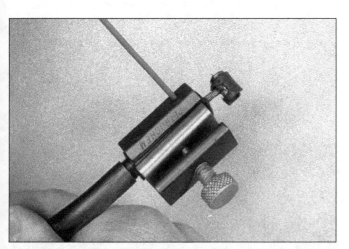

14.3 Lubricating a cable with a pressure lube adapter (make sure the tool seats around the inner cable)

14.5a Use a grease gun with a flexible hose to lubricate the friction points with grease nipples ...

and lubricate the choke cable. The choke lever should stay in the Off, halfway and On positions by itself.
2 Follow the cable from the lever to the starting enrichment valve on the carburetor. Check for kinks, bends, loose retainers or other problems and correct them as necessary.
3 Place the choke lever in the Off position (all the way down) **(see illustration)**.
4 Measure the freeplay between the choke lever and the cable housing **(see illustration 13.3)**. It should be within the range listed in this Chapter's Specifications.

Adjustment

5 Turn the choke adjusting nut at the carburetor counterclockwise until there's 1/4-inch of freeplay or more between the lever and cable housing **(see illustration 12.3a or 12.3b)**. Then turn the adjusting nut back until the freeplay is within the range listed in this Chapter's Specifications.

14 Lubrication - general

Refer to illustrations 14.3 and 14.5a through 14.5l

1 Since the controls, cables and various other components of an ATV are exposed to the elements, they should be lubricated periodically to ensure safe and trouble-free operation.

2 The throttle and brake levers, brake pedal and transmission shift lever should be lubricated frequently. In order for the lubricant to be applied where it will do the most good, the component should be disassembled. However, if chain and cable lubricant is being used, it can be applied to the pivot joint gaps and will usually work its way into the areas where friction occurs. If motor oil or light grease is being used, apply it sparingly as it may attract dirt (which could cause the controls to bind or wear at an accelerated rate). **Note:** *One of the best lubricants for the control lever pivots is a dry-film lubricant (available from many sources by different names).*
3 The throttle, choke and brake cables should be removed and treated with a commercially available cable lubricant which is specially formulated for use on vehicle control cables. Small adapters for pressure lubricating the cables with spray can lubricants are available and ensure that the cable is lubricated along its entire length **(see illustration)**. When attaching the cable to the lever, be sure to lubricate the barrel-shaped fitting at the end with multi-purpose grease.
4 To lubricate the cables, disconnect them at the lower end, then lubricate the cable with a pressure lube adapter **(see illustration 14.3)**. See Chapter 4 (throttle and choke cables) or Chapter 7 (brake cables).
5 To lubricate the steering and suspension [pivot points, use a grease gun equipped with a flexible hose **(see illustration)**. Many lubrication points have grease nipples **(see illustrations)**.

1-16 Chapter 1 Tune-up and routine maintenance

14.5b ... these include the swingarm pivot on each side (lower arrow) and the transmission output shaft (upper arrow) ...

14.5c ... the front chain adjuster eccentric on 4x4 and 6x6 models ...

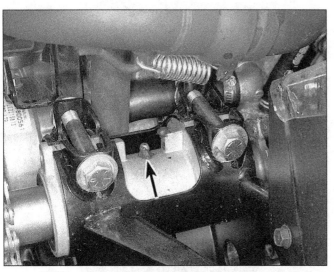

14.5d ... the center chain adjuster eccentric on 4x4 and 6x6 models ...

14.5e ... the rear chain adjuster eccentric on all models ...

14.5f ... The center rear chain adjuster on 6x6 models ...

14.5g ... the top of the steering column ...

Chapter 1 Tune-up and routine maintenance

14.5h ...the bottom of the steering column (left arrow) and the steering linkage inner tie-rod ends (right arrows)...

14.5i ...the steering linkage outer tie-rod end on each side...

14.5j ...the ball-joint at the bottom of each front suspension strut...

14.5k ...the front suspension pivot shafts...

14.5l ...and the rear suspension pivot shafts on shaft drive models

15 Engine oil/filter - change

4-stroke models

Refer to illustrations 15.5, 15.7, 15.8 and 15.12

1 Consistent routine oil and filter changes are the single most important maintenance procedure you can perform on a 4-stroke engine. The oil not only lubricates the internal parts of the engine, transmission and clutch, but it also acts as a coolant, a cleaner, a sealant, and a protectant. Because of these demands, the oil takes a terrific amount of abuse and should be replaced often with new oil of the recommended grade and type. Saving a little money on the difference in cost between a good oil and a cheap oil won't pay off if the engine is damaged.

2 Before changing the oil and filter, warm up the engine so the oil will drain easily. Be careful when draining the oil, as the exhaust pipe, the engine and the oil itself can cause severe burns.

3 Park the vehicle over a clean drain pan.

4 Remove the dipstick/oil filler cap to vent the crankcase and act as a reminder that there is no oil in the engine.

5 Being careful not to touch the hot exhaust components, place the

15.5 Most of the oil is drained by removing the oil tank drain plug (arrow)

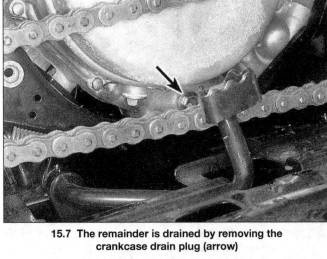

15.7 The remainder is drained by removing the crankcase drain plug (arrow)

15.8 Unscrew the oil filter with a filter wrench

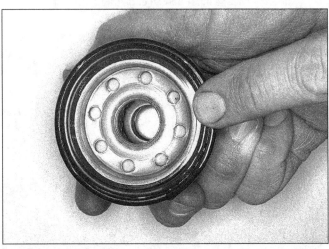

15.12 Apply a thin film of clean engine oil to the filter gasket

drain pan under the oil tank drain plug and remove the plug **(see illustration)**. You may want to wear gloves while unscrewing the plug the final few turns if the engine is really hot.

6 Allow the old oil to drain into the pan. It may be necessary to move the pan farther under the engine as the oil flow slows to a trickle. Inspect the old oil for the presence of metal shavings and chips.

7 To drain the remainder of the oil (about a cup), remove the drain plug on the right side of the crankcase near the bottom **(see illustration)**.

8 As the oil is draining, loosen the oil filter with a filter wrench **(see illustration)**. Once the filter is loose, use your hands to unscrew it from the block. Just as the filter is detached from the engine, immediately tilt the open end up to prevent the oil inside the filter from spilling out.
Warning: *The engine exhaust components may still be hot, so be careful.* If additional maintenance is planned for this time period, check or service another component while the oil is allowed to drain completely.

9 After all the oil has drained, wipe off the drain plugs with a clean rag. Even minute metal particles clinging to the plugs would immediately contaminate the new oil.

10 Inspect the drain plug sealing washers. If there's any doubt about their condition, replace them with new ones. Clean the area around the drain plug openings, reinstall the plugs and tighten them to the torques listed in this Chapter's Specifications.

11 With a clean rag, wipe off the oil filter mounting surface on the engine. If a residue of old oil is allowed to remain, it will smoke when the engine is heated up. It will also prevent the new filter from seating properly. Also make sure that the none of the old gasket remains stuck to the mounting surface. It can be removed with a scraper if necessary.

12 Compare the old filter with the new one to make sure they are the same type. Smear some engine oil on the rubber gasket of the new filter and screw it into place **(see illustration)**. Because overtightening the filter will damage the gasket, do not use a filter wrench to tighten the filter. Tighten it by hand until the gasket contacts the seating surface. Then seat the filter by giving it an additional 1/2-turn.

13 Before refilling the engine, check the old oil carefully. If the oil was drained into a clean pan, small pieces of metal or other material can be easily detected. If the oil is very metallic colored, then the engine is experiencing wear from break-in (new engine) or from insufficient lubrication. If there are flakes or chips of metal in the oil, then something is drastically wrong internally and the engine will have to be disassembled for inspection and repair.

14 If the inspection of the oil turns up nothing unusual, refill the crankcase to the proper level with the recommended oil and install the dipstick/filler cap. Start the engine and let it idle for one or two minutes. Shut it off, wait a few minutes, then check the oil level. If necessary, add more oil to bring the level up to the upper level mark on the dipstick. Check around the drain plugs and filter for leaks.

15 The old oil drained from the engine cannot be reused in its present state and should be disposed of. Check with your local refuse disposal company, disposal facility or environmental agency to see whether they will accept the oil for recycling. Don't pour used oil into drains or onto the ground. After the oil has cooled, it can be drained into a suitable container (capped plastic jugs, topped bottles, milk cartons, etc.) for transport to one of these disposal sites.

Chapter 1 Tune-up and routine maintenance

1-19

15.16 The 2-stroke oil filter and fuel filter (arrows) look alike; the oil filter is in the hose from the oil tank and the fuel filter is in the hose from the fuel tank

2-stroke models

Refer to illustration 15.16

16 Two-stroke models use an inline oil filter, similar to a fuel filter, in the line between the oil tank and oil pump **(see illustration)**. Periodic oil changes on two-strokes aren't necessary because the oil is blended with the gasoline and burned in the combustion chamber.

17 Check the new filter to make sure it's the right type. Don't substitute another type of filter. Check the oil hoses for damage or deterioration and replace them if any problems are found.

18 Remove the hose clamps and detach the old filter from the oil line. Position the new filter in the line, making sure the arrow indicating direction of oil flow points toward the oil pump.

19 Tighten the clamps. Run the engine and check for leaks.

16 Front gearcase oil (shaft drive models) - change

Refer to illustration 16.2

1 Place a drain pan beneath the differential.
2 Remove the oil filler plug, then the drain bolt and sealing washer **(see illustration)**. Let the oil drain for several minutes, until it stops dripping.
3 Clean the drain bolt and sealing washer. If the sealing washer is in good condition, it can be reused; otherwise, replace it.
4 Install the drain bolt. Tighten it securely, but don't overtighten and strip the threads.

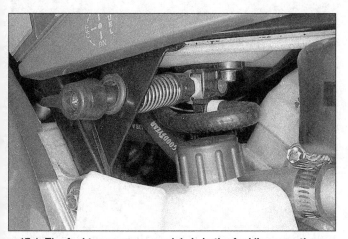

17.1 The fuel tap on some models is in the fuel line; on others (shown) it's on the underside of the fuel tank

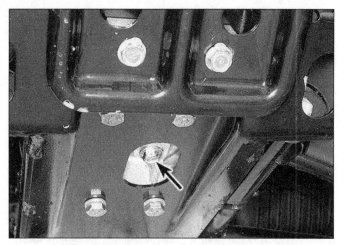

16.2 Front gearcase drain plug (shaft drive models)

5 Add oil of the type and amount listed in this Chapter's Specifications, then install the filler plug and tighten it securely. **Note:** *The specified amount is approximate. Oil level should be up to the bottom of the filler hole.*

6 Refer to Step 15 of Section 15 to dispose of the drained oil.

17 Fuel system - check and filter cleaning

Refer to illustrations 17.1 and 17.6

Warning: *Gasoline is extremely flammable, so take extra precautions when you work on any part of the fuel system. Don't smoke or allow open flames or bare light bulbs near the work area, and don't work in a garage where a natural gas-type appliance (such as a water heater or clothes dryer) is present. If you spill any fuel on your skin, wash it off immediately with soap and water. When you perform any kind of work on the fuel system, wear safety glasses and have a fire extinguisher suitable for class B type fires (flammable liquids) on hand.*

1 Check the carburetor, fuel tank, the fuel tap and the line for leaks and evidence of damage **(see illustration)**.
2 If carburetor gaskets are leaking, the carburetor should be disassembled and rebuilt (see Chapter 4).
3 If the fuel tap is leaking, tightening the screws may help. If leakage persists, the tap should be removed and replaced with a new one.
4 If the fuel lines are cracked or otherwise deteriorated, replace them with new ones.
5 Place the fuel tap lever in the Off position **(see illustration 15.1)**.
6 Undo the fuel filter clamps and remove the filter from the line **(see illustration 15.16 and the accompanying illustration)**.

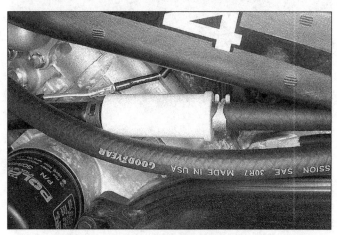

17.6 Here's a typical inline fuel filter used on 4-stroke models

Chapter 1 Tune-up and routine maintenance

18.2 Make sure the nuts that hold the exhaust pipe to the cylinder head are tight

18.3 Unscrew the plug or plugs from the underside of the muffler, then hold a rag against the muffler opening and rev the engine a few times to blow carbon out of the spark arrester

7   Install a new filter, making sure the arrow that indicates the direction of fuel flow points toward the carburetor.
8   After installation, run the engine and check for fuel leaks.
9   If the vehicle will be stored for a month or more, remove and drain the fuel tank. Also loosen the float chamber drain screw and drain the fuel from the carburetor **(see illustration 12.3b)**. If you're working on a 4-stroke model, the float chamber should be drained at the intervals listed in this Chapter's Specifications.

18 Exhaust system - inspection

Refer to illustrations 18.2 and 18.3
1   Periodically check the exhaust system for leaks and loose fasteners. If tightening the holder nuts at the cylinder head fails to stop any leaks, replace the gasket with a new ones (a procedure which requires removal of the system).
2   The exhaust pipe flange nuts at the cylinder head are especially prone to loosening, which could cause damage to the head. Check them frequently and keep them tight **(see illustration)**.
3   With the engine cold, remove the cleanout plug(s) from the underside of the muffler **(see illustration)**.
5   **Warning:** *Have an assistant sit on the vehicle and operate the controls during the next steps. You'll need to be near the muffler. Don't stand directly behind or in front of the vehicle. Make sure the selector lever is in Neutral or the vehicle will drive off when the engine is revved.*

Wear eye protection so you won't be injured by flying carbon chunks. Place the transmission in Neutral and rev the engine several times to expel carbon. If carbon is blown out, cover the end of the pipe with a rag. Have an assistant rev the engine several more times while you tap sharply on the pipe with a wrench or similar tool.
6   If you think there's still carbon in the muffler, back the vehicle up a slope so the rear end is about a foot higher than the front, then repeat Step 5.
7   If there's still carbon in the muffler, drive the vehicle up the slope so the front is about a foot higher than the rear, then repeat Step 5.

19 Ignition timing - check

Refer to illustrations 19.3, 19.6a and 19.6b
1   Perform this check with the engine at 68-degrees F. If necessary, let it cool overnight. Timing will retard approximately 2-degrees from the specified setting as the engine warms up.
2   Connect a tune-up tachometer and timing light to the engine, following the instructions supplied with the test equipment.
3   Remove the plug from the ignition timing check hole **(see illustration)**.
4   Place the range select lever in Neutral. **Warning:** *If the range select lever is in any position other than neutral, the vehicle will drive off when engine speed is increased in Step 5.*
5   Raise engine speed to the ignition timing check speed listed in

19.3 Unscrew the timing hole plug (arrow)

19.6a The threaded portion of the plug hole is cut into a V-shaped pointer (arrow) . . .

Chapter 1 Tune-up and routine maintenance

19.6b ... which should align with the center of the three timing marks stamped into the flywheel

20.2a Twist the spark plug cap back and forth to free it ...

this Chapter's Specifications.

6 Point the timing light at the timing marks **(see illustrations)**. The timing pointer should align with the center mark. If it doesn't, refer to Chapter 5 and adjust the position of the stator plate. Rotate the stator plate counterclockwise to advance timing or clockwise to retard it.

20 Spark plug - replacement

Refer to illustrations 20.2a, 20.2b, 20.6a and 20.6b

1 On some models, you'll need to remove the right body panel for access to the plug (see Chapter 8). On some 2-strokes, the plug is accessible through the right footwell.
2 Twist the spark plug cap to break it free from the plug, then pull it off **(see illustrations)**. If available, use compressed air to blow any accumulated debris from around the spark plug. Unscrew the plug with a spark plug socket.
3 Inspect the electrodes for wear. Both the center and side electrodes should have square edges and the side electrode should be of uniform thickness. Look for excessive deposits and evidence of a cracked or chipped insulator around the center electrode. Compare your spark plugs to the color spark plug reading chart. Check the threads, the washer and the ceramic insulator body for cracks and other damage.
4 If the electrodes are not excessively worn, and if the deposits can be easily removed with a wire brush, the plug can be regapped and

20.2b ... then pull it off the plug and unscrew the plug with a spark plug socket

reused (if no cracks or chips are visible in the insulator). If in doubt concerning the condition of the plug, replace it with a new one, as the expense is minimal.
5 Cleaning the spark plug by sandblasting is permitted, provided you clean the plug with a high flash-point solvent afterwards.
6 Before installing a new plug, make sure it is the correct type and

20.6a Spark plug manufacturers recommend using a wire type gauge when checking the gap - if the wire doesn't slide between the electrodes with a slight drag, adjustment is required

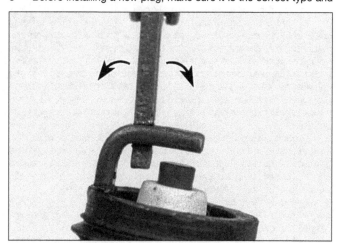

20.6b To change the gap, bend the side electrode only, as indicated by the arrows, and be very careful not to crack or chip the ceramic insulator surrounding the center electrode

Chapter 1 Tune-up and routine maintenance

21.6 A compression gauge with a threaded fitting for the spark plug hole is preferred over the type that requires hand pressure to maintain the seal

22.7a Align the timing pointer with the top dead center line on the flywheel (arrow) . . .

heat range. Check the gap between the electrodes, as it is not preset. For best results, use a wire-type gauge rather than a flat gauge to check the gap **(see illustration)**. If the gap must be adjusted, bend the side electrode only and be very careful not to chip or crack the insulator nose **(see illustration)**. Make sure the washer is in place before installing the plug.

7 Since the cylinder head is made of aluminum, which is soft and easily damaged, thread the plug into the head by hand. Slip a short length of hose over the end of the plug to use as a tool to thread it into place. The hose will grip the plug well enough to turn it, but will start to slip if the plug begins to cross-thread in the hole - this will prevent damaged threads and the accompanying repair costs.

8 Once the plug is finger tight, the job can be finished with a socket. If a torque wrench is available, tighten the spark plug to the torque listed in this Chapter's Specifications. If you do not have a torque wrench, tighten the plug finger tight (until the washer bottoms on the cylinder head) then use a wrench to tighten it an additional 1/4 turn. Regardless of the method used, do not over-tighten it.

9 Reconnect the spark plug cap.

21 Cylinder compression - check

Refer to illustration 21.6

1 Polaris 4-stroke engines are equipped with an automatic decompressor that opens an exhaust valve slightly when the engine is being cranked by the starter. This eases starting, but it affects compression readings (they change with engine cranking speed). For this reason, Polaris recommends a cylinder leakage test rather than a compression tests on 4-stroke models. A compression test can be done on 4-stroke engines, but it won't give the usual specific results (and Polaris doesn't provide compression specifications for 4-stroke engines). It will, however, indicate extreme high or low readings, which can point to specific problems.

2 Among other things, poor engine performance may be caused by leaking valves or incorrect valve clearances (4-strokes), a leaking head gasket, worn piston, rings and/or cylinder wall. A cylinder compression check will help pinpoint these conditions and can also indicate the presence of excessive carbon deposits in the cylinder head.

3 The only tools required are a compression gauge and a spark plug wrench. Depending on the outcome of the initial test, a squirt-type oil can may also be needed.

4 Start the engine and allow it to reach normal operating temperature, then remove the spark plugs (see Section 20, if necessary). Work carefully - don't strip the spark plug hole threads and don't burn your hands.

5 Disable the ignition by disconnecting the primary (low tension) wires from the coil (see Chapter 4). Be sure to mark the locations of the wires before detaching them.

6 Install the compression gauge in the spark plug hole **(see illustration)**. Hold or block the throttle wide open.

7 Crank the engine over a minimum of four or five revolutions (or until the gauge reading stops increasing) and observe the initial movement of the compression gauge needle as well as the final total gauge reading.

8 If the compression built up quickly and evenly, you can assume the engine upper end is in reasonably good mechanical condition. Worn or sticking piston rings and worn cylinders will produce very little initial movement of the gauge needle, but compression will tend to build up gradually as the engine spins over. Valve and valve seat leakage (4-strokes), or head gasket leakage (all models), is indicated by low initial compression which does not tend to build up.

9 To further confirm your findings, add a small amount of engine oil to the cylinder by inserting the nozzle of a squirt-type oil can through the spark plug hole. The oil will tend to seal the piston rings if they are leaking.

10 If the compression increases significantly after the addition of the oil, the piston rings and/or cylinder are definitely worn. If the compression does not increase, the pressure is leaking past the valves or the head gasket. Leakage past the valves may be due to insufficient valve clearances, burned, warped or cracked valves or valve seats or valves that are hanging up in the guides.

10 If compression readings are considerably higher than specified, the combustion chamber may be coated with excessive carbon deposits. It is possible (but not very likely) for carbon deposits to raise the compression enough to compensate for the effects of leakage past rings or valves. Refer to Chapter 2 Part B, remove the cylinder head and carefully decarbonize the combustion chamber.

11 Other possible causes of very high readings include a damaged decompressor and worn or damaged exhaust cam lobes. Refer to Chapter 2, Part B for decompressor and camshaft service procedures.

22 Valve clearances - check and adjustment

Refer to illustrations 22.7a, 22.7b, 22.9, 22.10 and 22.13

1 The engine must be completely cool for this maintenance procedure, so if possible let the machine sit overnight before beginning.

2 Refer to Section 7 and disconnect the cable from the negative terminal of the battery.

3 Remove the left and right side panels, the seat and the fuel tank (see Chapters 8 and 4).

4 Refer to Section 20 and remove the spark plug. This will make it easier to turn the engine.

Chapter 1 Tune-up and routine maintenance   1-23

22.7b ... and make sure the camshaft dowel (arrow) is straight up; there should now be some clearance on all four valves

22.9 The intake valve clearances can be measured one at a time, or both at once as shown here

22.10 Loosen the locknut and turn the adjusting screw to change the clearance

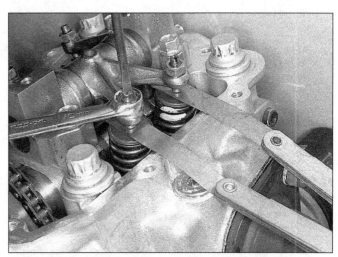

22.13 The exhaust valves must both be measured at once with a pair of feeler gauges, since they share a common rocker arm

5   Remove the valve cover and camshaft sprocket cover (see Chapter 2 Part B).
6   Remove the ignition timing hole plug **(see illustration 19.3)**.
7   Position the piston at Top Dead Center (TDC) on the compression stroke. To do this, turn the crankshaft (by pulling slowly on the recoil starter rope) until the mark on the rotor is aligned with the TDC mark inside the timing hole **(see illustration)**. Check the camshaft sprocket dowel position; it should be straight up **(see illustration)**. If it's straight down, the piston is at TDC on the exhaust stroke, so you'll need to turn the crankshaft one full turn to bring the dowel upward.
8   With the engine in this position, all four of the valves can be checked.

Intake valves

9   To check, insert a feeler gauge of the thickness listed in this Chapter's Specifications between the valve stem and rocker arm **(see illustration)**. Pull the feeler gauge out slowly - you should feel a slight drag. If there's no drag, the clearance is too loose. If there's a heavy drag, the clearance is too tight.
10 If the clearance is incorrect, loosen the adjuster locknut with a box wrench. Turn the adjusting screw with a screwdriver until the correct clearance is achieved, then tighten the locknut. Polaris recommends tightening valve adjuster locknuts with a flank drive torque wrench adapter, mounted at 90-degrees to the torque wrench handle **(see illustration)**.

11 After adjusting, recheck the clearance with the feeler gauge to make sure it wasn't changed when the locknut was tightened.
12 Now measure the other intake valve, following the same procedure you used for the first valve. Make sure to use a feeler gauge of the specified thickness.

Exhaust valves

13 Adjustment is the same as for intake valves, but two feeler gauges must be used since the valves share a common rocker arm **(see illustration)**. Both feeler gauges must be in position between the valves stems and rocker arms when adjustments are made and when the locknuts are tightened.

All valves

14 With all of the clearances within the Specifications, install the timing hole plug, valve cover and camshaft sprocket cover.
15 Install the fuel tank and body panels. Reconnect the cable to the negative terminal of the battery.

23 Fasteners - check

1   Since vibration of the machine tends to loosen fasteners, all nuts, bolts, screws, etc. should be periodically checked for proper tightness.

1-24 Chapter 1 Tune-up and routine maintenance

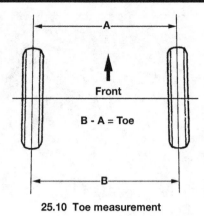

25.10 Toe measurement

25.11 Hold and turn the tie rod by placing an open end wrench on the flat (A); loosen the locknut at inner and outer ends (B) so the tie rod can be turned

Also make sure all cotter pins or other safety fasteners are correctly installed.
2   Pay particular attention to the following:
   Spark plug
   Engine oil drain plug
   Oil filter
   Gearshift lever
   Brake pedal
   Engine and transmission mount bolts
   Shock absorber mount bolts
   Front axle nuts
   Rear axle nuts
   Skid plate bolts
3   If a torque wrench is available, use it along with the torque specifications at the beginning of this, or other, Chapters.

24 Suspension - check

1   The suspension components must be maintained in top operating condition to ensure rider safety. Loose, worn or damaged suspension parts decrease the vehicle's stability and control.
2   Lock the front brake and push on the handlebars to compress the front shock absorbers several times. See if they move up-and-down smoothly without binding. If binding is felt, the shocks should be disassembled and inspected as described in Chapter 6.
3   Check the tightness of all front suspension nuts and bolts to be sure none have worked loose.
4   Inspect the rear shock absorber for fluid leakage and tightness of the mounting nuts and bolts. If leakage is found, the shock should be replaced.
5   Support the vehicle securely upright with its rear wheel off the ground. Grab the swingarm on each side, just ahead of the axle. Rock the swingarm from side to side - there should be no discernible movement at the rear. If there's a little movement or a slight clicking can be heard, make sure the swingarm pivot shaft is tight. If the pivot shaft is tight but movement is still noticeable, the swingarm will have to be removed and the bearings replaced as described in Chapter 6.
6   Inspect the tightness of the rear suspension nuts and bolts.

25 Steering system - inspection and toe adjustment

Inspection
1   This vehicle is equipped with bushings at the upper and lower ends of the steering shaft, which can become dented, rough or loose during normal use of the machine. In extreme cases, worn or loose parts can cause steering wobble that is potentially dangerous.
2   To check the bushings, block the rear wheels so the vehicle can't roll, jack up the front end and support it securely on jackstands.
3   Point the wheel straight ahead and slowly move the handlebars from side-to-side. Dents or roughness in the bearing or bushing will be felt and the bars will not move smoothly. **Note:** *Make sure any hesitation in movement is not being caused by the cables and wiring harnesses that run to the handlebars.*
4   If the handlebars don't move smoothly, or if they move horizontally, refer to Chapter 6 to remove and inspect the steering shaft bushing and bearing.

Toe adjustment
Refer to illustrations 25.10 and 25.11
5   Remove the front carrier and fender (see Chapter 7).
6   Roll the vehicle forward onto a level surface and stop it with the front wheels pointing straight ahead.
7   Make a mark at the front and center of each tire, even with the centerline of the front hub.
8   Measure the distance between the marks with a toe-in gauge or steel tape measure.
9   Have an assistant push the vehicle backward while you watch the marks on the tires. Stop pushing when the tires have rotated exactly one-half turn, so the marks are at the backs of the tires.
10   Again, measure the distance between the marks. Subtract the rear measurement from the front measurement to get toe-out **(see illustration)**.
11   If toe-out is not as listed in this Chapter's Specifications, hold each tie-rod with a wrench on the flats and loosen the locknuts **(see illustration)**. Turn the tie-rods an equal amount to change toe-out. When toe-out is set correctly, tighten the locknuts to the torque listed in this Chapter's Specifications.

26 Drive chain and sprockets - check, adjustment and lubrication

1   A neglected drive chain won't last long and can quickly damage the sprockets. Routine chain adjustment isn't difficult and will ensure maximum chain and sprocket life.
Caution: *The tension of the rear chain on models with four tires (not 6x6 models) changes as the swingarm moves up and down. Pay special attention to the pre-adjustment steps; otherwise the chain may over-tighten during operation and damage the transmission, chain or sprockets.*

Screw-type chain adjusters
Rear chain adjustment
2   Early models use a screw-type chain adjuster at the rear. To

Chapter 1 Tune-up and routine maintenance 1-25

26.11 Compress the rear suspension so the rear axle, swingarm pivot and countershaft (arrows) are in a straight line, then adjust chain tension

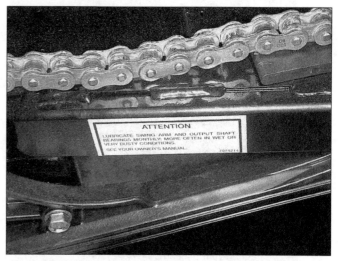

26.12 Measure rear chain tension along the top run

Eccentric chain adjusters

Rear chain adjustment (except 6x6)

Refer to illustrations 26.11, 26.12, 26.13a and 26.13b

11 Before checking the adjustment, you'll need to compress the rear suspension until the transmission output shaft, the swingarm pivot and the rear axle are in a straight line **(see illustration)**. This is where the chain reaches its tightest point. One way to do this is to wrap a cargo strap in a loop around the rear axle and the lower tube of the rear bumper, then tighten the loop to compress the suspension. Lay a straightedge (a long carpenter's level, a straight board, etc.) along the three points to make sure they're in a straight line.

12 To check the adjustment, roll the machine forward slightly to create slack in the top run of the chain. Loosen or remove the chain guard if necessary for access (see Chapter 6). Pull up and push down on the top run of the chain and measure the slack midway between the two sprockets **(see illustration)**.

13 If the measurement is not as listed in this Chapter's Specifications, adjust the chain. Loosen the locking bolts and nuts on the rear eccentric **(see illustration)**. Insert a punch through the hole in the rear sprocket into the eccentric **(see illustration)**. With the punch in the hole, roll the vehicle slightly forward or back to change the chain tension.

14 With the tension set correctly, tighten the locking bolts to the torque listed in this Chapter's Specifications. Recheck the setting after you tighten the bolts to make sure the tension didn't change.

adjust this type, loosen the three pinch bolts at the adjuster bracket (one above and two below).

3 Early models except Cyclone have a single adjuster bolt; tighten or loosen this to change the adjustment, then tighten the adjuster and pinch bolts to the torque listed in this Chapter's Specifications.

4 Cyclone models have two adjuster nuts at the rear of the adjuster and a jam nut at the lower left of the adjuster. Loosen the jam nut, tighten or loosen the adjuster nuts as needed, then tighten the jam nut. Tighten the pinch bolts.

Center and front chain adjustment (1987 and 1988 4x4 models)

5 Adjust the center chain first, then the front chain.
6 Loosen the lockbolt in the adjuster slide at the center sprocket housing.
7 Loosen the locknuts on the adjuster rod (it's next to the lower run of the front chain, and connects the center housing to a stationary bracket).
8 Adjust the length of the adjuster rod, then tighten the locknuts.
9 At the front sprocket, loosen the two lockbolts (there's one on each side of the axle housing).
10 Loosen or tighten the nut on the adjuster rod, attached to the top of the axle housing. Once the front chain tension is set correctly, tighten the lockbolts.

26.13a Loosen the locknuts (arrows) and bolts so the adjusting eccentric can be rotated

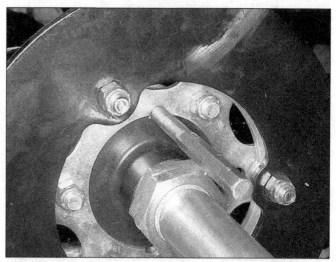

26.13b Insert a punch through a sprocket hole into the eccentric, then roll the vehicle slightly to change tension

Chapter 1 Tune-up and routine maintenance

26.16 There's an adjusting bolt and locknut on each side of the center rear eccentric on 6x6 models

26.20 Remove the center chain guard and measure tension along the top run of the chain

15 If the chain guard was removed, reinstall it, leaving at least 1/8-inch clearance between it and the chain.

Middle and rear axle chain adjustment (6x6)

Refer to illustration 26.16

16 The middle axle chain is adjusted before the rear axle chain. The adjustment is the same as for other models, described above, but you don't need to compress the suspension **(see illustration)**. It's important to note, though, that the specified chain play is considerably greater than for other chains to allow for changing tension as the suspension moves through its travel.

17 The rear axle chain adjustment is the same as for four-wheeled models, described above.

Center and front chain adjustment (4x4 and 6x6 models)

Refer to illustrations 26.20, 26.21, 26.22, 26.23a and 26.23b

18 Adjust the center chain first, then the front chain. Adjusting the center chain changes the adjustment of the front chain, so the front chain needs to be adjusted afterward.
19 Remove the center and front chain guards (see Chapter 6).
20 Measure chain tension along the top run of the chain **(see illustration)**.
21 Loosen the lockbolts on the center eccentric **(see illustration)**.
22 Insert a punch between two sprocket teeth into the adjustment hole in the center eccentric **(see illustration)**. Roll the vehicle slightly forward or back to change the adjustment. Once it's set correctly, tighten the lockbolts, and remove the punch.
23 The front chain can be inspected through a window in the chain

26.21 Loosen the lockbolts and nuts (arrows) so the center eccentric can be rotated

guard **(see illustration)**. The chain is adjusted in the same way as the center chain **(see illustration)**. After adjustment, tighten the eccentric lockbolts to the torque listed in this Chapter's Specifications, then recheck the chain adjustment to make sure it didn't change while you were tightening the bolts.

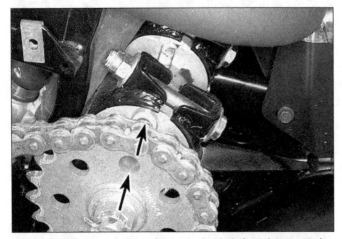

26.22 Insert a punch through a sprocket hole into the eccentric (arrows), then roll the vehicle to adjust the chain

26.23a The front chain tension can be checked through this window in the chain guard

Chapter 1 Tune-up and routine maintenance

26.23b Loosen the locknut (arrow) and bolt so the front eccentric can be rotated

27.2 Transmission drain plug (arrow)

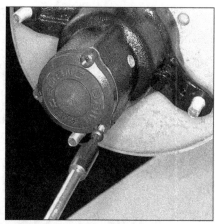

29.2a If the hub cover has screws like this one, remove them with a Torx bit

27 Transmission oil - change

Refer to illustration 27.2

1 Place a drain pan beneath the transmission.
2 Remove the oil dipstick, then the drain plug and sealing washer **(see illustration)**. Let the oil drain for several minutes, until it stops dripping.
3 Clean the drain bolt and sealing washer. If the sealing washer is in good condition, it can be reused; otherwise, replace it.
4 Install the drain plug. Tighten it securely, but don't overtighten and strip the threads.
5 Add oil of the type and amount listed in this Chapter's Specifications, then install the filler plug and tighten it securely. **Note:** *The specified amount is approximate. Oil level should be up to the bottom of the filler hole.*
6 Refer to Step 15 of Section 15 to dispose of the drained oil.

28 Counterbalancer fluid - change

1 Place a drain pan under the counterbalancer drain plug **(see illustration 6.2)**.
2 Remove the filler plug or dipstick **(see illustrations 3.19 and 3.20a)**.
3 Unscrew the drain plug and let the fluid drain **(see illustration 6.2)**.
4 Clean the drain plug and sealing washer. If the sealing washer is in good condition, it can be reused; otherwise, replace it.
5 Install the drain plug. Tighten it securely, but don't overtighten and strip the threads.
6 Add oil of the type and amount listed in this Chapter's Specifications, then install the filler plug or dipstick and tighten it securely. **Note:** *The specified amount is approximate. Oil level should be up to the bottom of the filler hole (350) or in the crosshatched area of the dipstick (400).*
7 Refer to Step 15 of Section 15 to dispose of the drained oil.

29 Front hub fluid - change

Refer to illustrations 29.2a, 29.2b and 29.3

1 Jack up the front end and support it securely on jackstands. Remove the front wheels.
2 Place a drain pan beneath the hub and remove the cap. The caps on mid-1996 and later 4x4 and 6x6 models are secured by three Torx screws **(see illustration)**. On all other models, insert a flat-bladed screwdriver in the removal notches and pry the cap off **(see illustration)**.
3 Check the cap O-ring and replace it if it's damaged or flattened **(see illustration)**.
4 Unscrew the hub filler plug **(see illustration 3.22)**.
5 Reinstall the cap and O-ring. If it's equipped with Torx screws, tighten them securely, but don't overtighten them and strip out the threads.
6 Refer to Section 3 to fill the hub.

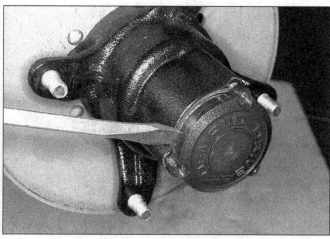

29.2b Pry the hub cover off; models without cover screws have prying notches

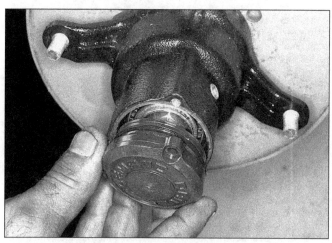

29.3 Replace the cover O-ring if it's flattened or deteriorated

1-28   Chapter 1 Tune-up and routine maintenance

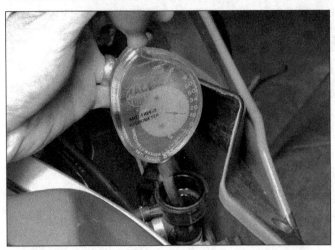

30.8 An antifreeze hydrometer is helpful in determining the condition of the coolant

31.1 Radiator drain tap (arrow)

30 Cooling system - inspection

Refer to illustration 30.8
Warning: *The engine must be cool before beginning this procedure.*
Note: *Refer to Section 3 and check the coolant level before performing this check.*

1   The entire cooling system should be checked carefully at the recommended intervals. Look for evidence of leaks, check the condition of the coolant, check the radiator for clogged fins and damage and make sure the fan operates when required.
2   Remove body panels as necessary for access to the cooling system components (see Chapter 8).
3   Examine each of the rubber coolant hoses along its entire length. Look for cracks, abrasions of other damage. Squeeze each hose at various points. They should feel firm, yet pliable, and return to their original shape when released. If they are dried or hard, replace them with new ones.
4   Check for evidence of leaks at each cooling system joint. Tighten the hose clamps to prevent future leaks. If coolant has been leaking from the joints of steel or aluminum coolant tubes, remove the tubes and replace the O-rings or gaskets (see Chapter 3).
5   Check the radiator for evidence of leaks or other damage. Leaks in the radiator leave telltale scale deposits or coolant stains on the outside of the core below the leak. If leaks are noted, remove the radiator (see Chapter 3) and have it repaired by a radiator shop or replace it with a new one. **Caution:** *Do not use a liquid leak stopping compound to try to repair leaks.*
6   Check the radiator fins for mud, dirt, weeds and insects, which may impede the flow of water through the radiator. If the fins are dirty, force water or low pressure compressed air through the fins from the backside. If the fins are bent or distorted, straighten them carefully with a screwdriver.
7   Remove the pressure cap by turning it counterclockwise until it reaches a stop. If you hear a hissing sound (indicating there is still pressure in the system) wait until it stops. Now, press down on the cap with the palm of your hand and continue turning the cap counterclockwise until it can be removed. Check the condition of the coolant in the system. If it is rust colored or if accumulations of scale are visible, drain, flush and refill the system with new coolant. Check the cap[gaskets for cracks and other damage. Have the cap tested by a dealer service department or replace it with a new one. Install the cap by turning it clockwise until it reaches the first stop, then push down on the cap and continue turning until it can turn no further.
8   Check the antifreeze content of the coolant with a coolant hydrometer **(see illustration)**. Sometimes coolant may look like it's in good condition, but might be too weak to offer adequate protection. If the hydrometer indicates a weak mixture, drain, flush tand refill the

cooling system (see Section 31).
9   Start the engine and let it reach normal operating temperature, then check for leaks again. As the coolant temperature increases, the fan should come on automatically and the temperature should begin to drop. If it doesn't, refer to Chapter 3 and check the fan and fan circuit carefully.
10   If the coolant level is consistently low, and no evidence of leaks can be found, have the entire system checked by a Polaris service department, ATV repair shop or service station.

31 Cooling system - draining, flushing and refilling

Warning: *Allow the engine to cool completely before performing this maintenance operation. Also, don't allow antifreeze to come in contact with your skin or painted surfaces of the vehicle. Rinse off spills immediately with plenty of water. Antifreeze is highly toxic if ingested. Never leave antifreeze lying around in an open container or in puddles on the floor; children and pets are attracted by its sweet smell and may drink it. Check with local authorities about disposing of used antifreeze, Many communities have collection centers which will see that antifreeze is disposed of safely. Antifreeze is also combustible, so don't store or use it near open flames.*

Draining

Refer to illustrations 31.1 and 31.2
1   Place a pan beneath the radiator drain valve **(see illustration)**.
2   Open the radiator drain tap. Coolant should dribble out at first, then flow freely after you remove the radiator cap **(see illustration)**.

2-stroke models

Refer to illustration 31.3
3   Remove the bleed plug from the top of the cylinder head **(see illustration)**.

4-stroke models

Refer to illustration 31.4
4   Disconnect the hose from the thermostat housing on the cylinder **(see illustration)**. Cap or plug the housing; it must be airtight so coolant won't flow out in the next step.

All models

Refer to illustrations 31.5a and 31.5b
5   If the water pump housing has a drain plug, remove it **(see illustration)**. If not, disconnect the outlet hose from the water pump **(see illustration)**. Point the end of the hose into a drain pan.
6   Let the coolant drain into the pan. If you're working on a 4-stroke, uncap the thermostat housing so the coolant can flow.

Chapter 1 Tune-up and routine maintenance

31.2 The radiator cap is at the top center of the radiator

31.3 Open the bleed plug (arrow) on 2-stroke models so the coolant can displace air as it's added

Flushing

7   Flush the system with clean water by inserting a garden hose into the radiator filler neck. Allow the water to run through the system until it is clear when it exits the radiator and engine. If the radiator is extremely corroded, remove it (see Chapter 3) and have it cleaned at a radiator shop.

8   Check the gasket on the water pump drain plug (if the machine has one). Replace it with a new one if necessary. Clean the drain hole, then install the drain plug and tighten it securely, but don't overtighten it and strip the threads.

9   Reconnect the water pump hose if it was disconnected. If you're working on a 4-stroke, don't reconnect the thermostat housing hose yet.

Refilling

10   Fill the coolant reservoir with the proper coolant mixture (see this Chapter's Specifications).

11   If you're working on a 2-stroke, slowly pour coolant into the radiator. The bleed plug on top of the cylinder head must be removed so the rising coolant can expel air from the engine. When coolant flows from the plug hole, reinstall the plug.

12   If you're working on a 4-stroke, slowly add coolant to the radiator until it flows from the thermostat housing. When that happens, reconnect the thermostat housing hose.

13   Fill the radiator the rest of the way.

31.4 Undo the hose from the thermostat housing (arrow)

14   Run the engine briefly with the radiator cap off (don't let it warm up to operating temperature). Shut the engine off, recheck coolant level in the radiator and top it up if necessary.

16   Install the radiator cap, then warm up the engine to normal operating temperature. Shut the engine off.

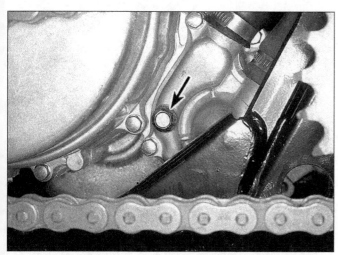

31.5a If there's a drain plug on the water pump (arrow), remove it . . .

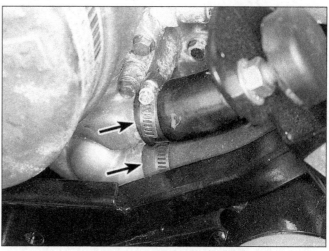

31.5b . . . if not, undo the water pump lower hose (arrow)

17 Check coolant level in the reservoir and top it off if necessary.

18 **Warning:** *Never remove the radiator cap with the engine warm. The coolant is scalding hot and under pressure. Opening the cap too soon will let it spray out forcefully and burn you.* Let the engine cool, then remove the radiator cap and check the coolant level. Add coolant if necessary.

32 PVT - check and clean

1 The Polaris Variable Transmission (PVT) should be inspected at the specified intervals. Check the following:
   a) *Belt tension and width*
   b) *Pulley offset*
   c) *Pulley faces and clutch components for wear and damage*
   d) *Cover seal condition*
   e) *Air ducts (they should be free of obstructions)*

2 Refer to Chapter 2, Part C for service procedures.

3 If water gets into the PVT air ducts, place the transmission in Neutral and rev the engine slightly above idle (not at high speed) for a few minutes. This will blow air through the PVT cover to dry out the water. Let the engine idle, shift the transmission into a forward range and operate the vehicle to make sure the belt doesn't slip.

4 If the belt slips, place the transmission back in neutral and rev the engine some more. Once the belt stops slipping, operate the vehicle in low range (if equipped) or at low speed to dry out any remaining water.

33 Brake fluid - change

At the specified intervals, pump all the old brake fluid out of the hydraulic system and replace it with new fluid. Refer to brake bleeding procedures in Chapter 7 for details.

Chapter 2 Part A
Two-stroke engines

Contents

	Section
Counterbalancer (350 and 400 models) - removal, inspection and installation	15
Crankcase - disassembly and reassembly	16
Crankcase - inspection and servicing	17
Crankcase pressure and vacuum - check	5
Crankshaft and connecting rod - removal, inspection and installation	18
Cylinder - removal, inspection and installation	10
Cylinder head - removal, inspection and installation	8
Engine - removal and installation	6
Engine disassembly and reassembly - general information	7
General information	1
Major engine repair - general note	4
Oil pump - check valve test, cable adjustment and pump test	12
Oil pump - removal, inspection, installation and bleeding	13
Operations possible with the engine in the frame	2
Operations requiring engine removal	3
Piston and rings - removal, inspection and installation	11
Recoil starter - removal, inspection and installation	14
Recommended start-up and break-in procedure	19
Reed valve - removal, inspection and installation	9

Specifications

Cylinder head warpage limit	Not specified
Reed stopper height	0.350 inch
Cylinder	
Bore	
250 models (244 cc)	2.835 inches
300 models (283 cc)	2.935 inches
350 models (352 cc)	3.152 inches
400 models (379 cc)	3.270 inches
Taper and out-of-round limits	Not specified
Piston clearance	
250 models (244 cc)	
1985 through 1990	
Standard	0.0014 to 0.0028 inch
Limit	0.006 inch
1991-on	
Standard	0.0011 to 0.0028 inch
Limit	0.006 inch
300 models (283 cc)	
Standard	0.0012 to 0.0026 inch
Limit	0.006 inch
350 models (352 cc)	
Standard	0.0024 to 0.0037 inch
Limit	0.006 inch
300 models (379 cc)	
Standard	0.0023 to 0.0037 inch
Limit	0.006 inch
Piston diameter measuring point (above bottom of piston)	3/8-inch

Ring end gap
    250 models (244 cc)
        1985 through 1990 .... 0.008 to 0.016 inch
        1991-on .... 0.009 to 0.018 inch
    300 models (283 cc) .... 0.012 to 0.022 inch
    350 models (352 cc) .... 0.008 to 0.016 inch
    400 models (379 cc) .... 0.007 to 0.015 inch
Crankshaft
    Connecting rod side clearance
        Standard .... 0.016 to 0.020-inch
        Limit .... 0.036 inch
    Runout limit .... 0.005 inch
    End play .... 0.008 to 0.016 inch

Torque specifications
ft-lbs (unless otherwise noted)

Cylinder head bolts .... 18 to 20
Cylinder base bolts .... 25 to 29
Crankcase bolts
    6 mm .... 72 to 96 inch-lbs
    8 mm .... 17 to 18
Drive clutch bolt 40
All 6mm bolts not listed .... 72 to 96 inch-lbs
Flywheel
    250 and 300 .... 44 to 62
    350 and 400 .... 29 to 44
Crankshaft slotted nut (350 and 400) .... 29 to 44

1 General information

The engine/transmission unit on all models is a single-cylinder two-stroke design. The engine/transmission assembly is constructed from aluminum alloy. The crankcase is divided vertically.

The cylinder, piston, crankshaft bearings and connecting rod lower end bearing are lubricated by the fuel, which is a mixture of gasoline and two-stroke oil. The transmission (and counterbalancer, if equipped) is lubricated by four-stroke engine oil, which is contained in a sump within the crankcase.

2 Operations possible with the engine in the frame

The components and assemblies listed below can be removed without having to remove the engine from the frame. If, however, a number of areas require attention at the same time, removal of the engine is recommended.

Cylinder and piston
Oil pump (250 and 400)
Counterbalancer (if equipped)
Water pump
Recoil starter (if equipped)

3 Operations requiring engine removal

It is necessary to remove the engine/transmission assembly from the frame and separate the crankcase halves to gain access to the following components:

Crankshaft and connecting rod
Oil pump (350)
Connecting rod
Crankshaft
Crankcase bearings

4 Major engine repair - general note

1 It is not always easy to determine when or if an engine should be completely overhauled, as a number of factors must be considered.

2 High mileage is not necessarily an indication that an overhaul is needed, while low mileage, on the other hand, does not preclude the need for an overhaul. Regular maintenance is probably the single most important consideration. This is especially true if the bike is used in competition. An engine that has regular and frequent oil changes, as well as other required maintenance, will most likely give many hours of reliable service. Conversely, a neglected engine, or one which has not been broken in properly, may require an overhaul very early in its life.

3 Poor running that can't be accounted for by seemingly obvious causes (fouled spark plug, leaking head or cylinder base gasket, worn piston ring, carburetor problems) may be due to leaking crankshaft seals. In two-stroke engines, the crankcase acts as a suction pump to draw in fuel mixture and as a compressor to force it into the cylinder. If the crankcase seals are leaking, the pressure drop will cause a loss of performance.

4 If the engine is making obvious knocking or rumbling noises, the connecting rod and/or main bearings are probably at fault.

5 A top-end overhaul consists of replacing the piston and ring and inspecting the cylinder bore. The cylinder on some models can be bored for an oversize piston if necessary; on others, the cylinder and piston must be replaced with new ones if they're worn.

6 A lower-end engine overhaul generally involves inspecting the crankshaft, crankcase bearings and seals. Unlike four-stroke engines equipped with plain main and connecting rod bearings, there isn't much in the way of machine work that can be done to refurbish existing parts. Worn bearings and seals should be replaced with new ones. The crankshaft and connecting rod are permanently assembled, so if one of these components (or the connecting rod lower end bearing) needs to be replaced both must be. While the engine is being overhauled, other components such as the carburetor can be rebuilt also. The end result should be a like-new engine that will give as many trouble-free hours as the original.

7 Before beginning the engine overhaul, read through all of the related procedures to familiarize yourself with the scope and requirements of the job. Overhauling an engine is not all that difficult, but it is time consuming. Plan on the vehicle being tied up for a minimum of two (2) weeks. Check on the availability of parts and make sure that any necessary special tools, equipment and supplies are obtained in advance.

8 Most work can be done with typical shop hand tools, although a number of precision measuring tools are required for inspecting parts to determine if they must be replaced. Often a dealer service depart-

6.13 Remove the ground cable at the rear of the engine (if equipped)

6.15 Remove the ground cable and engine mount at the top front of the engine . . .

ment or repair shop will handle the inspection of parts and offer advice concerning reconditioning and replacement. As a general rule, time is the primary cost of an overhaul so it doesn't pay to install worn or sub-standard parts.

9 As a final note, to ensure maximum life and minimum trouble from a rebuilt engine, everything must be assembled with care in a spotlessly clean environment.

5 Crankcase pressure and vacuum - check

This test can pinpoint the cause of otherwise unexplained poor running. It can also prevent piston seizures by detecting air leaks that can cause a lean mixture. It requires special equipment, but can easily be done by a Polaris dealer or other motorcycle shop. If you regularly work on two-stroke engines, you might want to consider purchasing the tester for yourself (or with a group of other riders). You may also be able to fabricate the tester.

The test involves sealing off the intake and exhaust ports, then applying vacuum and pressure to the spark plug hole with a hand vacuum/pressure pump, similar to the type used for brake bleeding and automotive vacuum testing.

First, remove the carburetor and exhaust system. Block off the carburetor opening with a rubber plug, clamped securely in position. Place a rubber sheet (cut from a tire tube or similar material) over the exhaust port and secure it with a metal plate.

Apply air pressure to the spark plug hole with the vacuum/pressure pump. Check for leaks at the crankcase gasket, intake manifold, reed valve gasket, cylinder base gasket and head gasket. Also check the seal at the alternator end of the crankshaft. If the leaks are large, air will hiss as it passes through them. Small leaks can be detected by pouring soapy water over the suspected area and looking for bubbles.

After checking for air leaks, apply vacuum with the pump. If vacuum leaks down quickly, the crankshaft seals are leaking.

6 Engine - removal and installation

Note: *Engine removal and installation should be done with the aid of an assistant to avoid damage or injury that could occur if the engine is dropped. A hydraulic floor jack should be used to support and lower the engine if possible (they can be rented at low cost).*

Removal

Refer to illustrations 6.13, 6.15, 6.16 and 6.17,
1 Disconnect the negative cable from the battery.
2 Drain the coolant (see Chapter 1).
3 Remove the seat, both side covers and the fuel tank cover (see

6.16 . . . one at the left rear . . .

Chapter 8).
4 Remove the fuel tank, air cleaner housing, carburetor and exhaust system (see Chapter 4). The carburetor cables can be left connected.
5 Disconnect the spark plug wire (see Chapter 1).
6 Remove the oil pump, leaving the cables connected, or disconnect the oil pump cables (see Section 12).
7 Label and disconnect the ignition pulse generator wires (refer to Chapter 5 for component location if necessary). Detach the wires from their retainers. Disconnect the wire for the coolant temperature sender (if equipped).
8 If you're working on a 4wd or 6wd vehicle, remove the auxiliary brake adjusting bolt and the brake actuator arm (see Chapter 7).
9 If you're working on a 4wd or 6wd vehicle, remove the center drive chain (see Chapter 6).
10 Remove the PVT inner cover (see Chapter 6).
11 Remove the starter motor (see Chapter 8).
12 Detach the transmission linkage rods from the shift select lever (see Chapter 2 Part C). Tie them up out of the way so they won't interfere with engine removal.
13 Disconnect the ground cable at the rear of the engine (if equipped) **(see illustration)**.
14 Support the engine with a jack, using a block of wood between the jack and the engine to protect the crankcase.
15 Remove the upper engine mount and the ground cable **(see illustration)**.
16 Remove the nut from the rear engine mount **(see illustration)**.

6.17 ... and one at the front

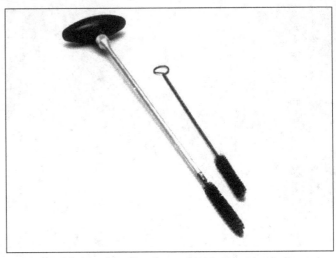

7.2 A selection of brushes is required for cleaning holes and passages in the engine components

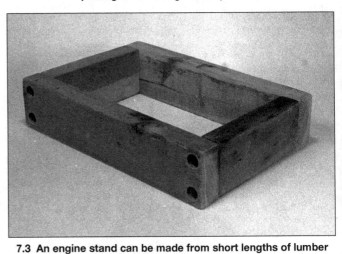

7.3 An engine stand can be made from short lengths of lumber and lag bolts or nails

17 Remove the front engine mount **(see illustration)**.
18 If you're working on a liquid cooled model, disconnect the coolant hoses from the water pump (see Chapter 3). You may need to raise the engine for removal clearance.
19 Have an assistant help you lift the engine out of the frame from the left side.
20 Slowly lower the engine to a suitable work surface.

Installation

21 Have an assistant help lift the engine into the frame so it rests on the jack and block of wood. Use the jack to align the mounting holes, then install the bolts, nuts and washers. Don't forget to install the ground cable(s).
22 The remainder of installation is the reverse of the removal steps, with the following additions:
   a) *Use new gaskets at all exhaust pipe connections.*
   b) *Adjust the throttle cable and transmission linkage following the procedures in Chapter 1.*
   c) *Fill the engine with coolant and bleed the cooling system (if equipped), also following the procedures in Chapter 1.*
   d) *If the engine was overhauled, refer to the break-in procedures at the end of this Chapter.*
   e) *Run the engine and check for oil, coolant or exhaust leaks.*

7 Engine disassembly and reassembly - general information

Refer to illustrations 7.2 and 7.3

1 Before disassembling the engine, clean the exterior with a degreaser and rinse it with water. A clean engine will make the job easier and prevent the possibility of getting dirt into the internal areas of the engine.
2 In addition to the precision measuring tools mentioned earlier, you will need a torque wrench and oil gallery brushes **(see illustration)**. Some new, clean engine oil of the correct grade and type (two-stroke oil, four-stroke oil or both, depending on whether it's a top-end or bottom-end overhaul), some engine assembly lube (or moly-based grease) and a tube of RTV (silicone) sealant will also be required.
3 An engine support stand made from short lengths of 2 x 4's bolted together will facilitate the disassembly and reassembly procedures **(see illustration)**. If you have an automotive-type engine stand, an adapter plate can be made from a piece of plate, some angle iron and some nuts and bolts.
4 When disassembling the engine, keep "mated" parts together that have been in contact with each other during engine operation. These "mated" parts must be reused or replaced as an assembly.
5 Engine/transmission disassembly should be done in the following general order with reference to the appropriate Sections.

   Remove the cylinder head
   Remove the cylinder
   Remove the piston
   Remove the water pump
   Remove the alternator rotor
   Separate the crankcase halves
   Remove the crankshaft and connecting rod

6 Reassembly is accomplished by reversing the general disassembly sequence.

8 Cylinder head - removal, inspection and installation

Caution: *The engine must be completely cool before beginning this procedure, or the cylinder head may become warped.*
Note: *This procedure is described with the engine in the frame. If the engine has been removed, ignore the steps which don't apply.*

Removal

Refer to illustrations 8.1, 8.5a, 8.5b and 8.7

1 If you're working on a liquid-cooled model, drain the cooling system (see Chapter 1). Disconnect the upper end of the coolant hose that connects the cylinder head to the cylinder **(see illustration)**.
2 Disconnect the spark plug wire (see Chapter 1).

Chapter 2 Part A Two-stroke engines

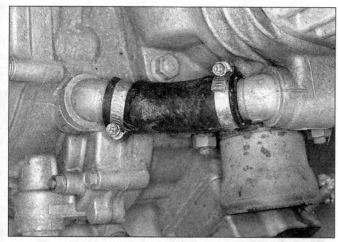

8.1 Disconnect the coolant hose from the cylinder head on liquid-cooled models

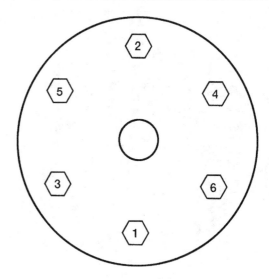

8.5a Cylinder head nut TIGHTENING sequence (250, 300 and 350)

8.5b Cylinder head nut TIGHTENING sequence (400)

8.7 The head gasket cutouts on 400 models spell UP; the word should be upright when the gasket is installed

3   Remove body components as necessary for access to the cylinder head (see Chapter 8).
4   Remove the carburetor and exhaust system (see Chapter 4).
5   Loosen the cylinder head nuts in two or three stages, in a criss-cross pattern **(see illustrations)**. Remove the nuts once they're all loose.
6   Lift the cylinder head off the cylinder. If the head is stuck, use a wooden dowel inserted into the spark plug hole to lever the head off. Don't attempt to pry the head off by inserting a screwdriver between the head and the cylinder - you'll damage the sealing surfaces.
7   Rotate the piston to the top of the cylinder or stuff a clean rag into the cylinder to prevent the entry of debris. Once this is done, remove the gasket from the cylinder **(see illustration)**.

Inspection

Refer to illustration 8.11

8   Check the cylinder head gasket and the mating surfaces on the cylinder head and cylinder for leakage, which could indicate warpage.
9   Clean all traces of old gasket material from the cylinder head and cylinder. Be careful not to let any of the gasket material fall into the cylinder or coolant passages.
10 Inspect the head very carefully for cracks and other damage. If cracks are found, a new head will be required.
11 Using a precision straightedge and a feeler gauge, check the head gasket mating surface for warpage. Lay the straightedge across the head, intersecting the head bolt holes, and try to slip a feeler gauge

8.11 Check for head warpage with a straightedge and feeler gauge in the directions shown

Chapter 2 Part A Two-stroke engines

8.12 The head gasket tab (arrow) goes toward the front of the engine

9.2 Unbolt the reed valve body and remove it from the cylinder . . .

under it, on either side of the combustion chamber **(see illustration)**. The feeler gauge thickness should be the same as the cylinder head warpage limit listed in this Chapter's Specifications. If the feeler gauge can be inserted between the head and the straightedge, the head is warped and must either be machined or, if warpage is excessive, replaced with a new one.

Installation

Refer to illustration 8.12

12 Lay the new gasket in place on the cylinder, making sure it's installed in the correct direction. On 350L models, the wide side of the metal ring around the gasket hole should be downward. The small hole in the edge of the gasket should be below the opening where the coolant hose fitting meets the cylinder head. On 400 models, the word UP cut in the gasket should be upright and the tab on the edge of the gasket should be on the exhaust side of the cylinder, on top of the tab cast into the cylinder **(see illustration 8.7 and the accompanying illustration)**. Never reuse the old gasket and don't use any type of gasket sealant.

13 Carefully lower the cylinder head over the studs.

14 Install the cylinder head nuts and tighten them evenly, in a criss-cross pattern, to the torque listed in this Chapter's Specifications.

15 The remainder of installation is the reverse of the removal steps. Be sure to refill the cooling system (see Chapter 1).

9 Reed valve - removal, inspection and installation

Removal

Refer to illustrations 9.2 and 9.3

1 Remove the carburetor (see Chapter 4).

2 Unbolt the carburetor intake tube from the cylinder **(see illustration)**.

3 Take off the intake tube and its O-ring. Pull the reed valve out of the cylinder and remove the gasket **(see illustration)**.

Inspection

Refer to illustrations 9.5 and 9.6

4 Check the reed valve for obvious damage, such as cracked or broken reeds or stoppers. Also make sure there's no clearance between the reeds and the edges where they make contact with the seats.

5 Check the clearance between the reeds and reed stoppers **(see illustration)**.

6 If necessary, remove the screws and take the assembly apart **(see illustration)**. The reed valve must be replaced as an assembly if any problems are found.

Installation

7 Installation is the reverse of the removal steps, with the following

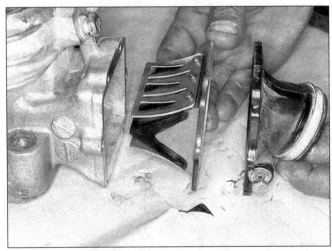

9.3 . . . then pull out the reed valve and remove the gasket

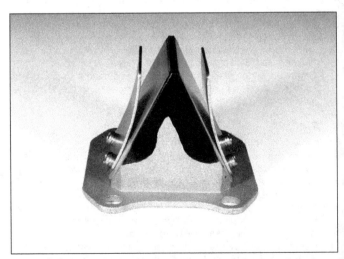

9.5 Measure the gap between the reed stoppers and reeds

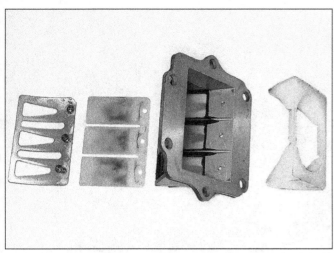

9.6 Remove the screws to detach the reed stoppers from the reed valve; the plastic insert comes with 400 models and can be retrofitted to 350 models

10.2a Loosen one of the cylinder base nuts on one side of the engine . . .

additions:
a) Use a new gasket between the reed valve assembly and cylinder.
b) Use a new O-ring between the carburetor intake tube and the reed valve assembly.
c) Tighten the intake tube bolts in a criss-cross pattern.

10 Cylinder - removal, inspection and installation

Removal

Refer to illustrations 10.2a, 10.2b and 10.4

1 Remove the cylinder head (see Section 8). Make sure the piston is positioned at the top of its stroke.
2 Loosen the four nuts securing the cylinder to the crankcase in a criss-cross pattern, then remove them **(see illustrations)**.
3 Lift the cylinder straight up off the piston. If it's stuck, tap around its perimeter with a soft-faced hammer. Don't attempt to pry between the cylinder and the crankcase, as you'll ruin the sealing surfaces.
4 Stuff clean shop rags around the piston and remove the gasket and all traces of old gasket material from the surfaces of the cylinder and the crankcase **(see illustration)**.
5 Some models have the oil check valve in the base of the cylinder (see Section 12).

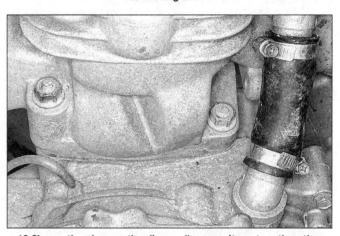

10.2b . . . then loosen the diagonally opposite nut on the other side, then loosen the remaining two nuts in a criss-cross pattern

Inspection

Refer to illustration 10.6

6 Check the top surface of the cylinder for warpage, using the same method as for the cylinder head (see Section 8). Measure along the sides, across the stud holes **(see illustration)**.

10.4 The arrowhead mark should point to the alternator side of the engine

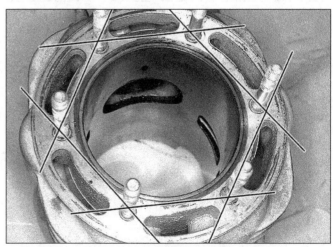

10.6 Measure cylinder surface warpage with a straightedge and feeler gauge in the directions shown

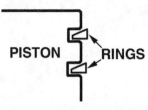

11.7a The beveled side of each ring faces upward

7 Check the cylinder walls carefully for scratches and score marks. Make sure the small decompressor port (if equipped) is unclogged **(see illustration 10.6)**.
8 Using the appropriate precision measuring tools, check the cylinder's diameter at the top, center and bottom of the cylinder bore, parallel to the crankshaft axis (see Chapter 2 Part B). Next, measure the cylinder's diameter at the same three locations across the crankshaft axis. Compare the results to this Chapter's Specifications.
9 If the cylinder walls are tapered, out-of-round, worn beyond the specified limits, or badly scuffed or scored, you can have the cylinder rebored and honed by a dealer service department or a motorcycle repair shop. If a rebore is done, oversize pistons and rings will be required as well. Polaris recommends against using the commonly available flexible hone; instead, a rigid-type hone which can enlarge the cylinder to the next oversize while maintaining the correct cylinder taper is recommended.
10 As an alternative, if the precision measuring tools are not available, a dealer service department or repair shop will make the measurements and offer advice concerning servicing of the cylinder.

Installation

11 Lubricate the piston with plenty of clean 2-stroke engine oil.
12 Install a new cylinder base gasket over the studs.
13 Install the cylinder over the studs and carefully lower it down until the piston crown fits into the cylinder liner. Push down on the cylinder, making sure the piston doesn't get cocked sideways, until the bottom of the cylinder liner slides down past the piston rings. Be sure not to rotate the cylinder, as this may snag the piston rings on the exhaust ports. A wood or plastic hammer handle can be used to gently tap the cylinder down, but don't use too much force or the piston will be damaged.
14 The remainder of installation is the reverse of the removal steps.

11 Piston and rings - removal, inspection and installation

1 The piston is attached to the connecting rod with a piston pin that's a slip fit in the piston and connecting rod needle bearing. It is secured by circlips that fit into grooves in the piston.
2 Before removing the piston from the rod, stuff a clean shop towel into the crankcase hole, around the connecting rod. This will prevent the circlips from falling into the crankcase if they are inadvertently dropped.

Removal

3 The piston should have an F mark or arrow on its crown that goes toward the magneto side of the engine **(see illustration 10.4)**. If this mark is not visible due to carbon buildup, scribe an arrow into the piston crown before removal.
4 The remainder of removal is the same as for 4-stroke pistons (Chapter 2 Part B).

Inspection

5 This is the same as for 4-stroke pistons, except that a needle bearing is used inside the connecting rod small end. If the piston pin wobbles in the bearing, and the pin is not worn, replace the bearing.

11.7b Center the ring gaps on the dowel in each ring groove (arrows)

Installation

Refer to illustrations 11.7a and 11.7b

6 Install the piston with its arrow mark toward the alternator side of the engine. Lubricate the pin and the rod needle bearing with 2-stroke oil of the type listed in the Chapter 1 Specifications.
7 Ring installation is basically the same as for 4-stroke engines (see Chapter 2 Part B). The beveled side of each ring faces upward **(see illustration)**. Center each of the ring end gaps on the dowel in the upper edge of the ring land **(see illustration)**.

12 Oil pump - check valve test, cable adjustment and pump test

1 The check valve on all models is in the inlet line from the pump. It lets oil flow into the engine, but keeps crankcase pressure out of the oil line. It's mounted on the oil pump, on the engine cylinder or in the banjo fitting at the cylinder, depending on model.

Check valve test

250, 350 and 400 models

Refer to illustrations 12.2a and 12.2b

2 If you're working on a 250, 350 or 400 model, locate the check valve. It's mounted on the cylinder or in the banjo fitting **(see illustrations)**.

12.2a The oil check valve on some models is located on the cylinder (arrow) . . .

Chapter 2 Part A Two-stroke engines

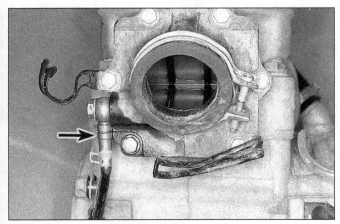

12.2b ... while on others it's in the banjo fitting (arrow)

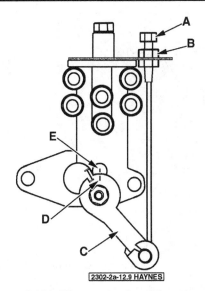

12.10a Oil pump details - 250 engine

- A Cable adjuster
- B Locknut
- C Oil pump lever
- D Idle index
- E Static line

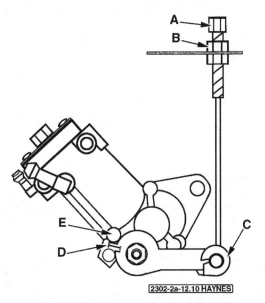

12.10b Oil pump details - 300 engine

- A Cable adjuster
- B Locknut
- C Oil pump
- D Lever stop
- E Stopper post

12.13a On Type II 400 models, remove the oil pump cover...

3 Disconnect the line from the pump. Attach a pressure/vacuum pump (Mity Vac or equivalent) to the line fitting on the check valve.
4 Apply pressure (not vacuum) to the check valve. It should open at 2 to 5 psi. If not, replace it.

300 models

5 Locate the check valve. It's mounted on the oil pump.
6 Disconnect the line from the pump. Attach a pressure/vacuum pump (Mity Vac or equivalent) to the line fitting on the check valve.
7 Apply vacuum (not pressure) to the check valve. It should open at 3 to 7 inches of mercury. If not, replace it.

Cable adjustment

8 Check the idle speed and throttle lever freeplay (see Chapter 1). Make any necessary adjustments before adjusting the oil pump.

1985 through 1996 250, 1994 and 1995 300 models

9 Push the throttle lever very lightly just far enough to take up the lever freeplay. While an assistant holds the throttle lever in this position, check the position of the alignment marks on the pump lever and pump body. If they're not aligned with each other, loosen the locknut, turn the cable adjuster to align the marks and tighten the locknut.

1997 and later 250, 1996 and later 300 models

Refer to illustrations 12.10a and 12.10b

Warning: *Place the transmission in Neutral before doing this procedure so the vehicle doesn't accidentally drive off.*

9 Run the engine at idle speed.
10 Operate the throttle lever. The lever on the oil pump should start to move just as engine speed increases **(see illustrations)**. When the throttle lever is released, the stopper arm built into the pump lever should rest against the stopper post cast into the oil pump body.
11 If necessary, loosen the locknut and turn the cable adjuster so the pump lever begins to move at the same time as the throttle piston **(see illustration 12.10a or 12.10b)**. Tighten the locknut and recheck the adjustment.
12 Recheck idle speed and throttle lever freeplay to make sure they didn't change when the pump lever was adjusted.

350 and 400 models

Refer to illustrations 12.13a, 12.13b, 12.14 and 12.15

Warning: *Place the transmission in Neutral before doing this procedure so the vehicle doesn't accidentally drive off.*

13 If you're working on a 400 with a Type II pump, remove the cover **(see illustrations)**.

12.13b ... and inspect the gasket

12.14 Oil pump details (400 Type II; 400 Type I and 350 similar)

A   Index lines
B   Pump mounting screws
C   Pump housing screws (400 Type II only)

14 With the engine idling, move the throttle lever until engine speed just starts to increase. At this point, the index marks on the cable pulley and pump body should be aligned with each other **(see illustration)**.

15 If the marks aren't aligned, loosen the cable adjuster locknut **(see illustration)**. Turn the adjuster until the marks align when they're supposed to, then tighten the locknut.

Oil pump test

Refer to illustrations 12.19a and 12.19b

16 Drain the gasoline from the fuel tank and replace it with a 40:1 mixture of gasoline and 2-stroke oil. **Caution:** *This test requires running the engine with the oil pump disconnected. Don't run it on plain gasoline or the engine may seize.*

17 Test the check valve and bleed the oil pump as described above.

18 Disconnect the oil inlet line from the engine.

19 Start the engine and let it idle. Twist the cable pulley or lever on the oil pump to the full open position and loosen the bleed screw **(see illustrations)**. Oil should pump from the bleed screw every few seconds.

20 If oil doesn't flow, check the oil lines and oil tank vent line for clogging or pinching. Check the inline filter in the oil lines for clogging (see Chapter 1). If the lines and filter are clear, the oil pump itself is probably defective.

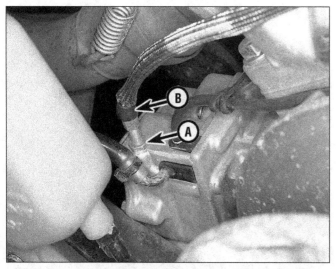

12.15 Loosen the locknut (A) and turn the cable adjuster (B) to align the marks

12.19a Oil pump bleed screw (250 models)

12.19b Rotate the pulley as shown and loosen the bleed screw (arrow)

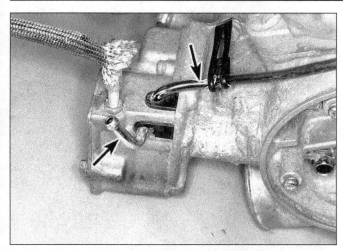

13.1 Disconnect the oil inlet and outlet lines from the pump fittings (arrows) (400 Type II shown)

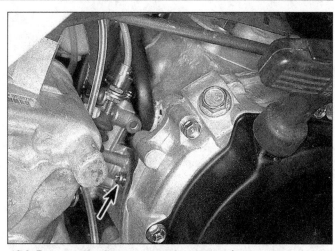

13.3 Remove the oil pump mounting screws (outer screw shown, inner screw hidden)

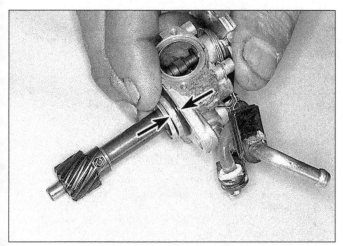

13.5a Measure the depth from the pump shoulder to the mounting surface on the pump . . .

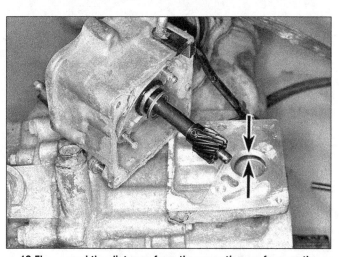

13.5b . . . and the distance from the mounting surface on the engine to the pump bushing

13 Oil pump - removal, inspection, installation and bleeding

Removal

Refer to illustrations 13.1 and 13.3

1 Disconnect and plug the oil inlet and outlet lines **(see illustration)**.
2 If you're working on a 400 with a Type II pump, remove the cover **(see illustration 12.13)**.
3 Remove the pump mounting screws and take the pump off **(see illustration 12.14 and the accompanying illustration)**.

Inspection

4 Check the pump for visible wear and damage. Individual pump parts aren't available, so you'll need to replace the pump as an assembly if it's worn.

Installation

Refer to illustrations 13.5a and 13.5b

5 If the pump, its shims, its bushing or the crankcase have been replaced, you'll need to measure the pump installed clearance. Using a vernier caliper, measure the depth from the pump shoulder to the pump mating surface, and from the crankcase mating surface to the bushing in the oil pump bore **(see illustrations)**. The difference between the two measurements is the installed clearance. If the existing shims will bring the clearance within the specified range, use them; otherwise, add or reduce shims as needed to obtain the correct clearance.
6 The remainder of installation is the reverse of the removal steps. Bleed the pump after installation as described below.

Oil pump bleeding

7 Make sure the engine oil reservoir is full. If you're working on a 400 model with a Type II oil pump, remove the cover **(see illustration 12.13a)**.
8 With the engine off, open the pump bleed screw exactly one turn (don't remove the screw). Let oil flow from around the screw for ten seconds, then tighten the screw.
9 Run the engine at idle. Hold the lever on the oil pump in the full open position for 20 seconds on all except 3550 models, or 1-1/2 minutes on 350 models. This will remove any remaining air from the system.
10 Release the lever. Reinstall the oil pump cover (if removed).

14 Recoil starter - removal, inspection and installation

Warning: *The spring in the recoil starter may fly out suddenly and cause severe eye injury. Wear eye protection, a face shield and heavy gloves during these procedures.*

14.2a Remove the cover screws (this is a 250) . . .

14.2b . . . and this is a 400

Early 250 models

1 If you're working on an early 250 model, refer to Section 22 in Part B of this Chapter. The design of the recoil starter on early 250 models is very similar to that used for four-stroke engines, except that a center nut is used in place of the center bolt and washer used on four-strokes.

Later 250 and all 300, 350 and 400 models

Refer to illustrations 14.2a and 14.2b

2 Unbolt the recoil starter cover from the engine **(see illustrations)**. Hold the cover while the last bolt is removed, then let the cover rotate to release the tension of the recoil spring.
4 Remove the cover once the spring tension has been released. If you're only planning to replace the pull rope, you can do it now without removing any more parts.

Rope replacement

5 Unwind the rope from the reel and pull it out. Separate the other end of the rope from the handle.
6 Pass the new rope through the hole in the cover, then through the reel. Tie a knot in the end.
7 Position the flywheel cover and reel on the engine, then turn the cover clockwise to wind up the rope. When the rope is wound all the way into the cover, turn the cover three more turns clockwise to preload the spring and install the cover bolts or screws.

Disassembly

8 Remove the flywheel housing from the engine.
9 Unhook the pawl return spring and take it out.
10 Support the recoil starter in a vise, with the corners of the vise jaws gripping the sides of the ratchet pawl bracket. Wrap a strap wrench around the outside of the rope reel and unscrew it from the threaded shaft on the ratchet pawl bracket.
11 Detach the ratchet pawl bracket and pawl, the recoil cup (1993 and earlier only), the friction ring and the friction spring from the flywheel housing. Leave the spring retainer in place unless you plan to remove the spring.

Inspection

12 Check all parts for wear and damage and replace any that have visible problems.
13 If you're installing a recoil spring, hook its outer end into the housing groove. Wind the spring in counterclockwise, toward the center of the spring. Lubricate the spring with cable lube or lightweight, low-temperature grease, then install the spring retaining plate.
14 Lubricate the center bushing and install it, engaging its slot with the inner end of the recoil spring.

15 Lubricate the center seal in the flywheel housing.
16 Make sure the friction spring is on the friction ring. Install them, engaging the friction spring with the retaining plate tab and noting the direction of the friction ring tab.
17 Align the pin on the ratchet pawl bracket with the hole in the square drive and install the ratchet pawl bracket, making sure it aligns with the center bushing. Hold the ratchet pawl bracket in place with one hand, turn the assembly over and thread the reel onto the shaft of the pawl bracket.
18 Place the ratchet pawl bracket in the corners of a vise (as in Step 10) and tighten the reel securely, using your hands only (not the strap wrench).
19 Pass the rope through the handle hole in the cover and into the reel, then tie a knot in the end. Pass the other end through the handle and knot it.
20 Place the cover over the reel, then place the cover and reel on the flywheel housing and turn the cover clockwise to wind up the rope. Turn the cover three more turns to preload the spring, then install the cover screws or bolts.
21 Hook the pawl return spring to the ratchet pawl bracket.
22 Pull the rope to test for correct assembly.
23 Apply a coat of Loctite 515 Gasket Eliminator to the mating surfaces of the flywheel housing and crankcase, then install the flywheel housing.

15 Counterbalancer (350 and 400 models) - removal, inspection and installation

Removal

Refer to illustrations 15.3a, 15.3b, 15.4a, 15.4b, 15.6, 15.7a, 15.7b and 15.9

1 Remove the engine from the vehicle and remove the recoil starter (see Sections 6 and 14).
2 Remove the starter motor, alternator rotor and stator (see Chapter 5).
3 Unscrew the slotted nut from the crankshaft with Polaris tool 2870967 or equivalent **(see illustrations)**. The nut has left-hand threads (unscrews clockwise). **Note:** *If you don't have the special tool, you may be able to get the crankcase cover off without removing the nut. However, this will probably damage the cover seal.*
4 Unbolt the crankcase cover **(see illustration)**. Tap the cover gently with a soft-faced mallet to free it from the crankcase, then lift it off and locate the dowels **(see illustration)**.
5 Remove the water pump impeller and housing from the counterbalancer (see Chapter 3).

15.3a Engage the special wrench with the slots in the nut . . .

15.3b . . . and turn it with a breaker bar

15.4a Remove the cover bolts and nuts (arrows) . . .

15.4b . . . and take the cover off for access to the water pump and counterbalancer

6   Turn the crankshaft so the alignment marks on the gears are directly opposite each other **(see illustration)**.
7   If you haven't already removed the slotted nut from the crankshaft, do it now (see Step 3). Remove the O-ring, washer bearing and collar from the crankshaft **(see illustrations)**. Pull the gear off the

crankshaft. If it won't come by hand, carefully pry it off with a pair of screwdrivers.
8   Unbolt the counterbalancer bracket, but don't remove it yet **(see illustration 15.6)**.
9   Attach a puller (Polaris tool 2870968 or equivalent) to the counter-

15.6 Align the counterbalancer marks (center arrows) and unbolt its bracket (upper and lower arrows)

15.7a Remove the crankshaft end components (O-ring shown)

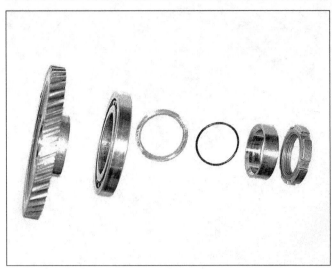

15.7b Crankshaft end components include the nut, collar, O-ring, guide washer, bearing and gear

15.9 This special tool is used to pull the counterbalancer

15.12a Check the counterbalancer gear and bearing for wear or damage

15.12b The opposite counterbalancer bearing may remain in the crankcase bore

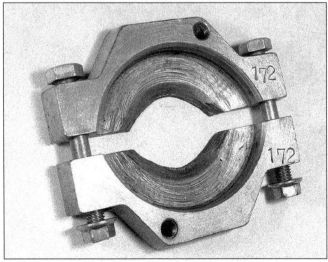

15.13 If the bearing stays on the counterbalancer, remove it with a bearing splitter (and a press if necessary)

balancer shaft **(see illustration)**.

10 Heat the counterbalancer bearing areas on the outside of the case. A powerful hand-held hair dryer is the safest way to do this. A small propane torch can be used without damaging the crankcase, provided the flame is kept at least one inch from the case and it isn't applied for more than two minutes. However, using a torch may heat oil and gasoline residue on the cases to the point of being highly explosive, then the flame may set an explosion off.

11 With the cases heated, tighten the puller nut to withdraw the counterbalancer from the crankcase.

Inspection

Refer to illustrations 15.12a, 15.12b and 15.13

12 Check the counterbalancer and its bearing for obvious wear and damage. The bearing may have come out with the counterbalancer; if it didn't, look inside the crankcase and inspect the bearing inside the far side of the case **(see illustrations)**.

13 If the bearings are loose, rough or noisy when rotated, replace them. Use a puller or a press and bearing splitter to remove the bearing from the counterbalancer **(see illustration)**. If the far side bearing has remained in the crankcase, remove it with a slide hammer fitted with puller jaws.

Chapter 2 Part A Two-stroke engines

16.8a Unbolt the rear engine mount bracket from the crankcase

16.8b Remove the Woodruff keys (arrows) before separating the crankcase halves

16.9 Crankcase bolt TIGHTENING sequence (bolts 6 and 7 used on 350/400 models only)

16.10a Pry the crankcase apart at this pry point . . .

Installation

14 Installation is the reverse of the removal steps, with the following additions:
   a) *Make sure the counterbalancer bracket is in place before installing the counterbalancer in the crankcase.*
   b) *Heat the counterbalancer bearing areas of the crankcase, then reinstall the counterbalancer with a press.*

16 Crankcase - disassembly and reassembly

1 To examine and repair or replace the crankshaft, connecting rod, bearings and transmission components, the crankcase must be split into two parts.

Disassembly

Refer to illustrations 16.8a, 16.8b, 16.9, 16.10a, 16.10b, 16.11a and 16.11b

2 Remove the engine from the vehicle (see Section 6).
3 Remove the carburetor (see Chapter 3).
4 Remove the alternator (see Chapter 4).
5 Remove the cylinder head, cylinder and piston (see Sections 8, 10 and 11).
6 Remove the oil pump (see Section 13).
7 Remove the recoil starter (see Section 14).
8 Check carefully to make sure there aren't any remaining components that attach the halves of the crankcase together, such as the rear engine mount bracket **(see illustration)**. Remove the Woodruff keys from the crankshaft if you haven't already done so **(see illustration)**.
9 Loosen the crankcase bolts evenly in two or three stages, then remove them **(see illustration)**.
10 Heat the crankcase in the bearing support area around the

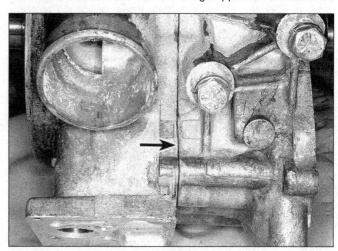

16.10b . . . and this one

16.11a Shims (arrow) are used to adjust crankshaft end play

16.11b Make sure the dowels (arrows) are in place before assembly

crankshaft. **Warning:** *Before heating the case, wash it thoroughly with soap and water so no explosive fumes are present. Also, don't use a flame to heat the case.* Carefully tap on each end of the crankshaft to ease the crankcase apart and separate the halves **(see illustrations)**. Don't pry against the mating surfaces or they'll develop leaks.

11 Locate the crankshaft shims and the crankcase dowels **(see illustrations)**.

12 Refer to Sections 17 and 18 for information on the internal components of the crankcase.

Reassembly

Refer to illustration 16.21

13 Before reassembling the case halves, refer to Chapter 2, Part B and check the crankshaft bearing end play.

14 Remove all traces of old sealant from the crankcase mating surfaces with a sharpening stone or similar tool. Be careful not to let any fall into the case as this is done and be careful not to damage the mating surfaces.

15 With the crankshaft end seals removed, heat the PVT side of the crankcase to expand it, then lower the crankshaft into the case. Make sure the connecting rod is aligned with its opening in the top of the crankcase.

16 Check to make sure the dowel pins are in place in their holes in the crankcase mating surface **(see illustration 16.11b)**.

17 Coat one of the crankcase mating surfaces with Loctite 515 Gasket Eliminator or equivalent.

18 Heat the alternator half of the crankcase, then carefully place it onto the PVT crankcase half. While doing this, make sure the crankshaft bearing fits into its bore in the alternator half. Also make sure the connecting rod is properly aligned with its opening in the top of the crankcase.

19 Install the crankcase bolts and tighten them so they are just snug. Then tighten them evenly in two or three stages to the torque listed in this Chapter's Specifications.

20 Turn the crankshaft to make sure it turns freely.

21 Install the crankshaft end seals **(see illustration)**.

22 The remainder of assembly is the reverse of disassembly.

17 Crankcase - inspection and servicing

1 Separate the crankcase and remove the crankshaft (see Sections 16 and 18).

2 Clean the crankcase halves thoroughly with new solvent and dry them with compressed air. All oil passages should be blown out with compressed air and all traces of old gasket should be removed from the mating surfaces. **Caution:** *Be very careful not to nick or gouge the*

16.21 The crankshaft end seals are critical to 2-stroke performance

crankcase mating surfaces or leaks will result. Check both crankcase halves very carefully for cracks and other damage.

3 Replace the oil seals whenever the crankcase is disassembled **(see illustration 16.21)**. The crankshaft seals are critical to the performance of two-stroke engines, so they should be replaced whenever the crankcase is disassembled, even if they look perfectly alright.

4 If any damage is found that can't be repaired, replace the crankcase halves as a set.

5 Assemble the case halves (see Section 16) and check to make sure the crankshaft and the transmission shafts turn freely.

18 Crankshaft and connecting rod - removal, inspection and installation

Crankshaft

Removal

1 The crankshaft is removed as part of the crankcase disassembly procedure (Section 16).

Inspection

2 Clean the bearings on the ends of the crankshaft. Lubricate them with clean 2-stroke oil, then spin the outer race of each bearing with a finger. If the bearings don't turn smoothly, they'll need to be replaced.

3   Slip an extra (third) bearing onto the PVT end of the crankshaft and place it in V-blocks. Turn the crankshaft and measure runout with dial indicator. If it exceeds the amount listed in this Chapter's Specifications, it may be possible to straighten the crankshaft. This is a specialized job that should be done by a Polaris dealer or a machine shop familiar with two-stroke crankshafts.
4   Slip a feeler gauge between the lower end of the connecting rod and crankshaft to measure the clearance (Chapter 2 Part B). If it's beyond the range listed in this Chapter's Specifications, replace the crankshaft and connecting rod as an assembly.
5   Rock the small end of the connecting rod from side-to-side and check for play. Polaris doesn't provide specifications for play in this situation, but if there's noticeable play, the connecting rod big-end bearing is probably worn and the crankshaft and connecting rod should be replaced as an assembly.
6   Check the crankshaft and splines for visible wear or damage, such as step wear of the splines or scoring. If any of these conditions are found, replace the crankshaft and connecting rod as an assembly.

Shim selection

7   This is similar to 4-stroke crankshafts, described in Chapter 2 Part B. Measure the width of all three ball bearings while they're installed on the crankshaft (two on the PVT end and one on the alternator end). Measure the depth of the bearing bore in each crankcase half and add these together. Select shims that will give a crankshaft end play within the range listed in this Chapter's Specifications.

Installation

8   Refer to Section 16 for the remainder of the crankshaft installation and case assembly procedure.

19   Recommended start-up and break-in procedure

1   This procedure should be followed each time the piston and rings, cylinder, crankshaft or crankshaft bearings are replaced. Make sure the transmission and controls, especially the brakes, function properly before riding the machine.
2   Pre-mixing oil and gasoline is normally not necessary on these models, since it's done automatically by the oil injection system. However, the first tank of fuel after an overhaul should be pre-mixed with 2-stroke engine oil at a ratio of one part oil to 40 parts gasoline.
3   Refer to Section 13 and bleed the oil pump. Fill the oil reservoir with the specified 2-stroke oil.
4   Start the engine and warm it up completely before operating the machine.
5   Operate the machine normally, but avoid sustained full-throttle operation for the first two tanks of fuel. After the first tank is used up, check the oil reservoir. Oil level should have dropped, indicating that the oil pump is working.
6   If the vehicle is liquid cooled, check coolant level (see Chapter 1).

Notes

Chapter 2 Part B
Four-stroke engines

Contents

	Section		Section
Cam chain tensioner - removal and installation	8	General information	1
Counterbalancer - removal and installation	20	Initial start-up after overhaul	23
Crankcase - disassembly and reassembly	17	Major engine repair - general note	4
Crankcase components - inspection and servicing	19	Oil pump - removal, inspection and installation	18
Crankshaft and connecting rod - removal, inspection and installation	21	Operations possible with the engine in the frame	2
Cylinder head and valves - disassembly, inspection and reassembly	12	Operations requiring engine removal	3
		Piston - removal, inspection and installation	14
		Piston rings - installation	15
Cylinder head covers - removal, inspection and installation	7	Recoil starter - removal and installation	22
Cylinder head - removal and installation	10	Recommended break-in procedure	24
Cylinder - removal, inspection and installation	13	Rocker arms and camshaft - removal, inspection and installation	9
Engine - removal and installation	5		
Engine disassembly and reassembly - general information	6	Valves/valve seats/valve guides - servicing	11
External oil lines, check valve and tank – removal and installation	16		

Specifications

General
Bore
    425 .... 3.4606 inch
    500 .... 3.622 inch
Stroke
    425 .... 2.758 inch
    500 .... 2.953 inch
Displacement
    425 .... 425 cc
    500 .... 498 cc

Rocker arms
Rocker shaft outside diameter .... 0.8656 to 0.8661 inch
Rocker arm inside diameter .... 0.8669 to 0.8678 inch
Shaft-to-arm clearance
    Standard .... 0.00008 to 0.0021 inch
    Limit .... 0.0039 inch

Camshaft

Lobe height	
Standard	1.2884 to 1.2924 inch
Limit	1.2766 inch
Journal diameter	1.4935 to 1.4941 inch
Bearing inside diameter	1.4963 to 1.4970 inch
Bearing clearance	
Standard	0.0022 to 0.0035 inch
Limit	0.0039 inch
Camshaft runout	Not specified
Cam chain tensioner spring length	2.320 inch
Cam chain free length (20 pitches)	5.407 inch

Cylinder head, valves and valve springs

Cylinder head warpage limit	0.002 inch
Cylinder head height	3.870 inch
Valve stem runout	not specified
Valve stem diameter	
Intake	0.2343 to 0.2348 inch
Exhaust	0.2341 to 0.2346 inch
Valve guide inside diameter (intake and exhaust)	0.2362 to 0.2367 inch
Valve guide protrusion above head	0.689 to 0.709 inch
Valve stem oil clearance	
Standard	
Intake	0.0014 to 0.0024 inch
Exhaust	0.0016 to 0.0026 inch
Limit (intake and exhaust)	0.0059 inch
Valve seat width	
Intake	
Standard	0.028 inch
Limit	0.055 inch
Exhaust	
Standard	0.039 inch
Limit	0.071 inch
Valve spring free length	
Painted orange	
Standard	1.7342 inch
Limit	1.656 inch
Painted yellow	
Standard	1.654 inch
Limit	1.575 inch
Valve spring bend limit	0.075 inch
Cylinder	
Surface warpage limit	0.002 inch
Bore diameter	
425	3.4606 to 3.4614 inch
500	3.6216 to 3.6224 inch
Taper and out-of-round limits	0.002 inch
Overbore limit	0.020 inch
Pistons and pins	
Piston diameter	
425	
Standard	3.4596 to 3.4600 inch
0.010 overbore	3.4695 to 3.4699 inch
0.020 overbore	3.4793 to 3.4797 inch
500	
Standard	3.6206 to 3.6210 inch
0.010 overbore	3.6304 to 3.6310 inch
0.020 inch	3.6403 to 3.6407 inch
Diameter measurement point	0.2 inch from bottom of skirt
Piston-to-cylinder clearance	
Standard	0.0006 to 0.0018 inch
Limit	0.0024 inch
Piston pin bore	
In piston	0.9055 to 0.9057 inch
In connecting rod	0.9058 to 0.9063 inch
Piston pin outer diameter	0.9053 to 0.9055 inch
Piston pin-to-piston clearance	0.0002 to 0.0003 inch
Piston pin-to-connecting rod clearance	0.0003 to 0.0010 inch

Ring side clearance
  Top
    Standard .... 0.0016 to 0.0031 inch
    Limit .... 0.0059 inch
  Second
    Standard .... 0.0012 to 0.0028 inch
    Limit .... 0.0059 inch
Ring end gap
  Standard (top, second, oil ring rails) .... 0.0079 to 0.0138 inch
  Limit (top and second) .... 0.039 inch
  Limit (oil ring rails) .... 0.059 inch

Oil pump and check valve

Check valve spring free length .... 1.450 inch
Outer rotor to body clearance
  Standard .... 0.001 to 0.003 inch
  Limit .... 0.004 inch
Inner to outer rotor clearance
  Standard .... 0.005 inch
  Limit .... 0.008 inch
Side clearance (rotors to straightedge)
  Standard .... 0.001 to 0.003 inch
  Limit .... 0.004 inch
Oil pump shaft end play in crankcase .... 0.008 to 0.016 inch

Counterbalancer

Shaft end play .... 0.008 to 0.016 inch

Crankshaft

Connecting rod side clearance
  Standard .... 0.0039 to 0.0256 inch
  Limit .... 0.0315 inch
Connecting rod big end radial clearance
  Standard .... 0.0004 to 0.0015 inch
  Limit .... 0.002 inch
Runout limit .... 0.0024 inch
End play .... 0.008 to 0.016 inch

Torque specifications

Cylinder head covers .... 72 inch-lbs
Cam chain tensioner
  Cap bolt .... 17 ft-lbs
  Mounting bolts .... 72 inch-lbs
Cam sprocket bolts .... 72 inch-lbs
Rocker assembly mounting bolts .... 108 inch-lbs
Rocker shaft locating bolts .... 72 inch-lbs
Cam chain rear side guide bolt .... 72 inch-lbs
Cylinder head bolts
  Small .... 72 inch-lbs
  Large
    First stage .... 22 ft-lbs
    Second stage .... 51 ft-lbs
    Third stage .... Loosen 1/2 turn
    Fourth stage .... Loosen another 1/2 turn
    Fifth stage .... Tighten to 11 ft-lbs
    Sixth stage .... Tighten 1/4 turn
    Seventh stage .... Tighten another 1/4 turn
Cam sprocket slotted nut on crankshaft .... 35 to 51 ft-lbs
Cylinder base bolts
  Small .... 72 inch-lbs
  Large .... 46 ft-lbs
External oil pipe union bolts .... 20 ft-lbs
Check valve .... 14 to 19 ft-lbs
Oil pump bolts .... 60 to 78 inch-lbs
Crankcase bolts .... 14 ft-lbs

1 General information

The engine is of the liquid-cooled, single-cylinder four-stroke design. The four valves are operated by an overhead camshaft which is chain driven off the crankshaft.

The crankcase incorporates a dry sump, pressure-fed lubrication system which uses a gear-driven rotor-type oil pump, an filter and a strainer screen. A one-way valve keeps oil stored in the tank from draining into the crankcase when the engine is off.

A counterbalancer, mounted in the crankcase and driven by the crankshaft, reduces engine vibration.

2 Operations possible with the engine in the frame

The components and assemblies listed below can be removed without having to remove the engine from the frame. If, however, a number of areas require attention at the same time, removal of the engine is recommended.

Transmission
Starter motor and starter clutch
Alternator rotor and stator
Cam chain tensioner
Camshaft
Rocker arm assembly
Cylinder head
Cylinder and piston

3 Operations requiring engine removal

It is necessary to remove the engine/transmission assembly from the frame and separate the crankcase halves to gain access to the following components:

Crankshaft and connecting rod
Oil pump
Counterbalancer

4 Major engine repair - general note

1 It is not always easy to determine when or if an engine should be completely overhauled, as a number of factors must be considered.
2 High mileage is not necessarily an indication that an overhaul is needed, while low mileage, on the other hand, does not preclude the need for an overhaul. Frequency of servicing is probably the single most important consideration. An engine that has regular and frequent oil and filter changes, as well as other required maintenance, will most likely give many miles of reliable service. Conversely, a neglected engine, or one which has not been broken in properly, may require an overhaul very early in its life.
3 Exhaust smoke and excessive oil consumption are both indications that piston rings and/or valve guides are in need of attention. Make sure oil leaks are not responsible before deciding that the rings and guides are bad. Refer to Chapter 1 and perform a cylinder compression check to determine for certain the nature and extent of the work required.
4 If the engine is making obvious knocking or rumbling noises, the connecting rod and/or main bearings are probably at fault.
5 Loss of power, rough running, excessive valve train noise and high fuel consumption rates may also point to the need for an overhaul, especially if they are all present at the same time. If a complete tune-up does not remedy the situation, major mechanical work is the only solution.
6 An engine overhaul generally involves restoring the internal parts to the specifications of a new engine. During an overhaul the piston rings are replaced and the cylinder walls are bored and/or honed. If a

5.6 Disconnect the breather hose (A), oil feed hose (B) and return hose (C)

rebore is done, then a new piston is also required. The crankshaft and connecting rod are permanently assembled, so if one of these components needs to be replaced both must be. Generally the valves are serviced as well, since they are usually in less than perfect condition at this point. While the engine is being overhauled, other components such as the carburetor and the starter motor can be rebuilt also. The end result should be a like-new engine that will give as many trouble-free miles as the original.
7 Before beginning the engine overhaul, read through all of the related procedures to familiarize yourself with the scope and requirements of the job. Overhauling an engine is not all that difficult, but it is time consuming. Plan on the vehicle being tied up for a minimum of two (2) weeks. Check on the availability of parts and make sure that any necessary special tools, equipment and supplies are obtained in advance.
8 Most work can be done with typical shop hand tools, although a number of precision measuring tools are required for inspecting parts to determine if they must be replaced. Often a dealer service department or repair shop will handle the inspection of parts and offer advice concerning reconditioning and replacement. As a general rule, time is the primary cost of an overhaul so it doesn't pay to install worn or sub-standard parts.
9 As a final note, to ensure maximum life and minimum trouble from a rebuilt engine, everything must be assembled with care in a spotlessly clean environment.

5 Engine - removal and installation

Note: *Engine removal and installation should be done with the aid of an assistant to avoid damage or injury that could occur if the engine is dropped. A hydraulic floor jack should be used to support and lower the engine if possible (they can be rented at low cost).*

Removal

Refer to illustrations 5.6, 5.13, 5.15a, 5.15b, 5.15c, 5.16 and 5.18

1 Drain the engine oil and coolant (see Chapter 1). Remove the oil filter from the engine and cover its opening with a clean rag.
2 Disconnect the negative battery cable from the engine (see Chapter 1).
3 Remove the seat, both side covers, the fuel tank, the rear cargo rack and the rear body panel (see Chapter 8).
4 Remove the fuel tank, carburetor, exhaust system and air cleaner housing (see Chapter 4). Plug the carburetor intake opening with rags.
5 Disconnect the spark plug wire (see Chapter 1).
6 Disconnect the engine breather hose and oil tank hoses from the engine **(see illustration)**. Remove the oil tank (Section 16).

Chapter 2 Part B Four-stroke engines

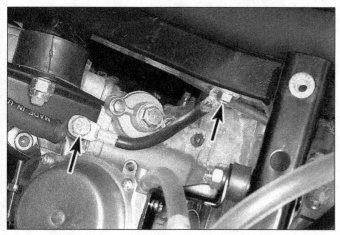

5.13 Disconnect the upper ground cable (arrows)

7   If you're working on a 4wd or 6wd model, remove the center drive chain guard and the drive chain (see Chapter 6).

8   If you're working on a 4wd or 6wd model with an auxiliary footbrake, remove the brake actuator arm (see Chapter 7).

9   Remove the Polaris Variable Transmission (PVT) and its air ducts (see Chapter 2 Part C).

9   Look at the vehicle and decide whether you want to remove the alternator rotor and stator (Chapter 5). This isn't strictly necessary, but will provide extra clearance to remove the engine.

10 Label and disconnect the following wires (refer to Chapter 5 for component location if necessary):

   Ignition pulse generator and alternator (if you didn't remove the alternator)
   Coolant temperature and neutral switches

11 Remove the starter motor (see Chapter 5).

12 Remove the coolant reservoir tank and disconnect the hose from the thermostat housing on the engine (see Chapter 3).

13 Remove the ground cable at the top of the engine **(see illustration)**.

14 Support the engine securely from below.

15 Remove the engine mounting bolts, nuts and brackets at the top right and lower front **(see illustrations)**. Mark the locations of the studs in the frame with a felt pen or paint.

16 Unscrew the nut on the rear engine mount stud to the end of the stud, but don't remove it completely **(see illustration)**.

17 Disconnect the coolant hoses from the water pump (see Chapter). Tie the ends of the hoses up so they don't drip coolant.

18 Have an assistant help you lift the engine. As you lift, disengage the rear engine mount. Remove the engine to the left side of the vehicle **(see illustration)**.

19 Slowly lower the engine to a suitable work surface.

Installation

20 Check the rubber engine supports for wear or damage and replace them if necessary before installing the engine.

21 Slip the stud on the rear engine mount into the slot in its bracket. Install the engine mounting brackets, bolts and nuts, aligning the marks made on removal. Tighten the fasteners temporarily.

22 Install the PVT and align the PVT clutches (Chapter 2 Part C).

23 Tighten the engine mounting bolts and nuts securely.

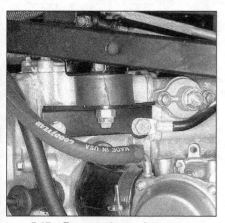

5.15a Remove the engine mount at the top right . . .

5.15b . . . and at the bottom left . . .

5.15c . . . the bottom left mount on some models is at the front of the engine

5.16 Thread the rear motor mount nut to the end of the stud, but don't remove it

5.18 Remove the engine to the left

2B-6 Chapter 2 Part B Four-stroke engines

7.3 Cylinder head upper cover bolts

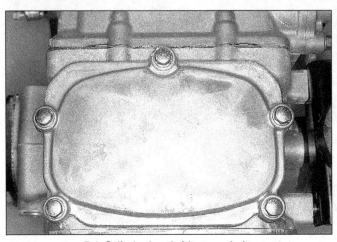

7.4 Cylinder head side cover bolts

24 The remainder of installation is the reverse of the removal steps, with the following additions:
 a) Use new gaskets at all exhaust pipe connections.
 b) Adjust the throttle cable and reverse selector cable following the procedures in Chapter 1.
 c) Fill the engine oil tank with oil until it flows form the feed hose, then attach the feed hose to the engine (Section 16). After the hoses are connected, fill the engine oil to the mark on the dipstick (see Chapter 1).
 d) Fill and bleed the cooling system as described in Chapter 1.
 e) Adjust the transmission linkage (see Chapter 2 Part C).
 f) Run the engine and check for leaks.

6 Engine disassembly and reassembly - general information

1 General disassembly and cleaning tools and methods are the same as for 2-stroke engines (see Chapter 2 Part A).
2 Engine/transmission disassembly should be done in the following general order with reference to the appropriate Sections.

 Remove the cylinder head cover and rocker assembly
 Remove the cam chain tensioner and camshaft
 Remove the cylinder head
 Remove the cylinder
 Remove the piston
 Remove the alternator rotor
 Remove the Polaris Variable Transmission (PVT)
 Separate the crankcase halves
 Remove the oil pump
 Remove the counterbalancer, crankshaft and connecting rod

3 Reassembly is accomplished by reversing the general disassembly sequence.

7 Cylinder head covers - removal, inspection and installation

1 There are two covers bolted to the cylinder head, one on top for access to the rocker arms and one on the side for access to the cam sprocket. **Note:** *The covers can be removed with the engine in the frame. If the engine has been removed, ignore the steps which don't apply.*

Removal
Refer to illustrations 7.3 and 7.4
2 Remove the fuel tank (see Chapter 4).
3 Unbolt the rocker cover from the engine **(see illustration)**. Tap it loose and lift it off. If it's stuck, don't attempt to pry it off - tap around the sides of it with a plastic hammer to dislodge it.
4 Unbolt the camshaft and cover from the engine **(see illustration)**. Again, tap it loose if it's stuck - don't pry it.

Inspection
Refer to illustrations 7.5a and 7.5b
5 Peel away the cover gaskets and clean off any traces that remain **(see illustrations)**.
6 Check the covers for cracks, warpage and damaged gasket surfaces. Replace them if problems are found.

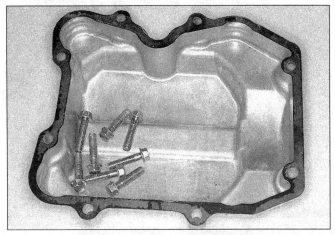

7.5a Cylinder head upper cover gasket

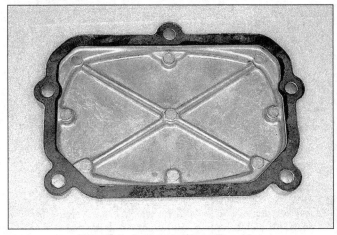

7.5b Cylinder head side cover gasket

Chapter 2 Part B Four-stroke engines

8.1 Loosen the tensioner cap bolt (center arrow); remove the mounting bolts (upper and lower arrows) . . .

8.2 . . . and pull the tensioner out of the engine

8.8 Lift the latch (arrow), compress the tensioner piston and release the latch before installing the tensioner

Installation

7   Installation is the reverse of removal. Use a new gasket. Tighten the bolts evenly to the torque listed in this Chapter's Specifications.

8   Cam chain tensioner - removal and installation

Removal

Refer to illustrations 8.1 and 8.2

1   Loosen the tensioner cap bolt **(see illustration)**.
2   Remove the tensioner mounting bolts and detach it from the cylinder block **(see illustration)**. **Caution:** *The tensioner piston locks in place as it extends. Once the tensioner bolts have been loosened, the tensioner piston must be reset before the bolts are tightened. If the bolts are loosened partway and then retightened without resetting the tensioner piston, the piston will be forced against the cam chain, damaging the tensioner or the chain.*
3   Remove the cap bolt and sealing washer from the tensioner body and wash them with solvent.

Inspection

4   Check the tensioner spring for signs of collapse, bending or breakage. Replace it if problems are found.
5   Check the tensioner for wear, especially at the plunger tip. Also make sure the ratchet lever and grooves aren't worn. The tensioner must be replaced as an assembly if there's a problem with any of the parts.

Installation

Refer to illustration 8.8

6   Clean all old gasket material from the tensioner body and engine.
7   Lubricate the friction surfaces of the components with moly-based grease.
8   Lift the tensioner latch, compress the tensioner piston into the body and release the latch to hold the piston in **(see illustration)**.
9   Install a new tensioner gasket on the cylinder. Position the tensioner body on the cylinder and install the bolts, tightening them to the torque listed in this Chapter's Specifications **(see illustration 8.1)**.
10 Install the cap bolt with a new sealing washer and tighten it to the torque listed in this Chapter's Specifications.

9   Rocker arms and camshaft - removal, inspection and installation

Note: *This procedure can be performed with the engine in the frame.*

Removal

Refer to illustrations 9.4a, 9.4b, 9.5a, 9.5b, 9.6a, 9.6b, 9.7a, 9.7b, 9.7c, 9.8a and 9.8b

1   Refer to the valve adjustment procedure in Chapter 1 and position the piston at top dead center (TDC) on its compression stroke.
2   Remove the cylinder head covers following the procedure given in Section 7.
3   Remove the camshaft chain tensioner (see Section 8).
4   Unbolt the rocker assembly and lift it off **(see illustrations)**.

9.4a Rocker assembly mounting bolts

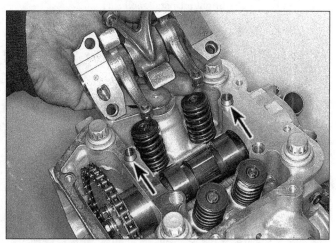

9.4b Lift the rocker assembly off the engine and locate its dowels

9.5a Remove the camshaft end cover bolts, the cover and gasket

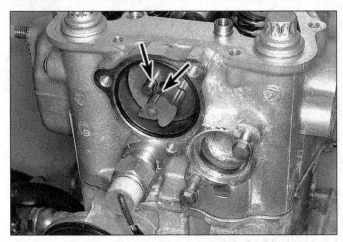

9.5b ... for access to the decompressor; on installation, the spring (right arrow) should be preloaded against the stop pin (left arrow) about 1/4 inch

Locate the dowels to make sure they don't fall into the engine.
5 Unbolt the camshaft end cover from the cylinder head **(see illustrations)**.
6 With the camshaft dowel in the straight-up position, remove the sprocket bolts **(see illustrations)**.
7 Lift the sprocket off the camshaft and disengage it from the chain **(see illustration)**. Pull up on the camshaft chain and carefully guide the camshaft out **(see illustration)**. With the chain still held taut, tie it to the engine with a piece of wire so it doesn't drop into the crankcase

(see illustration).
8 Pull the exhaust side (front) cam chain guide out of its slot **(see illustrations)**. To remove the intake side (rear) cam chain guide, it's necessary to remove the alternator and stator plate so the pivot bolt can be removed (see Chapter 5).
9 Cover the top of the cylinder head with a rag to prevent foreign objects from falling into the engine.

9.6a The camshaft dowel (arrow) must be upright on installation . . .

9.6b . . . if it rotates out of position while you undo the bolts, rotate it back

9.7a Disengage the sprocket from the chain . . .

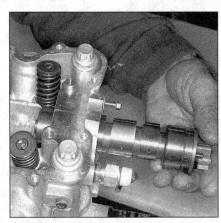

9.7b . . . pull the camshaft out, rotating as you pull . . .

9.7c . . . then tie the chain up so it doesn't drop down off the crankshaft sprocket

9.8a If you remove the stator plate, DO NOT forget the oil passage O-ring on assembly (arrow)

9.8b If you use the chain marks to set valve timing, the single bright link (upper arrow), the crankcase mark (lower arrow) AND the sprocket punch mark must all be aligned with each other

11 Slide the rocker arms off the shaft **(see illustration)**.
12 Measure the outer diameter of each rocker shaft and the inner diameter of the rocker arms with a micrometer and compare the measurements to the values listed in this Chapter's Specifications. If rocker arm-to-shaft clearance is excessive, replace the rocker arm or shaft, whichever is worn.
13 Coat the rocker shafts and rocker arm bores with moly-based grease. Install the rocker shafts and rocker arms in the cylinder head cover. There's a single rocker arm for each intake valve; a single rocker arm with two fingers operates the exhaust valves.

Camshaft

Refer to illustrations 9.14, 9.15a, 9.15b, 9.15c and 9.16

Note: *Before replacing the camshaft or cylinder head because of damage, check with local machine shops specializing in ATV or motorcycle engine work. In the case of the camshaft, it may be possible for cam lobes to be welded, reground and hardened, at a cost far lower than that of a new camshaft. If the bearing surfaces in the cylinder head or cover are damaged, it may be possible for them to be bored out to accept bearing inserts. Due to the cost of a new cylinder head it is recommended that all options be explored before condemning it as trash!*

14 Inspect the cam bearing surfaces of the cylinder head and the journal surfaces of the camshaft **(see illustration)**. Look for score marks, deep scratches and evidence of spalling (a pitted appearance).
15 Check the camshaft lobes for heat discoloration (blue appearance), score marks, chipped areas, flat spots and spalling **(see illustration)**. Measure the height of each lobe and the diameter of each bearing journal with a micrometer **(see illustration)** and compare the

9.11 Rocker assembly details

Inspection

Rocker arms and shaft

Refer to illustration 9.11

10 Check the rocker arms for wear at the cam contact surfaces and at the tips of the valve adjusting screws **(see illustration 9.4b)**. Try to twist the rocker arms from side-to-side on the shafts. If they're loose on the shafts or if there's visible wear, remove them as described below.

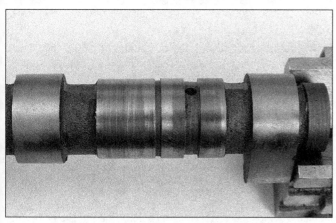

9.14 Check the cam bearing surfaces (arrows) for wear or damage

9.15a Check the cam lobes for wear - here's a good example of damage which will require replacement (or repair) of the camshaft

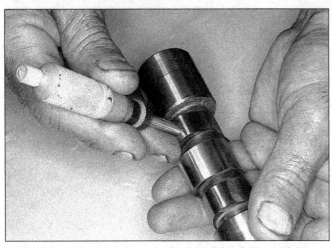

9.15b Use a micrometer to measure the height of the cam lobes . . .

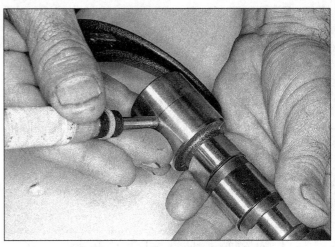

9.15c . . . and the journal diameters

results to the minimum lobe height listed in this Chapter's Specifications. If damage is noted or wear is excessive, the camshaft must be replaced. Also, be sure to check the condition of the rocker arms, as described above.

16 Check the decompressor for wear at the tip that fits inside the camshaft **(see illustration)**. Check the decompressor ball in the camshaft and make sure it isn't worn, damaged or frozen in place. If the ball is missing, replace the camshaft and decompressor.

17 Except in cases of oil starvation, the camshaft chain wears very little. If the chain has stretched excessively, which makes it difficult to maintain proper tension, replace it with a new one. To remove the chain from the crankshaft sprocket, remove the alternator rotor and stator plate (see Chapter 5).

18 Check the sprocket for wear, cracks and other damage, replacing it if necessary. If the sprocket is worn, the chain is also worn, and possibly the sprocket on the crankshaft **(see illustration 9.8b)**. If wear this severe is apparent, the entire engine should be disassembled for inspection.

19 Check the chain guides for wear or damage. If they are worn or damaged, replace them.

20 Check the thrust surface on the inside of the camshaft end cap for wear. If it's visibly worn, replace the end cap.

Installation

21 Make sure the cam bearing surfaces in the cylinder head and the shaft are clean.

22 If the decompressor spring isn't already on the shaft, install it **(see illustration 9.16)**. Coat the shaft with clean engine oil.

23 Either pull the actuator ball toward the outside of the camshaft or turn the camshaft so the actuator ball is on the bottom and held outward by gravity. With the ball in this position, slide the decompressor shaft into the camshaft as far as it will go. The spring should rest lightly against the stop pin **(see illustration 9.5b)**. **Note:** *When the spring is positioned correctly, the end against the stop in will be preloaded about 1/4 inch from the released position. If the spring is loaded this much plus one full turn, it won't allow the decompressor to release after the engine starts.*

24 Check to make sure the decompressor ball is held outward (away from the center of the camshaft) by the decompressor shaft. Also make sure the centrifugal weight on the end of the camshaft returns under spring pressure when rotated against the tension of the spring. The spring pressure should be very light.

25 Coat the cam bearing journals and lobes with molybdenum disulfide grease.

26 Slide the camshaft into the cylinder head and position its dowel straight up **(see illustration 9.6a)**.

27 There are two different ways to install the cam chain and align the valve timing marks, depending on whether or not the alternator is installed.

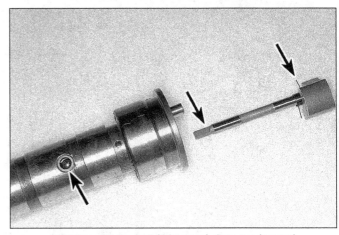

9.16 Inspect the ball (left arrow), flat area (center) and spring (right)

Cam chain installation (alternator installed)

28 Set the crankshaft to top dead center, using the timing mark on the alternator rotor and the pointer in the timing hole **(see illustrations 19.6a and 22.7a in Chapter 1)**.

29 Engage the sprocket with the timing chain. Place its two punch marks facing away from the engine and the dowel notch straight up **(see illustration 9.6a)**. **Note:** *"Straight up" means in a straight line with the centerlines of the crankshaft and camshaft.*

30 Place the sprocket on the camshaft. Recheck the crankshaft timing mark and camshaft dowel to make sure they're still at the correct positions.

Cam chain installation (alternator stator plate removed)

31 There are three bright plated links on the timing chain, one single link and one pair of links next to each other.

32 The single link aligns with a punch mark on the crankshaft sprocket **(see illustration 9.8b)**. The single link and the punch mark align with a cast protrusion on the crankcase casting below the crankshaft.

33 With the single link, punch mark and crankcase protrusion all aligned with each other, the pair of bright links should align with the two punch marks on the cam sprocket **(see illustration 9.6a)**. **Note:** *The timing chain marks won't line up again after the crankshaft is turned. If you're in doubt about whether the marks are lined up correctly, or if the bright links are hard to see on a used chain, use the method described in Steps 28 through 30. Place the alternator rotor and housing in position temporarily (they don't have to be tightened down) so you can see their marks.*

Chapter 2 Part B Four-stroke engines

10.5 Remove the two small head bolts (arrows)

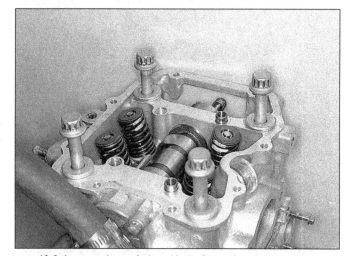

10.6 Loosen the main head bolts in a criss-cross pattern and remove the washers

Both methods

34 Apply non-permanent thread locking agent to the threads of the sprocket bolts. Install them and tighten to the torque listed in this Chapter's Specifications.
35 Rotate the engine back to the TDC position and recheck the timing marks as described above. If they aren't aligned, try removing the sprocket and repositioning it in the chain. Don't run the engine with the marks out of alignment or severe engine damage could occur.
36 Install the cam chain guides **(see illustration 9.8b)**.
37 Coat the cam lobes with moly-based grease or assembly lube.
38 Install the rocker assembly over the camshaft. Tighten the rocker assembly bolts evenly to the torque listed in this Chapter's Specifications.
39 Coat the mating surface of the camshaft end cap with Loctite 515 or 518 Gasket Eliminator or Three Bond 1215. Place a new O-ring on the cap and install it on the cylinder head, then tighten the bolts.
40 Install the cam chain tensioner and cylinder head covers (Sections 7 and 8).
41 Adjust the valve clearances (see Chapter 1).
42 The remainder of installation is the reverse of removal.

10 Cylinder head - removal and installation

Caution: *The engine must be completely cool before beginning this procedure, or the cylinder head may become warped.*

Removal

Refer to illustrations 10.5, 10.6, 10.7, 10.8a and 10.8b

1 Remove the fuel tank, carburetor and exhaust system (see Chapter 4).
2 Drain the cooling system (see Chapter 3).
3 Remove the upper engine mount (Section 5).
4 Remove the cylinder head covers, cam chain tensioner, rocker arms and camshaft (Sections 7, 8 and 9).
5 Remove the two small head bolts located inside the cover opening **(see illustration)**.
6 Loosen the main cylinder head bolts 1/8 turn at a time in a criss-cross pattern **(see illustration)**. Remove the bolts.
7 Lift the cylinder head off the cylinder **(see illustration)**. If the head is stuck, tap around the side of the head with a rubber mallet to jar it loose, or use two wooden dowels inserted into the intake or exhaust ports to lever the head off. **Caution:** *Don't attempt to pry the head off by inserting a screwdriver between the head and the cylinder block - you'll damage the sealing surfaces. Also, don't tap on thin areas of the casting or it may break.*

10.7 Lift the cylinder head off

8 Support the cam chain so it won't drop into the cam chain tunnel, and stuff a clean rag into the cam chain tunnel to prevent the entry of debris. Once this is done, remove the gasket and the two dowel pins from the cylinder **(see illustrations)**.

10.8a Remove the gasket . . .

10.8b ... and locate the dowels to make sure they don't fall into the engine

12.7a Install a valve spring compressor . . .

9 Check the cylinder head gasket and the mating surfaces on the cylinder head and cylinder for leakage, which could indicate warpage. Refer to Section 12 and check the flatness of the cylinder head.

10 Clean all traces of old gasket material from the cylinder head and cylinder. Be careful not to let any of the gasket material fall into the crankcase, the cylinder bore or the bolt holes.

Installation

11 Install the two dowel pins if they were removed, then lay the new gasket in place on the cylinder. Never reuse the old gasket and don't use any type of gasket sealant.

12 Carefully lower the cylinder head over the dowels. It's helpful to have an assistant support the camshaft chain with a piece of wire so it doesn't fall and become kinked or detached from the crankshaft. When the head is resting on the cylinder, wire the cam chain to another component to keep tension on it.

13 Install the steel washers on the bolts, then thread them lightly into their holes.

14 Tighten the four main cylinder head bolts in stages to the final setting listed in this Chapter's Specifications. **Caution:** *The cylinder head tightening sequence has several stages, including loosening stages, which must be followed exactly to prevent damage to the cylinder head and gasket.*

15 Install all parts removed for access, making sure that the cam chain front guide engages the notch in the crankcase **(see illustration 9.8b)**.

16 Change the engine oil (see Chapter 1).

11 Valves/valve seats/valve guides - servicing

1 Because of the complex nature of this job and the special tools and equipment required, servicing of the valves, the valve seats and the valve guides (commonly known as a valve job) is best left to a professional.

2 The home mechanic can, however, remove and disassemble the head, do the initial cleaning and inspection, then reassemble and deliver the head to a dealer service department or properly equipped vehicle repair shop for the actual valve servicing. Refer to Section 12 for those procedures.

3 The dealer service department will remove the valves and springs, recondition or replace the valves and valve seats, replace the valve guides, check and replace the valve springs, spring retainers and keepers (as necessary), replace the valve seals with new ones and reassemble the valve components.

4 After the valve job has been performed, the head will be in like-new condition. When the head is returned, be sure to clean it again very thoroughly before installation on the engine to remove any metal particles or abrasive grit that may still be present from the valve service operations. Use compressed air, if available, to blow out all the holes and passages.

12 Cylinder head and valves - disassembly, inspection and reassembly

1 As mentioned in the previous Section, valve servicing and valve guide replacement should be left to a dealer service department or vehicle repair shop. However, disassembly, cleaning and inspection of the valves and related components can be done (if the necessary special tools are available) by the home mechanic. This way no expense is incurred if the inspection reveals that service work is not required at this time.

2 To properly disassemble the valve components without the risk of damaging them, a valve spring compressor is absolutely necessary. If the special tool is not available, have a dealer service department or vehicle repair shop handle the entire process of disassembly, inspection, service or repair (if required) and reassembly of the valves.

Disassembly

Refer to illustrations 12.7a, 12.7b, 12.7c and 12.7d

3 Remove the carburetor intake tube from the cylinder head (see Chapter 4).

4 Before the valves are removed, scrape away any traces of gasket material from the head gasket sealing surface. Work slowly and do not nick or gouge the soft aluminum of the head. Gasket removing solvents, which work very well, are available at most ATV shops and auto parts stores.

5 Carefully scrape all carbon deposits out of the combustion chamber area. A hand held wire brush or a piece of fine emery cloth can be used once most of the deposits have been scraped away. Do not use a wire brush mounted in a drill motor, or one with extremely stiff bristles, as the head material is soft and may be eroded away or scratched by the wire brush.

6 Before proceeding, arrange to label and store the valves along with their related components so they can be kept separate and reinstalled in the same valve guides they are removed from (again, plastic bags work well for this).

7 Compress the valve spring(s) on the first valve with a spring compressor, then remove the keepers and the retainer from the valve assembly **(see illustrations)**. Do not compress the spring(s) any more than is absolutely necessary. Carefully release the valve spring compressor and remove the spring(s), spring seat and valve from the head **(see illustration)**. If the valve binds in the guide (won't pull through), push it back into the head and deburr the area around the keeper groove with a very fine file or whetstone **(see illustration)**.

Chapter 2 Part B Four-stroke engines

12.7b ... compress the spring and remove the keepers ...

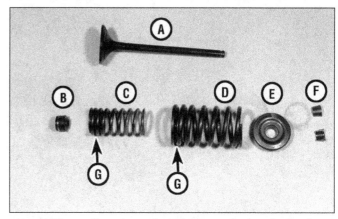

12.7c ... the valve spring retainer, springs and valve

A	Valve stem	E	Valve spring retainer
B	Oil seal	F	Keepers
C	Inner valve spring	G	Tightly wound coils
D	Outer valve spring		

8 Repeat the procedure for the remaining valves. Remember to keep the parts for each valve together so they can be reinstalled in the same location.

9 Once the valves have been removed and labeled, pull off the valve stem seals with pliers and discard them (the old seals should never be reused).

10 Next, clean the cylinder head with solvent and dry it thoroughly. Compressed air will speed the drying process and ensure that all holes and recessed areas are clean.

11 Clean all of the valve springs, keepers, retainers and spring seats with solvent and dry them thoroughly. Do the parts from one valve at a time so that no mixing of parts between valves occurs.

12 Scrape off any deposits that may have formed on the valve, then use a motorized wire brush to remove deposits from the valve heads and stems. Again, make sure the valves do not get mixed up.

Inspection

Refer to illustrations 12.14, 12.15, 12.16, 12.17, 12.18a, 12.18b, 12.19a and 12.19b

13 Inspect the head very carefully for cracks and other damage. If cracks are found, a new head will be required. Check the cam bearing surfaces for wear and evidence of seizure. Check the camshaft for wear as well (see Section 9).

14 Using a precision straightedge and a feeler gauge, check the head gasket mating surface for warpage. Lay the straightedge lengthwise, across the head and diagonally (corner-to-corner), intersecting the head bolt holes, and try to slip a feeler gauge under it, on either side of the combustion chamber **(see illustration)**. The feeler gauge thickness should be the same as the cylinder head warpage limit listed in this Chapter's Specifications. If the feeler gauge can be inserted between the head and the straightedge, the head is warped and must either be machined or, if warpage is excessive, replaced with a new one.

15 Examine the valve seats in each of the combustion chambers. If they are pitted, cracked or burned, the head will require valve service that is beyond the scope of the home mechanic. Measure the valve seat width **(see illustration)** and compare it to this Chapter's Specifications. If it is not within the specified range, or if it varies around its circumference, valve service work is required.

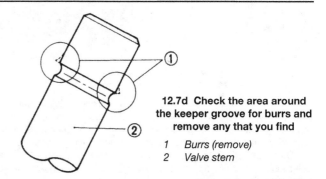

12.7d Check the area around the keeper groove for burrs and remove any that you find

1 Burrs (remove)
2 Valve stem

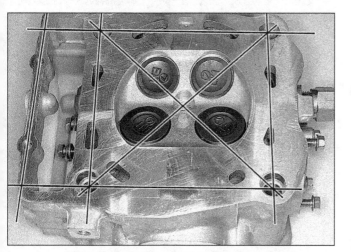

12.14 Measure the gasket surface flatness in the directions shown

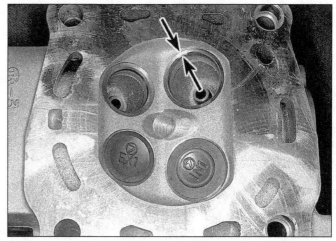

12.15 Measure the valve seat width

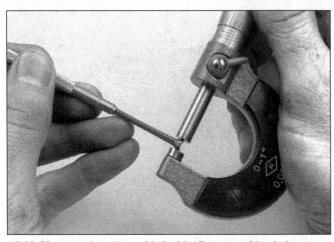

12.16 Measure the valve guide inside diameter with a hole gauge, then measure the gauge with a micrometer

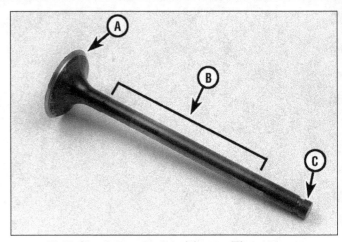

12.17 Check the valve face (A), stem (B) and keeper groove (C) for wear and damage

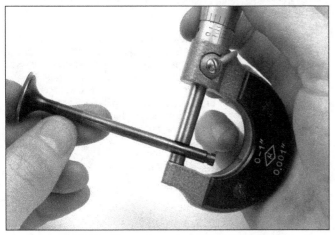

12.18a Measuring valve stem diameter

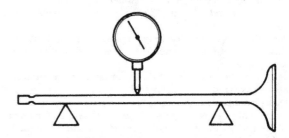

12.18b Check the valve stem for bends with a V-block (or V-blocks, as shown here) and a dial indicator

16 Clean the valve guides to remove any carbon buildup, then measure the inside diameters of the guides (at both ends and the center of the guide) with a small hole gauge and a micrometer **(see illustration)**. Record the measurements for future reference. The guides are measured at the ends and at the center to determine if they are worn in a bell-mouth pattern (more wear at the ends). If they are, guide replacement is an absolute must.

17 Carefully inspect each valve face for cracks, pits and burned spots. Check the valve stem and the keeper groove area for cracks

(see illustration). Rotate the valve and check for any obvious indication that it is bent. Check the end of the stem for pitting and excessive wear and make sure the edge of the stem at its top end is beveled (not worn to a sharp edge). The presence of any of the above conditions indicates the need for valve servicing.

18 Measure the valve stem diameter **(see illustration)**. If the diameter is less than listed in this Chapter's Specifications, the valves will have to be replaced with new ones. Also check the valve stem for bending. Set the valve in a V-block with a dial indicator touching the middle of the stem **(see illustration)**. Rotate the valve and look for a reading on the gauge (which indicates a bent stem). If the stem is bent, replace the valve.

19 Check the end of each valve spring for wear and pitting. Measure the free length **(see illustration)** and compare it to this Chapter's

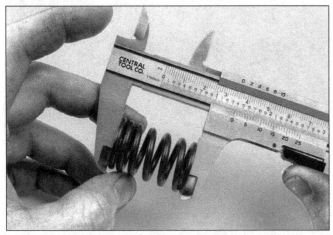

12.19a Measuring the free length of the valve springs

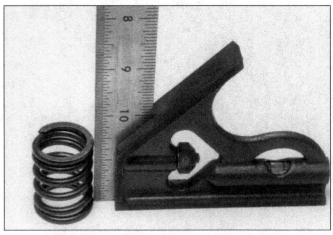

12.19b Checking a valve spring for squareness

12.23 Apply the lapping compound very sparingly, in small dabs, to the valve face only

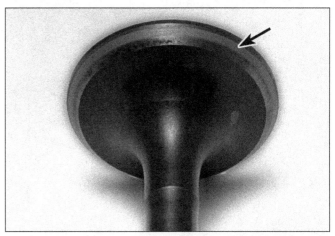

12.24 After lapping, the valve face should exhibit a uniform, unbroken contact pattern (arrow)

Specifications. Any springs that are shorter than specified have sagged and should not be reused. Stand the spring on a flat surface and check it for squareness **(see illustration)**.
20 Check the spring retainers and keepers for obvious wear and cracks. Any questionable parts should not be reused, as extensive damage will occur in the event of failure during engine operation.
21 If the inspection indicates that no service work is required, the valve components can be reinstalled in the head.

Reassembly

Refer to illustrations 12.23, 12.24, 12.26 and 12.27
22 If the valve seats have been ground, the valves and seats should be lapped before installing the valves in the head to ensure a positive seal between the valves and seats. This procedure requires coarse and fine valve lapping compound (available at auto parts stores) and a valve lapping tool. If a lapping tool is not available, a piece of rubber or plastic hose can be slipped over the valve stem (after the valve has been installed in the guide) and used to turn the valve.
23 Apply a small amount of coarse lapping compound to the valve face **(see illustration)**, then slip the valve into the guide. **Note:** *Make sure the valve is installed in the correct guide and be careful not to get any lapping compound on the valve stem.*
24 Attach the lapping tool (or hose) to the valve and rotate the tool between the palms of your hands. Use a back-and-forth motion rather than a circular motion. Lift the valve off the seat and turn it at regular intervals to distribute the lapping compound properly. Continue the lapping procedure until the valve face and seat contact area is of uniform width and unbroken around the entire circumference of the valve face and seat **(see illustration)**. Once this is accomplished, lap the valves again with fine lapping compound.
25 Carefully remove the valve from the guide and wipe off all traces of lapping compound. Use solvent to clean the valve and wipe the seat area thoroughly with a solvent soaked cloth. Repeat the procedure for the remaining valves.
26 Lay the spring seat in place in the cylinder head, then install new valve stem seals on both of the guides **(see illustration)**. Use an appropriate size deep socket to push the seals into place until they are properly seated. Don't twist or cock them, or they will not seal properly against the valve stems. Also, don't remove them again or they will be damaged.
27 Coat the valve stems with assembly lube or moly-based grease, then install one of them into its guide. Next, install the spring seat, springs and retainers, compress the springs and install the keepers. **Note:** *Install the springs with the tightly wound coils at the bottom (next to the spring seat). When compressing the springs with the valve spring compressor, depress them only as far as is absolutely necessary to slip the keepers into place. Apply a small amount of grease to the keepers* **(see illustration)** *to help hold them in place as the pressure is released from the springs. Make certain that the keepers are securely locked in*

12.26 Push the oil seals onto the valve guide (arrows); note the different appearance of the intake and exhaust seals

their retaining grooves.
28 Support the cylinder head on blocks so the valves can't contact the workbench top, then very gently tap each of the valve stems with a soft-faced hammer. This will help seat the keepers in their grooves.
29 Once all of the valves have been installed in the head, check for proper valve sealing by pouring a small amount of solvent into each of the valve ports. If the solvent leaks past the valve(s) into the combustion chamber area, disassemble the valve(s) and repeat the lapping procedure, then reinstall the valve(s) and repeat the check. Repeat the procedure until a satisfactory seal is obtained.

12.27 A small dab of grease will help hold the keepers in place on the valve while the spring compressor is released

Chapter 2 Part B Four-stroke engines

13.2 Lift the front cam chain guide (arrow) out of the engine

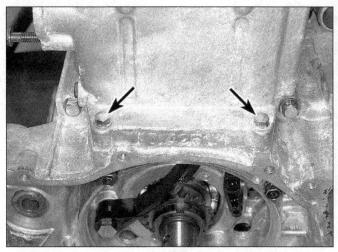

13.3a Remove the two bolts that attach the cylinder to the crankcase (arrows) . . .

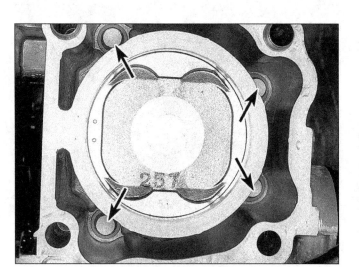

13.3b . . . and reach into the water jacket with a 12-point socket and extension to remove the four large bolts

13.4 Lift the cylinder off . . .

13 Cylinder - removal, inspection and installation

Removal

Refer to illustrations 13.2, 13.3a, 13.3b, 13.4 and 13.5

1 Following the procedure given in Section 10, remove the cylinder head. Make sure the crankshaft is positioned at top dead center (TDC).
2 Lift out the cam chain front guide **(see illustration)**.
3 Remove two small bolts that secure the base of the cylinder to the crankcase **(see illustration)**. Reach down inside the water jacket with a 12-socket and extension and remove the four main bolts **(see illustration)**. Loosen the bolts a little at a time in a criss-cross pattern. **Caution:** *Don't try to remove the bolts with a six-point socket or the heads will round off.*
4 Lift the cylinder straight up to remove it **(see illustration)**. If it's stuck, tap around its perimeter with a soft-faced hammer. Don't attempt to pry between the cylinder and the crankcase, as you'll ruin the sealing surfaces.
5 Locate the dowel pins (they may have come off with the cylinder or still be in the crankcase) **(see illustration)**. Be careful not to let these drop into the engine. Stuff rags around the piston and remove the gasket and all traces of old gasket material from the surfaces of the cylinder and the crankcase.

Inspection

Refer to illustrations 13.6 and 13.8

Caution: *Don't attempt to separate the liner from the cylinder.*

6 Check the top surface of the cylinder for warpage, using the same method as for the cylinder head (see Section 12). Measure along the sides and diagonally across the stud holes **(see illustration)**.

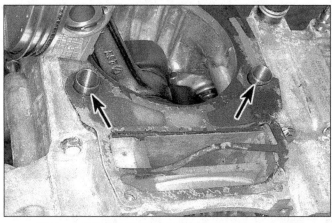

13.5 . . . and note the location of the dowels (arrows)

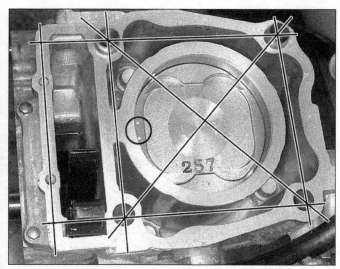

13.6 Check the cylinder top surface for warpage in the directions shown; if the directional arrow on the piston isn't visible, make your own marks (circled)

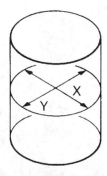

13.8 Measure the cylinder diameter in two directions, at top, center and bottom of the ring travel

13.16 If you're experienced and very careful, the cylinder can be installed over the rings without a ring compressor, but a compressor is recommended

7   Check the cylinder walls carefully for scratches and score marks.
8   Using the appropriate precision measuring tools, check the cylinder's diameter at the top, center and bottom of the cylinder bore, parallel to the crankshaft axis **(see illustration)**. Next, measure the cylinder's diameter at the same three locations across the crankshaft axis. Compare the results to this Chapter's Specifications. If the cylinder walls are tapered, out-of-round, worn beyond the specified limits, or badly scuffed or scored, have the cylinder rebored and honed by a dealer service department or an ATV repair shop. If a rebore is done, oversize pistons and rings will be required as well. **Note:** *Polaris supplies pistons in two oversizes.*
9   As an alternative, if the precision measuring tools are not available, a dealer service department or repair shop will make the measurements and offer advice concerning servicing of the cylinder.
10   If it's in reasonably good condition and not worn to the outside of the limits, and if the piston-to-cylinder clearance can be maintained properly (see Section 14), then the cylinder does not have to be rebored; honing is all that is necessary.
11   To perform the honing operation you will need the proper size flexible hone with fine stones as shown in Maintenance techniques, tools and working facilities at the front of this book, or a "bottle brush" type hone, plenty of light oil or honing oil, some shop towels and an electric drill motor. Hold the cylinder block in a vise (cushioned with soft jaws or wood blocks) when performing the honing operation. Mount the hone in the drill motor, compress the stones and slip the hone into the cylinder. Lubricate the cylinder thoroughly, turn on the drill and move the hone up and down in the cylinder at a pace which will produce a fine crosshatch pattern on the cylinder wall with the crosshatch lines intersecting at approximately a 60-degree angle. Be sure to use plenty of lubricant and do not take off any more material than is absolutely necessary to produce the desired effect. Do not withdraw the hone from the cylinder while it is running. Instead, shut off the drill and continue moving the hone up and down in the cylinder until it comes to a complete stop, then compress the stones and withdraw the hone. Wipe the oil out of the cylinder and repeat the procedure on the remaining cylinder. Remember, do not remove too much material from the cylinder wall. If you do not have the tools, or do not desire to perform the honing operation, a dealer service department or vehicle repair shop will generally do it for a reasonable fee.
12   Next, the cylinder must be thoroughly washed with warm soapy water to remove all traces of the abrasive grit produced during the honing operation. Be sure to run a brush through the bolt holes and flush them with running water. After rinsing, dry the cylinder thoroughly and apply a coat of light, rust-preventative oil to all machined surfaces.

Installation

Refer to illustration 13.16

13   Lubricate the cylinder bore with plenty of clean engine oil. Apply a thin film of moly-based grease to the piston skirt.
14   Install the dowel pins, then lower a new cylinder base gasket over them **(see illustration 13.5)**.
15   Attach a piston ring compressor to the piston and compress the piston rings. A large hose clamp can be used instead - just make sure it doesn't scratch the piston, and don't tighten it too much.
16   Install the cylinder over the studs and carefully lower it down until the piston crown fits into the cylinder liner **(see illustration)**. While doing this, pull the camshaft chain up, using a hooked tool or a piece of stiff wire. Push down on the cylinder, making sure the piston doesn't get cocked sideways, until the bottom of the cylinder liner slides down past the piston rings. A wood or plastic hammer handle can be used to gently tap the cylinder down, but don't use too much force or the piston will be damaged.
17   Remove the piston ring compressor or hose clamp, being careful not to scratch the piston.
18   The remainder of installation is the reverse of the removal steps.

14   Piston - removal, inspection and installation

1   The piston is attached to the connecting rod with a piston pin that is a slip fit in the piston and rod. The piston pin is secured by a circlip at each end.
2   Before removing the piston from the rod, stuff a clean shop towel into the crankcase hole, around the connecting rod. This will prevent the circlips from falling into the crankcase if they are inadvertently dropped.

Chapter 2 Part B Four-stroke engines

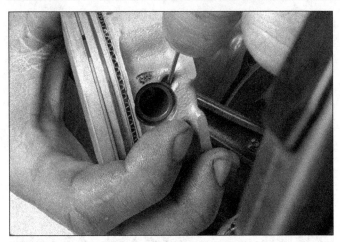

14.3 Wear eye protection and pry the circlip out of its groove with a pointed tool

14.4a Push the piston pin partway out, then pull it the rest of the way

14.4b The piston pin should come out with hand pressure - if it doesn't, you can use a special removal tool . . .

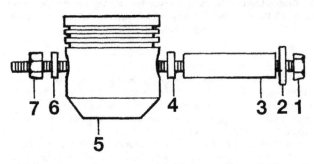

14.4c . . . or make your own tool from readily available parts

1   Bolt
2   Washer
3   Pipe (A)
4   Padding (A)
5   Piston
6   Washer (B)
7   Nut (B)
A   Large enough for piston pin to fit inside
B   Small enough to fit through piston pin bore

Removal

Refer to illustrations 14.3, 14.4a, 14.4b and 14.4c

3   The piston should have an arrow on its crown that points toward the right (alternator) side of the engine **(see illustration 13.6)**. If this mark is not visible due to carbon buildup, make your own marks on the piston crown before removal. Support the piston and pry the circlip out with a pointed tool **(see illustration)**.

14.6 Remove the piston rings with a ring removal and installation tool

4   Push the piston pin out from the opposite end to free the piston from the rod **(see illustration)**. You may have to deburr the area around the groove to enable the pin to slide out (use a triangular file for this procedure). If the pin won't come out, you can use a piston pin removal tool **(see illustration)**. The tool can be made from a long bolt, a nut, a piece of tubing and washers **(see illustration)**.

Inspection

Refer to illustrations 14.6, 14.11, 14.13, 14.14, 14.16a, 14.16b and 14.16c

5   Before the inspection process can be carried out, the pistons must be cleaned and the old piston rings removed.
6   Using a piston ring removal and installation tool, carefully remove the rings from the pistons **(see illustration)**. Do not nick or gouge the pistons in the process.
7   Scrape all traces of carbon from the tops of the pistons. A hand-held wire brush or a piece of fine emery cloth can be used once the majority of the deposits have been scraped away. Do not, under any circumstances, use a wire brush mounted in a drill motor to remove deposits from the pistons; the piston material is soft and will be eroded away by the wire brush.
8   Use a piston ring groove cleaning tool to remove any carbon deposits from the ring grooves. If a tool is not available, a piece broken off the old ring will do the job. Be very careful to remove only the carbon deposits. Do not remove any metal and do not nick or gouge the sides of the ring grooves.
9   Once the deposits have been removed, clean the pistons with

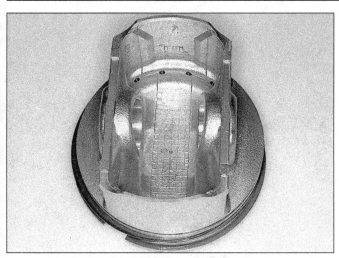

14.11 Make sure the internal oil holes are clear

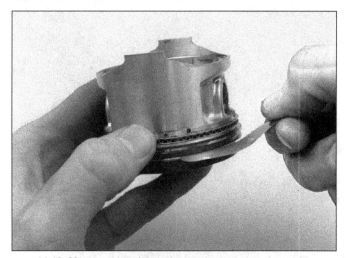

14.13 Measure the piston ring-to-groove clearance with a feeler gauge

solvent and dry them thoroughly. Make sure the oil return holes below the oil ring grooves are clear.

10 If the pistons are not damaged or worn excessively and if the cylinders are not rebored, new pistons will not be necessary. Normal piston wear appears as even, vertical wear on the thrust surfaces of the piston and slight looseness of the top ring in its groove. New piston rings, on the other hand, should always be used when an engine is rebuilt.

11 Carefully inspect each piston for cracks around the skirt, at the pin bosses and at the ring lands. Make sure the internal oil holes are clear **(see illustration)**.

12 Look for scoring and scuffing on the thrust faces of the skirt, holes in the piston crown and burned areas at the edge of the crown. If the skirt is scored or scuffed, the engine may have been suffering from overheating and/or abnormal combustion, which caused excessively high operating temperatures. The oil pump should be checked thoroughly. A hole in the piston crown, an extreme to be sure, is an indication that abnormal combustion (pre-ignition) was occurring. Burned areas at the edge of the piston crown are usually evidence of spark knock (detonation). If any of the above problems exist, the causes must be corrected or the damage will occur again.

13 Measure the piston ring-to-groove clearance (side clearance) by laying a new piston ring in the ring groove and slipping a feeler gauge in beside it **(see illustration)**. Check the clearance at three or four locations around the groove. Be sure to use the correct ring for each groove; they are different. If the clearance is greater then specified, new pistons will have to be used when the engine is reassembled.

14 Check the piston-to-bore clearance by measuring the bore (see Section 13) and the piston diameter **(see illustration)**. Measure the piston across the skirt on the thrust faces at a 90-degree angle to the piston pin, at the specified distance up from the bottom of the skirt. Subtract the piston diameter from the bore diameter to obtain the clearance. If it is greater than specified, the cylinder will have to be rebored and a new oversized piston and rings installed. If the appropriate precision measuring tools are not available, the piston-to-cylinder clearance can be obtained, though not quite as accurately, using feeler gauge stock. Feeler gauge stock comes in 12-inch lengths and various thicknesses and is generally available at auto parts stores. To check the clearance, slip a piece of feeler gauge stock of the same thickness as the specified piston clearance into the cylinder along with appropriate piston. The cylinder should be upside down and the piston must be positioned exactly as it normally would be. Place the feeler gauge between the piston and cylinder on one of the thrust faces (90-degrees to the piston pin bore). The piston should slip through the cylinder (with the feeler gauge in place) with moderate pressure. If it falls through, or slides through easily, the clearance is excessive and a new piston will be required. If the piston binds at the lower end of the cylinder and is loose toward the top, the cylinder is tapered, and if tight spots are encountered as the piston/feeler gauge is rotated in the cylinder, the

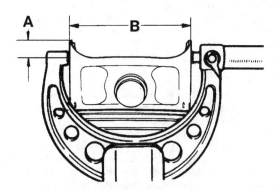

14.14 Measure the piston diameter with a micrometer

14.16a Slip the piston pin into the rod and try to rock it back-and-forth to check for looseness

cylinder is out-of-round. Be sure to have the cylinder and piston checked by a dealer service department or a repair shop to confirm your findings before purchasing new parts.

15 Apply clean engine oil to the pin, insert it into the piston and check for freeplay by rocking the pin back-and-forth **(see illustration 14.4a)**. If the pin is loose, a new piston and possibly new pin must be installed.

16 Repeat Step 15, this time inserting the piston pin into the connecting rod **(see illustration)**. If the pin is loose, measure the pin diam-

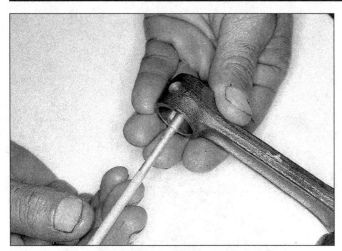

14.16b If there's any doubt, measure the rod bore with a hole gauge . . .

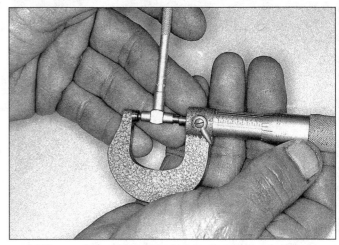

14.16c . . . then measure the hole gauge with a micrometer

eter and the pin bore in the rod (or have this done by a dealer or repair shop) **(see illustrations)**. A worn pin can be replaced separately; if the rod bore is worn, the rod and crankshaft must be replaced as an assembly.

17 Refer to Section 15 and install the rings on the pistons.

Installation

18 Install the piston with its arrow mark (or the marks you made on removal) toward the right side of the engine. Lubricate the pin and the rod bore with moly-based grease.

19 Install a new circlip in the groove in one side of the piston (don't reuse the old circlips). The gap in the circlip faces up or down. The offset tang on the circlip faces outward, away from the piston. Push the pin into position from the opposite side and install another new circlip. Compress the circlips only enough for them to fit in the piston. Make sure the clips are properly seated in the grooves.

15 Piston rings - installation

Refer to illustrations 15.2, 15.7a, 15.7b, 15.7c, 15.9 and 15.10

1 Before installing the new piston rings, the ring end gaps must be checked.

2 Insert the top (No. 1) ring into the bottom of the first cylinder and square it up with the cylinder walls by pushing it in with the top of the piston. The ring should be about one-half inch above the bottom edge of the cylinder. To measure the end gap, slip a feeler gauge between the ends of the ring **(see illustration)** and compare the measurement to the Specifications.

3 If the gap is larger or smaller than specified, double check to make sure that you have the correct rings before proceeding.

4 If the gap is too small, it must be enlarged or the ring ends may come in contact with each other during engine operation, which can cause serious damage.

5 Repeat the procedure for the second compression ring (ring gap is not specified for the oil ring rails or spacer).

6 Once the ring end gaps have been checked/corrected, the rings can be installed on the piston.

7 The oil control ring (lowest on the piston) is installed first. It is composed of three separate components. Slip the spacer into the groove, then install the upper side rail **(see illustrations)**. **Note:** *The tab on the end of the upper side rail fits in the notch above the oil ring groove.* Do not use a piston ring installation tool on the oil ring side rails as they may be damaged. Instead, place one end of the side rail into the groove between the spacer expander and the ring land. Hold it firmly in place and slide a finger around the piston while pushing the rail into the groove (taking care not to cut your fingers on the sharp edges). Next, install the lower side rail in the same manner.

8 After the three oil ring components have been installed, check to make sure that both the upper and lower side rails can be turned smoothly in the ring groove.

9 Install the no. 2 (middle) ring next. It can be readily distinguished from the top ring by its cross-section shape **(see illustration)**. Do not mix the top and middle rings.

15.2 Check the piston ring end gap with a feeler gauge at the top and bottom of the cylinder and compare the measurements

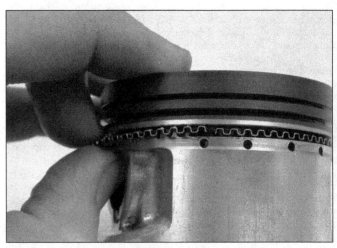

15.7a Installing the oil ring expander - make sure the ends don't overlap

Chapter 2 Part B Four-stroke engines

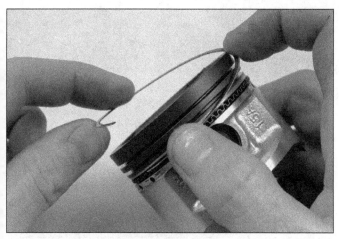

15.7b Installing an oil ring side rail - don't use a ring installation tool to do this

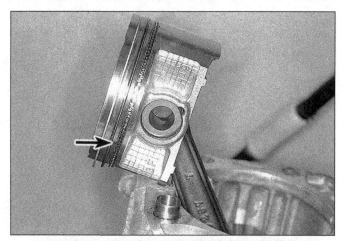

15.7c Place the tab of the oil ring top rail in the piston notch (arrow)

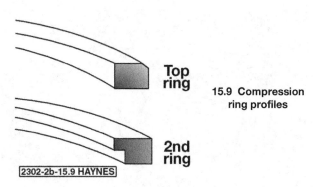

15.9 Compression ring profiles

10 To avoid breaking the ring, use a piston ring installation tool and make sure that the identification mark is facing up **(see illustration)**. Fit the ring into the middle groove on the piston. Do not expand the ring any more than is necessary to slide it into place.

11 Finally, install the no. 1 (top) ring in the same manner. Make sure the identifying mark is facing up. Be very careful not to confuse the top and second rings. Besides the different profiles, the top ring is narrower than the second ring.

12 Once the rings have been properly installed, stagger the end gaps, including those of the oil ring side rails.

a) The oil ring top rail is located by its tab, which fits in the piston notch.
b) The gap in the oil ring expander faces the front of the engine (exhaust side).
c) The gap in the lower oil ring rail goes on the opposite side of the expander gap from the top rail gap, and at least 30-degrees (1/12 turn) away from the expander gap.
d) The second compression ring gap goes toward the rear of the engine (carburetor side).
e) The top compression ring gap goes toward the front of the engine.

16 External oil lines, check valve and tank - removal and installation

Oil line and check valve
Removal
Refer to illustration 16.2

1 Remove the fuel tank and PVT (see Chapter 3 and Chapter 2, Part C).

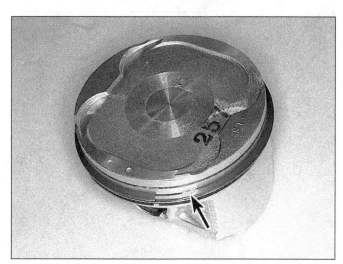

15.10 The manufacturer's marks on the compression rings face up (arrow)

16.2 The external oil lines and check valve (arrow) are at the left front of the engine

2 To remove the lines, remove the union bolt at the top and bottom of each line **(see illustration)**. Unscrew the check valve and remove its sealing washer, spring and plunger.

2B-22   Chapter 2 Part B   Four-stroke engines

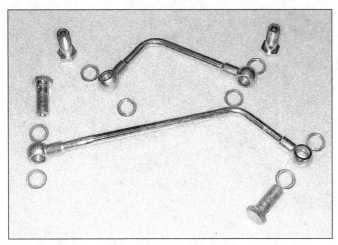

16.3 External oil line details

16.4 The oil tank is secured by Allen bolts in grommets (arrows)

16.5 Oil tank details

17.8 Loosen the case bolts (arrows) in two or three stages, in a criss-cross pattern

Installation

Refer to illustration 16.3

3 Installation is the reverse of the removal steps, with the following additions:

  a) *Replace the sealing washers at the union bolts and check valve whenever they're removed* **(see illustration)**.
  b) *Tighten the union bolts and check valve to the torque listed in this Chapter's Specifications.*

Oil tank removal and installation

Refer to illustrations 16.4 and 16.5

4 To remove the oil tank, disconnect its hoses from the engine, then undo its mounting bolts and take it off the frame **(see illustration)**.

5 Check the mounting grommets for deterioration and replace them as needed **(see illustration)**. To remove the upper hose, unscrew its union bolt and remove the sealing washers. To remove the lower hose, loosen its hose clamp and detach it from the fitting.

6 Installation is the reverse of removal. Use new sealing washers on the union bolt if it was removed. The lower hose from the oil tank connects to the upper fitting on the engine and vice versa.

17 Crankcase - disassembly and reassembly

1 To examine and repair or replace the crankshaft, connecting rod, bearings, counterbalancer, oil pump and the water pump mechanical seal, the crankcase must be split into two parts.

Disassembly

Refer to illustrations 17.8, 17.9 and 17.10

2 Remove the engine from the vehicle (see Section 5).
3 Remove the carburetor (see Chapter 4).
4 Remove the alternator rotor and the starter motor (see Chapter 5).
5 Remove the cylinder head cover, cam chain tensioner, cam chain, cylinder head, cylinder and piston (see Sections 7, 8, 9, 10, 13 and 14).
6 Remove the oil pipe (Section 16).
7 Check carefully to make sure there aren't any remaining components that attach the halves of the crankcase together.
8 Loosen the crankcase bolts in two or three stages, in a criss-cross pattern **(see illustration)**. Remove the bolts and label them; they are different lengths.
9 Carefully pry the crankcase apart and separate the halves **(see illustration)**. Don't pry against the mating surfaces or they'll develop leaks.
10 Remove the two crankcase dowels **(see illustration)**.
11 Refer to Sections 18 through 21 for information on the internal components of the crankcase.

Reassembly

12 Remove all traces of old gasket and sealant from the crankcase mating surfaces with a sharpening stone or similar tool. Be careful not to let any fall into the case as this is done and be careful not to damage the mating surfaces.
13 Check to make sure the two dowel pins are in place in their holes in the mating surface of the left crankcase half **(see illustration 17.10)**.

Chapter 2 Part B Four-stroke engines

17.9 Pry only at the pry points (arrow), not between the mating surfaces

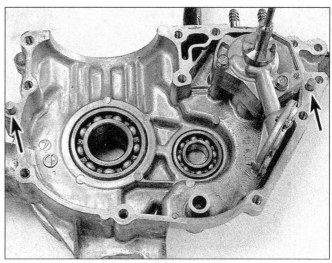

17.10 Note the locations of the case dowels (arrows)

18.2a Remove the shims (arrows) from the oil pump drive and driven gears

18.2b Remove the oil pump mounting bolts (A); on installation, be sure the dowel (B) engages the driven gear

14 Pour some engine oil over the oil pump gear and counterbalancer gears and the crankshaft bearing. Don't get any oil on the crankcase mating surface.
15 Install a new gasket on the crankcase mating surface.
16 Carefully place the right crankcase half onto the left crankcase half. While doing this, make sure the oil pump shaft, crankshaft and balancer fit into their bearings in the right crankcase half.
17 Install the crankcase bolts in the correct holes and tighten them so they are just snug. Then tighten them in two or three stages, in a criss-cross pattern, to the torque listed in this Chapter's Specifications.
18 Turn the crankshaft to make sure it turns freely.
19 The remainder of installation is the reverse of removal.

18 Oil pump - removal, inspection and installation

Removal

Refer to illustrations 18.2a, 18.2b and 18.2c

1 Remove the engine and separate the crankcase halves (Sections 5 and 17).
2 Remove the thrust washers and oil pump gears **(see illustration)**. Remove the mounting bolts, then pull the pump out of the engine **(see illustrations)**.

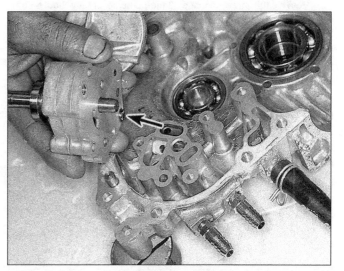

18.2c Pull the pump off the crankcase; the single screw (arrow) holds the pump together

2B-24   Chapter 2 Part B   Four-stroke engines

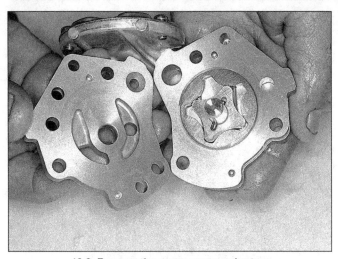

18.3 Remove the pump cover and rotors

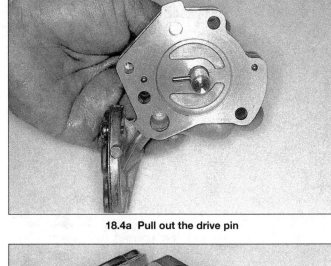

18.4a Pull out the drive pin

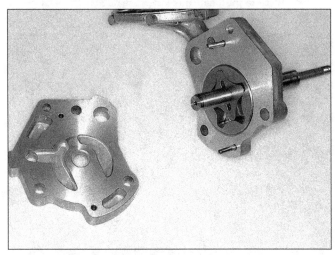

18.4b Take off the center cover, second inner rotor and drive pin

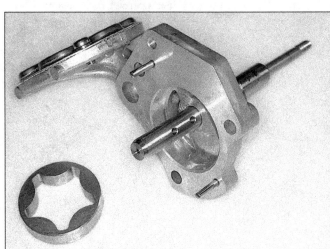

18.4c Remove the second outer rotor

Inspection

Refer to illustrations 18.3, 18.4a, 18.4b, 18.4c, 18.4d, 18.6a, 18.6b, 18.6c, 18.7a and 18.7b

3   Remove the screw and take off the pump cover **(see illustration)**.

4   Remove the rotors, drive pins and shaft **(see illustrations)**.
5   Wash all the components in solvent, then dry them off. Check the pump body, the rotors and the cover for scoring and wear. If any damage or uneven or excessive wear is evident, replace the pump. If you are rebuilding the engine, it's a good idea to install a new oil pump.

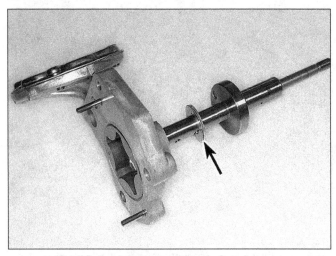

18.4d Pull out the pump drive shaft and remove the washer (arrow)

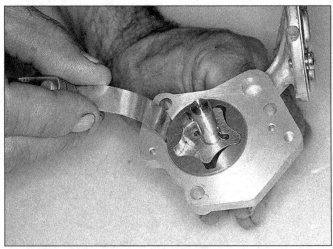

18.6a Measure the clearance between the outer rotor and body . . .

Chapter 2 Part B Four-stroke engines

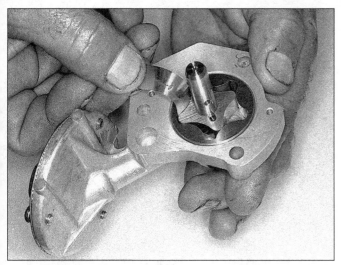

18.6b ... between the inner and outer rotors ...

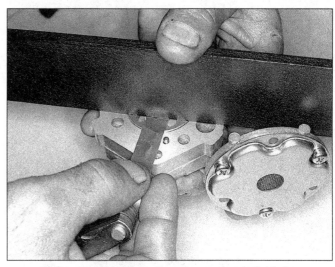

18.6c ... and between the rotors and a straightedge laid across the pump body

6 Place the rotors in the pump body. Measure the clearance between the outer rotor and body, and between the inner and outer rotors, with a feeler gauge **(see illustration)**. Also measure the rotor tip clearance **(see illustration)**. Do this for both sets of rotors. Place a straightedge across the pump body and rotors and measure the gap with a feeler gauge **(see illustration)**. If any of the clearances are beyond the limits listed in this Chapter's Specifications, replace the pump.

7 Remove the strainer screws and take the strainer off the pump **(see illustrations)**.

8 Reassemble the pump by reversing the disassembly steps, with the following additions:

a) *Before installing the cover, pack the cavities between the rotors with petroleum jelly - this will ensure the pump develops suction quickly and begins oil circulation as soon as the engine is started.*
b) *Make sure the dowels are in position.*
c) *Be sure to reinstall both drive pins.*

Installation

9 Installation is the reverse of removal, with the following additions:

a) *Make sure the oil cover screw is in position* **(see illustration 18.2b)**.
b) *Tighten the oil pump mounting bolts to the torque listed in this Chapter's Specifications.*

19 Crankcase components - inspection and servicing

Refer to illustrations 19.3a, 19.3b and 19.4

1 Separate the crankcase and remove the following:
   a) *Oil pump*
   b) *Counterbalancer*
   c) *Crankshaft and main bearings*

2 Clean the crankcase halves thoroughly with new solvent and dry them with compressed air. All oil passages should be blown out with compressed air and all traces of old gasket sealant should be removed from the mating surfaces. **Caution:** *Be very careful not to nick or gouge the crankcase mating surfaces or leaks will result. Check both crankcase sections very carefully for cracks and other damage.*

3 Check the bearings in the case halves **(see illustration)**. If they don't turn smoothly, replace them. For bearings that aren't accessible from the outside, a blind hole puller will be needed for removal **(see illustration)**. Drive the remaining bearings out with a bearing driver or a socket having an outside diameter slightly smaller than that of the bearing outer race. Before installing the bearings, allow them to sit in the freezer overnight, and about fifteen-minutes before installation, place the case half in an oven, set to about 200-degrees F, and allow it to heat up. The bearings are an interference fit, and this will ease installation. **Warning:** *Before heating the case, wash it thoroughly with soap*

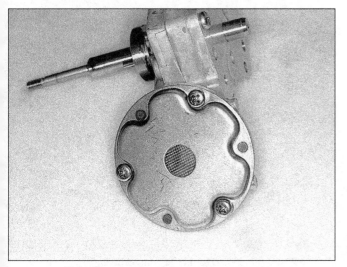

18.7a Remove the pickup screen screws ...

18.7b ... and take off the screen

19.3a Right side case bearings

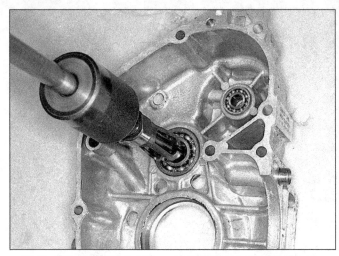

19.3b Bearings that aren't accessible from both sides can be removed with a blind hole puller

and water so no explosive fumes are present. Also, don't use a flame to heat the case. Install the ball bearings with a socket or bearing driver that bears against the bearing outer race. Install the needle roller bearing with a shouldered drift that fits inside the bearing to keep it from collapsing while the shoulder applies pressure to the outer race.

4   The water pump mechanical seal and the impeller shaft oil seal are behind one of the small case bearings **(see illustration 19.3a)**. Once the bearing is removed, pry the oil seal out of the bore, then drive the mechanical seal out from the inside of the case toward the outside. Tap a new oil seal in from the inside of the case, but install the mechanical seal from the outside after the crankcase has been reassembled **(see illustration)**.

5   If any damage is found that can't be repaired, replace the crankcase halves as a set.

6   Assemble the case halves (see Section 17) and check to make sure the crankshaft and the transmission shafts turn freely.

20 Counterbalancer - removal and installation

Refer to illustration 20.2

1   Remove the engine and separate the crankcase halves (Sections 5 and 17).

2   Turn the crankshaft and counterbalancer and check for the alignment marks **(see illustration)**. They must be aligned with other on

installation.

3   Lift the counterbalancer out of the crankcase. The bearing is supposed to fit tightly in the crankcase, so you may have to rock it slightly back and forth to free it.

4   Check the balancer drive and driven gears for worn, chipped or missing teeth. If either gear has problems, replace the balancer shaft and both gears as a set (the drive gear can be removed from the crankshaft).

5   Spin the bearings by hand and check for rough, loose or noisy movement. If you can hear or feel a problem, or if there's visible damage (scratches, chips, balls missing, blue-tinted areas), replace the bearings.

6   If you're replacing the counterbalancer, its gears, bearings or the crankcase, you'll need to select new counterbalancer shims. The method is basically the same as for selecting crankshaft shims (Section 21).

7   Installation is the reverse of removal. Lubricate the bearings with clean engine oil.

21 Crankshaft and connecting rod - removal, inspection and installation

Refer to illustration 21.3

Note: *Some of the procedures in this Section require special tools. If*

19.4 Tap the water pump mechanical seal back into position after the crankcase halves are assembled

20.2 The punch marks on the counterbalancer drive and driven gears (arrows) must be aligned on assembly

Chapter 2 Part B Four-stroke engines

21.3 Press the crankshaft out of the crankcase

21.4 Check the connecting rod side clearance with a feeler gauge

you don't have them, have the crankshaft removed, inspected and installed by a Polaris dealer or ATV shop.
1   Remove the engine and separate the crankcase halves (Sections 5 and 17).
2   Remove the counterbalancer (Section 20).
3   Support the crankcase on a press plate and press the crankshaft out **(see illustration)**.

Inspection

Refer to illustrations 21.4, 21.5 and 21.7

4   Measure the side clearance between the connecting rod and crankshaft with a feeler gauge **(see illustration)**. If it's more than the limit listed in this Chapter's Specifications, the crankshaft can be disassembled for thrust bearing replacement. This is a highly specialized job, which even some Polaris dealerships may not be equipped to handle. Check with your dealer or ATV shop and compare the cost of repair with the cost of a new crankshaft assembly.
5   Check the crankshaft end threads and Woodruff key slots, the cam chain sprocket, the ball bearing at the sprocket end of the crankshaft and the bearing journals for visible wear or damage **(see illustration)**. If you need to replace the sprocket or ball bearing, a cogged socket will be needed to unscrew the sprocket retaining nut (see Chapter 2 Part A).
6   Set the crankshaft in a lathe or a pair of V-blocks. Set up a dial indicator to contact each end in turn and measure the runout. If it's beyond the limit listed in this Chapter's Specifications, replace the crankshaft.
7   With the crankshaft supported in the V-blocks, place the dial indicator pointer against the big end of the connecting rod **(see illustration)**. Move the connecting rod up-and-down against the pointer to check the big end radial play. If it's beyond the limit listed in this Chapter's Specifications, the crankshaft can be disassembled for bearing replacement.

Shim selection

Refer to illustrations 21.9 and 21.10

8   Whenever the crankshaft is removed and installed, shims must be selected to adjust crankshaft end play.
9   Measure the width of the crankshaft between the two surfaces that its ball bearings rest against **(see illustration)**. On the cam

21.5 Inspect the components on the cam sprocket end of the crankshaft

21.7 Check the connecting rod radial clearance with a dial indicator

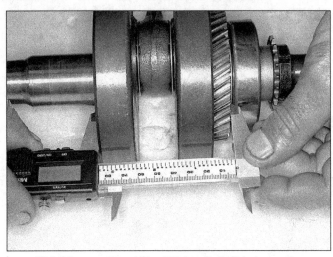

21.9 Measure the width of the crankshaft between the bearing seating surfaces . . .

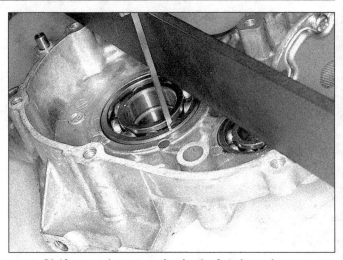

21.10 . . . and measure the depth of each crankcase half to the bearing surface

sprocket end, this surface is the side of the balancer drive gear; on the other end, it's the side of the crankshaft throw. Write this measurement down.

10 Lay a straightedge across one of the case halves. Measure from the straightedge to the bearing surface that contacts the crankshaft **(see illustration)**. Write this measurement down. **Note:** *If you measure from the top of the straightedge to the bearing, you'll need to subtract the width of the straightedge from your measurement. If you measure from the bottom of the straightedge to the bearing, this won't be necessary.*

11 Make the same measurement on the other crankcase half and write it down.

12 Add the measurements taken in Steps 9 and 10 together and write down the total.

13 Subtract the measurement you took in Step 8 from the total you wrote down in Step 11. The result is crankshaft end play. If it's more than the amount listed in this Chapter's Specifications, you'll need to reduce it by adding a shim at the gear end of the crankshaft, between the gear and the ball bearing. This involves unscrewing the sprocket nut with a special tool, removing the sprocket and removing the bearing (you may have to press the bearing off). The thickness of the shim must be enough to bring the crankshaft end play within the Specifications.

Installation

Refer to illustrations 21.14 and 21.15

14 If the crankshaft bearing was removed from the left side of the crankcase (Polaris Variable Transmission side), install it (see Section 19). Thread a puller adapter (Polaris tool 2871283 or equivalent) into the left end of the crankshaft **(see illustration)**.

15 Place the end of the crankshaft into the bearing. Install the puller on the outside of the case half, over the end of the crankshaft, and use it to pull the crankshaft into the bearing until it seats **(see illustration)**.

16 If you removed the cam sprocket and bearing to add a new shim, install the new bearing in the right side of the crankcase.

22 Recoil starter - removal, inspection and installation

Warning: *The spring in the recoil starter may fly out suddenly and cause severe eye injury. Wear eye protection, a face shield and heavy gloves during these procedures.*

Removal

Refer to illustrations 22.1, 22.5a, 22.5b and 22.5c

1 Unbolt the recoil starter housing from the engine and take it off

21.14 Thread the special tool adapter into the end of the crankshaft . . .

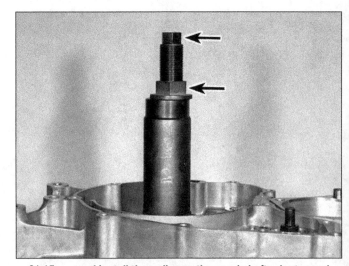

21.15 . . . and install the puller on the crankshaft adapter and crankcase half; hold the shaft (upper arrow) with a wrench and turn the nut to pull the crankshaft into the bearing

22.1 Unbolt the recoil housing and take it off

22.5a Center bolt and washer (A) and rope installation notch (B)

(see illustration).
2 Pull the rope out all the way and tie a knot in it so it won't slip back into the housing. You'll need to undo the knot soon, so make it loose.
3 Pry the cover off the starter pull handle. Pull the knot out of the handle, untie it and slip the rope out of the handle.
4 Hold the rope and untie the loose knot made in Step 3. Let the recoil spring slowly pull the rope into the housing until the spring tension is released.
5 Remove the center nut or bolt and washer, then lift off the friction plate, clip, ratchet pawl and spring **(see illustrations)**.
6 Carefully lift the reel away from the spring, making sure the spring stays in the recoil housing. **Warning:** *The spring is still under tension and will fly out if it is pulled out of the recoil housing.*
7 The spring can stay in the recoil housing unless it needs to be replaced. To remove it, place a hand on the spring to hold it in the housing. Grip the center with pliers and slowly pull the spring out, one coil at a time.

Inspection

8 Check all parts for wear and damage and replace any that have visible problems. Check the spring for breakage or small cracks. Check for wear where the ends of the spring contact the reel and housing.

Installation

Refer to illustrations 22.9, 22.14a and 22.14b
9 If the old spring was removed from the housing and you're rein-

22.5b Lift off the friction plate and clip . . .

stalling it, hook its outer end into the housing groove **(see illustration)**. Wind the spring in counterclockwise, toward the center of the spring. Lubricate the spring with cable lube or lightweight, low-temperature grease.
10 If you're installing a new spring, hook its outer end into the housing as described in Step 9. Hold the spring down in the housing and

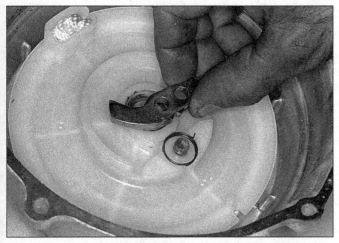

22.5c . . . and lift out the ratchet pawl and its spring

22.9 The outer end of the spring engages this notch in the housing (arrows)

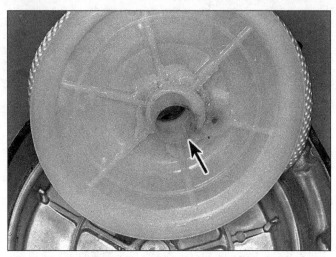

22.14a There's a notch in the center bushing for the spring's inner end (arrow) . . .

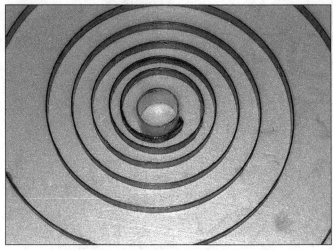

22.14b . . . the spring engages it like this (spring removed for clarity)

cut the retaining wire.

11 Insert the rope through its hole in the housing into the groove of the reel. Run it through the hole in the reel and tie a tight knot that's big enough to keep the rope from pulling out of the reel. Place the outer end of the rope through the handle and tie a knot in the end so the rope won't pull back through the handle.

12 Wind the rope around the reel counterclockwise (viewed from the inside of the housing). When it's wound all the way up, wedge it into the notch on the outer edge of the reel.

13 Lubricate the center bushing with light grease.

14 Install the reel onto the spring, making sure the center bushing slot engages the hook on the inner end of the spring **(see illustrations)**. Hold the reel down and wind it counterclockwise about six turns, until the notch with the rope wedged into aligns with the rope hole in the housing **(see illustration 22.5a)**.

15 Pull the rope out of the notch and let the reel unwind, pulling the rope with it. If there's too much slack in the rope, repeat Steps 12 and 14, but wind the reel one or more additional turns.

16 Install the return spring and the ratchet pawl **(see illustrations 22.5c and 22.5b)**.

17 Install the friction plate and clip **(see illustration 22.5a)**. Tighten the bolt to the torque listed in this Chapter's Specifications.

18 Install the recoil starter housing, using a new gasket. On models where the stator wire grommet passes through the edge of the housing, remove the grommet and coat it with silicone sealant.

23 Initial start-up after overhaul

1 Make sure the engine oil level is correct, then remove the spark plug from the engine. Place the engine kill switch in the Off position and disconnect the primary wires from the coil.

2 Turn on the key switch and crank the engine over with the starter several times to build up oil pressure. Reinstall the spark plug, connect the wires and turn the switch to On.

3 Make sure there is fuel in the tank, then operate the choke.

4 Start the engine and allow it to run at a moderately fast idle until it reaches operating temperature. **Caution:** *If the oil temperature light doesn't go off, or it comes on while the engine is running, stop the engine immediately.*

5 Check carefully for oil leaks and make sure the transmission and controls, especially the brakes, function properly before road testing the machine. Refer to Section 24 for the recommended break-in procedure.

6 Upon completion of the road test, and after the engine has cooled down completely, recheck the valve clearances (see Chapter 1).

24 Recommended break-in procedure

1 Any rebuilt engine needs time to break-in, even if parts have been installed in their original locations. For this reason, treat the machine gently for the first few miles to make sure oil has circulated throughout the engine and any new parts installed have started to seat.

2 Even greater care is necessary if the cylinder has been rebored or a new crankshaft has been installed. In the case of a rebore, the engine will have to be broken in as if the machine were new. This means greater use of the transmission (use Low range when necessary if the machine is so equipped) and a restraining hand on the throttle for the first few operating days. There's no point in keeping to any set speed limit - the main idea is to vary the engine speed, keep from lugging the engine and to avoid full-throttle operation. These recommendations can be lessened to an extent when only a new crankshaft is installed. Experience is the best guide, since it's easy to tell when an engine is running freely.

3 If a lubrication failure is suspected, stop the engine immediately and try to find the cause. If an engine is run without oil, even for a short period of time, irreparable damage will occur.

4 Change the engine oil and filter after 20 hours or 500 miles of operation or one month, whichever comes first. After the first oil change, change oil at the regular intervals specified in Chapter 1.

Chapter 2 Part C
Polaris Variable Transmission (PVT)

Contents

	Section		Section
Air ducts - removal and installation	2	Drive belt - width and play check	5
Clutches, inner cover and seal - removal and installation	7	General information	1
Drive and driven clutches - disassembly, inspection and reassembly	8	Outer cover and seal - removal, inspection and installation	3
Drive belt - removal, inspection and installation	4	Pulleys - alignment and offset check	6

Specifications

Drive belt deflection	1-1/8 inch
Drive belt width	
New	1-3/16 inch
Wear limit	1-1/8 inch

Tightening torques

Drive clutch bolt	40 ft-lbs
Driven clutch bolt	17 ft-lbs
Drive clutch cover bolts	90 inch-lbs
Drive clutch spider	200 ft-lbs

1 General information

All of the vehicles covered in this manual use the Polaris Variable Transmission (PVT). This consists of a drive clutch, drive belt and driven clutch, all mounted in a sealed housing on the left side of the machine. The PVT connects the engine to the transmission. The transmission on these models is a separate unit from the PVT. It allows the operator to select forward range (high or low on some models) or reverse range. Transmission service is covered in Part D of this Chapter.

The PVT's drive and driven clutches control variable-diameter pulleys. The drive clutch is mounted on the end of the engine crankshaft. The driven clutch is mounted on the transmission input shaft.

At low engine speeds, the drive clutch pulley has a small diameter and the driven clutch pulley has a large diameter. As engine speed increases, the drive pulley diameter gets larger and the driven pulley diameter gets smaller. This changes the effective gear ratio of the PVT.

The drive clutch changes its pulley diameter in response to changes in engine speed. At idle, the spring in the drive clutch forces the pulley halves apart, so the drive belt is not gripped and no power is transferred to the driven clutch. As engine speed increases, the drive clutch spins faster. This causes the three shift weights (centrifugal

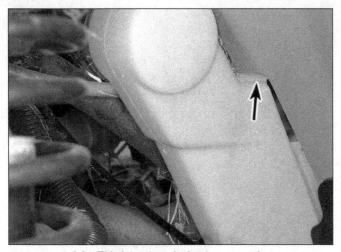

2.2a This inlet duct design has mounting bolts behind the fender (arrow) . . .

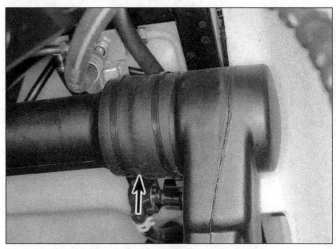

2.2b . . . in some versions, an extension of the inlet duct (arrow) passes across to the right side

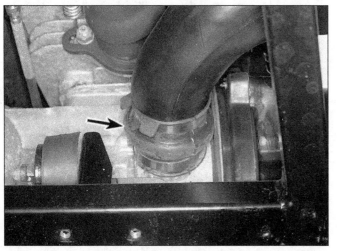

2.3a This type of inlet duct is secured to the front of the engine by a rubber coupling (arrow) . . .

2.3b . . . and this version is attached to the side of the engine near the starter

levers) in the drive clutch to push the sliding pulley half toward the fixed pulley half. As a result, the drive belt is gripped by the pulley and power is transferred to the driven clutch. As engine speed increases still more, the shift weights push the drive clutch's sliding pulley half closer to the fixed pulley half. This makes the pulley groove narrower, so the drive belt rides higher in the pulley groove (closer to the outer edge of the pulley). This causes the driven pulley to be rotated more times for each rotation of the drive pulley (higher gearing).

Since the drive belt doesn't change length, the driven clutch pulley must become smaller in diameter as the drive clutch pulley becomes larger. To make this happen, the driven clutch pulley changes diameter in response to the load placed on it by the drive belt. At low engine speeds, the spring in the driven clutch pushes the pulley halves together, which causes the drive belt to ride higher in the groove (closer to the outer edge of the pulley). This is the equivalent of a low gear in a conventional transmission. As engine speed increases, more power is applied to the drive belt. The belt forces the driven pulley halves apart, causing the belt to ride lower in the pulley groove (closer to the center of the pulley). This is the equivalent of a higher gear in a conventional transmission. Two things control the rate at which the driven pulley halves are forced apart; the tension of the driven clutch spring and the angle of the helical cam that connects the centers of the two halves.

The PVT is air cooled. The cooling air is drawn into the housing by fins on the fixed half of the drive clutch pulley. The air enters a duct

beneath the fuel tank and exits through another duct.

The system is designed so that at full throttle, engine rpm is maintained at a setting that produces peak power output. Springs, shift weights and driven clutch cams are specifically designed for each model. They must be replaced with the exact equivalents if new ones are needed. In addition, the clutch components are balanced as a unit, so installing used parts from another clutch will probably cause vibration.

2 Air ducts - removal and installation

Refer to illustrations 2.2a, 2.2b, 2.3a, 2.3b, 2.4a and 2.4b

1 If the vehicle's frame is made of round-section tubing, there is no inlet duct. The inlet air is routed through the engine case and one of the frame tubes.

2 Some models use a white or black plastic intake duct mounted on the left side of the machine forward of the front fender **(see illustration)**. The duct on liquid-cooled models is molded in one piece with the coolant reservoir. On some versions of this design, the duct is attached to a tube that passes across the front of the vehicle to the right side **(see illustration)**. The duct is attached to the engine case near the PVT housing by a short section of hose and two hose clamps.

3 The inlet duct on all other models is a length of round-section tubing attached to the engine case near the PVT housing by a rubber coupling and two hose clamps **(see illustrations)**.

2.4a The inlet duct is secured to the PVT housing by a rubber coupling . . .

2.4b . . . and is bolted on the top (shown) or at the front of the duct

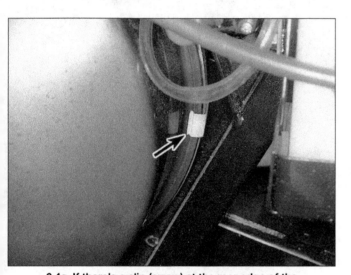

3.4a If there's a clip (arrow) at the rear edge of the PVT housing, remove it . . .

3.4b . . . remove the bolt(s) at the front of the housing . . .

4   The outlet duct is attached to the top of the PVT housing by a rubber coupling and two hose clamps **(see illustration)**. The duct is also bolted to the machine, either along to the top **(see illustration)** or at the front.
5   To remove an air duct, loosen its hose clamps, remove its mounting bolts and take it off.
6   Clean away any debris clogging the ducts. Check the ducts for cracks and the couplings for cracks, brittleness or deteriorated rubber. Replace any parts that have problems.

3 Outer cover and seal - removal, inspection and installation

Refer to illustrations 3.4a, 3.4b, 3.4c and 3.5

1   The seals in the outer cover and inner cover are important because they keep water from leaking in and cooling air from leaking out. A wet or overheated PVT belt will slip on the pulleys.
2   Remove the seat, rear fenders and left footrest (see Chapter 8).
3   Remove the outlet air duct (Section 2).
4   Remove the clip (if equipped) and the cover bolts **(see illustrations)**.

3.4c . . . and at the top (shown) and bottom of the housing

3.5 Pull the seal out of its groove (arrow)

4.1 Draw an arrow on the belt indicating its direction of rotation

4.2 Work the belt out of the driven pulley groove

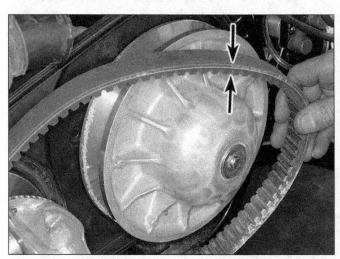

4.3 Check the belt for separated plies (upper arrow) and worn or broken cogs (lower arrow)

5 Carefully pull the seal out of its groove **(see illustration)**. Check for cracks, hardening and deterioration of the rubber. Also look for pinched spots caused by the cover being tightened while the seal is out of position. If you find any problems, replace the seal.
6 Installation is the reverse of the removal steps.

4 Drive belt - removal, inspection and installation

Refer to illustrations 4.1, 4.2, 4.3 and 4.4

1 Remove the outer cover (see Section 2). Draw an arrow on the belt with chalk or white paint to indicate its direction of rotation **(see illustration)**.
2 With the engine off, there will be enough slack in the belt to slip it off the driven pulley **(see illustration)**. You can then lift it out of the drive pulley groove and over the drive clutch.
3 Check the belt for worn or broken cogs and for separation of the plies **(see illustration)**. Check along the surface that contacts the pulleys for wear, especially for cupped spots that indicate severe slippage. Also check for burn marks caused by belt slippage. If you find any problems, replace the belt. If the belt has been slipping, check the outer cover O-ring and the inner cover seal for leaks that could allow the entry of water.
4 Check the friction surface on each pulley for wear, scoring or corrosion **(see illustration)**. If there's a problem, replace the complete front or rear clutch as a unit (Section 5).
5 Installation is the reverse of removal. If you're reinstalling the old belt, make sure its arrow faces the direction of rotation. If you're installing a new belt, look for letters and numbers along one edge. These should face away from the engine when the belt is installed.
6 Check belt play and adjust if necessary (Section 6).

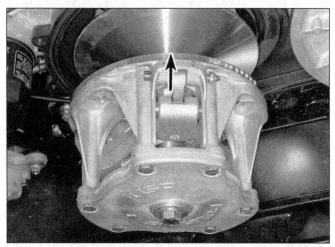

4.4 Check the pulley friction surfaces (arrow) for wear, scoring or corrosion

Chapter 2 Part C Polaris Variable Transmission (PVT) 2C-5

5.3 Push down on the belt and measure its play

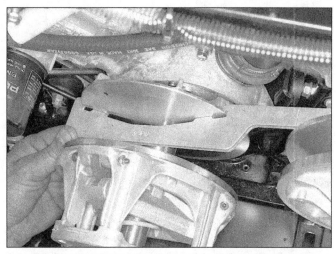

6.1a The factory alignment tool fits on the drive pulley like this . . .

6.1b . . . and across the back of the driven pulley like this

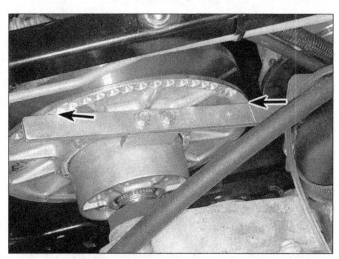

6.1c The extension should touch the driven pulley at the rear (left arrow) and leave a gap at the front (right arrow

5 Drive belt - width and play check

Refer to illustration 5.3

1 Remove the outer cover (Section 3).

2 Measure the width of the belt with a steel ruler or vernier caliper and compare it to the value listed in this Chapter's Specifications. Check at several places along the belt. If it's significantly narrower at any one point, or if it's worn to less than the minimum specification, replace it (Section 4).

3 Press down on the belt midway between the pulleys on the top side and measure its play **(see illustration)**. Press with a finger hard enough to tension the belt lightly. If the play is not within the range listed in this Chapter's Specifications, add or remove shims between the inner and outer halves of the driven pulley (Section 8). Removing shims (placing the pulley halves closer together) tightens the belt and decreases play. Adding shims (placing the pulley halves farther apart) loosens the belt and increases play.

6 Pulleys - alignment and offset check

Refer to illustrations 6.1a, 6.1b and 6.1c

1 Check the alignment of the drive and driven pulleys. The recommended method is to place a Polaris alignment tool (part no. 2870654) across the drive pulley, with its extension contacting the inner surface of the driven pulley **(see illustrations)**. The rear of the extension should touch the rear edge of the driven pulley, and there should be a small gap between the extension and the front edge of the driven pulley **(see illustration)**. Measure the gap with stacked-up feeler gauges. If it's not within the range listed in this Chapter's Specifications, adjust it as described below. **Note:** *If you don't have the correct Polaris tool, you can approximate the alignment measurement, but not the offset, using a straightedge. If you have to do this, check the belt frequently for uneven wear and have a dealer check the alignment and offset as soon as possible.*

2 If you only need to make a small adjustment, look for shims between the frame and the engine mount at the lower left front of the engine. Add shims to increase the gap, or remove shims to decrease it.

3 If you need to make a major adjustment (or if you need to decrease the gap and there aren't any shims to remove), loosen all of the engine mounts (see Chapter 2A or 2B). Move the engine to the right or left, which will move the drive pulley, changing the gap. Tighten the engine mounts to the torque listed in this Chapter's Specifications.

4 If you don't have the factory tool, you can approximate the measurement by placing a straightedge behind the drive pulley so it extends past the driven pulley. Measure the distances from the straightedge to the front and rear edges of the driven pulley; the difference between the two measurements should be the same as the gap listed in this Chapter's Specifications. If it isn't, adjust it as described in Step 3.

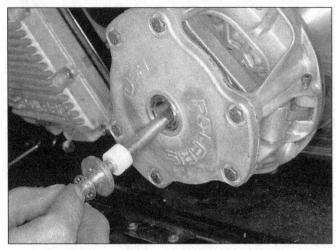

7.2 Remove the drive clutch bolt, lockwasher, washer and bushing

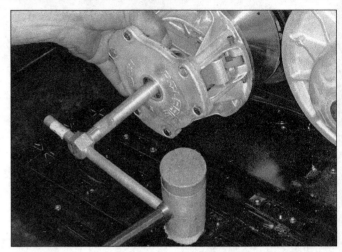

7.3a Install the removal tool and tap it with a hammer to free the drive clutch

7.3b Pull the drive clutch off the crankshaft

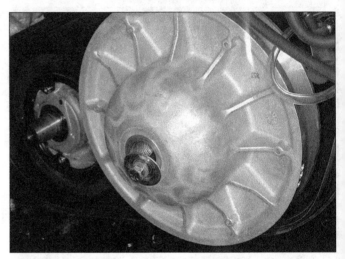

7.5a Remove the driven clutch bolt, lockwasher and washer . . .

5 Once you've adjusted the alignment, check the offset. To do this, install the Polaris measuring tool as described in Step 1. It should touching the rear edge of the driven pulley **(see illustration 6.1c)**. If not, adjust it by removing or adding shims between the driven pulley and inner cover (Section 7).

7 Clutches, inner cover and seal - removal and installation

1 Remove the outer cover and drive belt (Sections 3 and 4).

Drive clutch

Refer to illustrations 7.2, 7.3a and 7.3b

2 Remove the drive pulley bolt, lockwasher, washer and bushing **(see illustration)**.
3 Install a removal tool (Polaris part no. 2870506 or equivalent) **(see illustration)**. Tap the tool handle with a hammer to break the drive pulley loose from the crankshaft, then pull it off **(see illustration)**.
4 Installation is the reverse of the removal steps, with the following additions:
  a) Clean the taper on the end of the crankshaft and its bore inside the drive pulley before installing the pulley.
  b) Tighten the drive pulley bolt to the torque listed in this Chapter's Specifications.

Driven clutch

Refer to illustrations 7.5a, 7.5b and 7.6

5 Remove the pulley bolt, lockwasher and washer **(see illustrations)**. Note that the washer's smaller diameter faces toward the pulley.

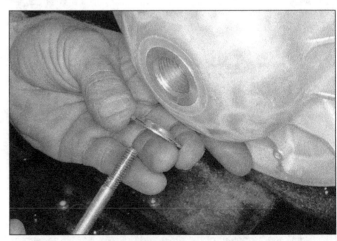

7.5b . . . the small diameter of the washer goes toward the driven clutch

Chapter 2 Part C Polaris Variable Transmission (PVT)

7.6 Pull the driven clutch off and remove the shims

7.8 These three screws secure the inner cover's seal retainer

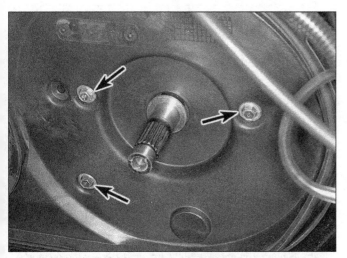

7.9 There's an O-ring and spacer on the shaft of each of these bolts, so don't remove them completely

8.2a The driven clutch holding fixture is designed for a slotted workbench or press plate . . .

6   Remove the driven clutch. If it won't come off the transmission input shaft easily, use the same tool as for the drive clutch **(see illustration 7.3a)**. Remove the shims and place them where they won't be lost **(see illustration)**. These are the shims used to adjust pulley offset (Section 6).

7   Installation is the reverse of the removal steps, with the following additions:
  a) *Install the same number of shims you removed.*
  b) *Check pulley offset (Section 6) and adjust it if necessary.*
  c) *Tighten the driven clutch bolt to the torque listed in this Chapter's Specifications.*

Inner cover and seal

Refer to illustrations 7.8 and 7.9

8   Remove the three screws that secure the inner cover and seal to the engine **(see illustration)**.

9   Loosen the three bolts (with lockwashers and washers) that secure the inner cover to the transmission **(see illustration)**. Loosen these bolts just enough to disengage them from the transmission, but don't remove them from the inner cover; there's a spacer and O-ring behind each of them.

10 Take the cover off and turn it over. Remove the O-ring and spacer from each of the three bolts. If the seal is deteriorated or damaged, carefully pry it out of the cover. Clean the seal groove in the cover thoroughly, then coat it with silicone sealant and press in a new seal, using a block of wood.

11 Installation is the reverse of the removal steps. Be sure to install a spacer, then an O-ring, on each of the three bolts.

8   Drive and driven clutches - disassembly, inspection and reassembly

1   Remove the clutch you plan to work on (Section 7).

Drive clutch

Disassembly

Refer to illustrations 8.2a, 8.2b, 8.3, 8.4a, 8.4b, 8.5a, 8.5b, 8.5c, 8.6a, 8.6b, 8.7, 8.8a, 8.8b, 8.9, 8.10a and 8.10b

2   Disassembly of the drive clutch requires a special holding fixture or its equivalent to secure the unit while the spider is unscrewed **(see illustrations)**. Because the spider is tightened to 200 ft-lbs, trying to unscrew it with improvised tools may damage it. You may be able to duplicate the factory tools if you have a well-equipped shop, but read through the procedure before trying it. **Caution:** *Whatever method you use, make sure it doesn't scratch the mild-steel friction surfaces of the pulley. The entire drive clutch must be replaced as a unit if one of the*

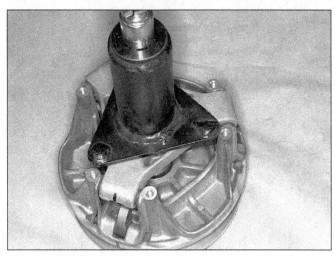

8.2b ... the spider adapter has a post to engage each spider arm and a square hole at the top for a torque wrench

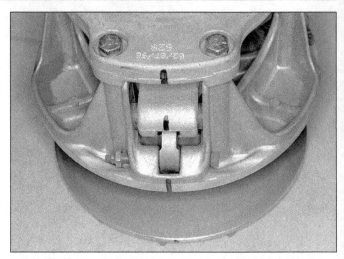

8.3 Mark the components so they can be reassembled in the same relative positions

8.4a Loosen the cover bolts in stages, in a criss-cross pattern

8.4b Two lengths of cover bolts are used, depending on model

pulley surfaces is damaged.

3   There should be "X" marks on the cover, spider and moving pulley half. Just in case they aren't there, mark the cover, the spider, the moving pulley half and the fixed pulley half with a felt pen so they can be reassembled in the same relationship **(see illustration)**. This is necessary to preserve the unit's balance.

4   Loosen the cover bolts in stages, in a criss-cross pattern, then remove them and take off the cover **(see illustrations)**. **Note:** *There are two cover designs, one with long bolts and one with short bolts. Make sure you use the correct type of bolt if you replace any of them.*

5   Remove the cover spring and the shim beneath it (if equipped) **(see illustrations)**. **Note:** *Before disassembling further, check each of*

8.5a Remove the spring . . .

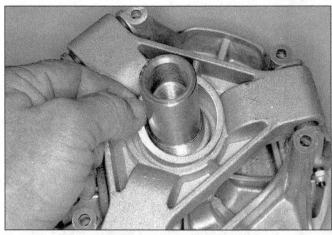

8.5b . . . and the shim beneath it (if equipped)

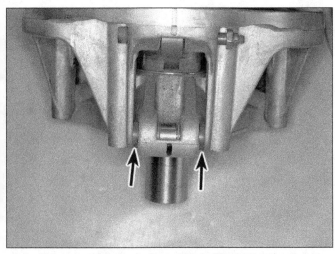

8.5c If there's clearance between the spider buttons and posts (arrows), install new buttons

8.6a Clamp the fixed pulley half and unscrew the spider . . .

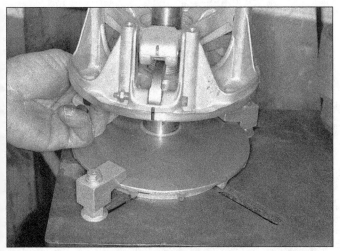

8.6b . . . and lift the clutch components off the fixed pulley half

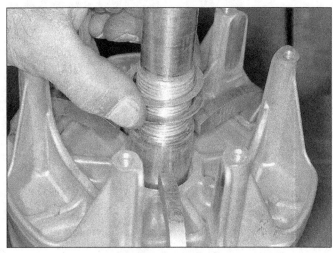

8.7 Remove the spacer shims

the spider arms for play at the buttons **(see illustration)**. If there's any clearance between the buttons and the posts next to them, replace the buttons as described later.

6 Place the unit in the holding fixture and unscrew the spider **(see illustration)**. Lift the clutch mechanism and moving pulley half off the fixed pulley half **(see illustration)**.

7 Remove the spacer shims from the clutch mechanism **(see illustration)**. Tie them together and place them where they won't be lost; it's important to reinstall the same number and thickness of shims.

8 Remove the nut from each of the shift weight pivot bolts **(see illustration)**. Pull out the bolt and remove the shift weight **(see illustration)**. Mark each of the shift weights with its position in the moving pulley half so it can be returned to its original position.

9 Check to make sure each roller in the spider turns freely **(see**

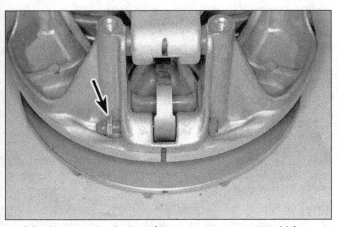

8.8a Unscrew the locknut (use a new one on assembly) . . .

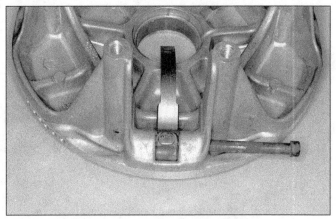

8.8b . . . and pull out the shift weight bolt

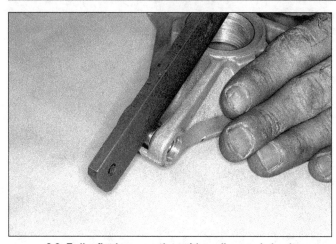

8.9 Roll a flat bar over the spider rollers and check for free movement

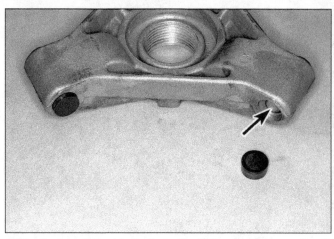

8.10a Remove the spider buttons, the O-ring or rubber washer (if equipped) and the pivot roller shaft (arrow)

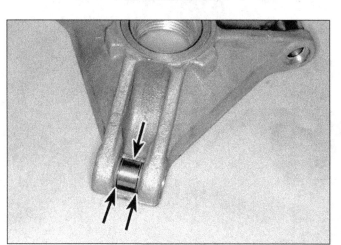

8.10b Remove the roller (upper arrow) and thrust washers (lower arrows)

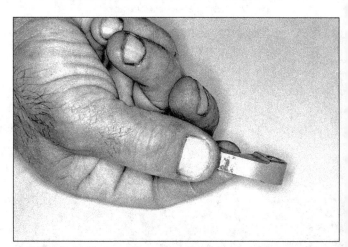

8.11a Replace the shift weights if they're corroded, like this, or cupped

illustration). If the roller turns roughly or not at all, remove and clean its components, then reassemble it and recheck its movement. Rollers that don't turn freely are a common cause of shift weight wear.

10 Remove the button and O-ring or washer from each end of each spider pivot (see illustration). The buttons shown in the illustration are the rubber-backed design introduced in mid-1995. These can be used on earlier models, provided they are used with solid, not hollow, roller pins. Note: *On some models, there's an O-ring inside the button on the driving side and a flat rubber washer inside the button on the trailing side of each spider pivot. Be sure to reinstall these in their correct locations.* Push out the roller pivot, then remove the roller and thrust washers (see illustration).

Inspection

Refer to illustrations 8.11a, 8.11b, 8.13a and 8.13b

11 Check the shift weight friction surfaces for wear, especially for cupping (see illustration). This is usually caused by sticking rollers. Replace all three shift weights, along with their nuts and bolts, if any problem is found with any one of them. Be sure to use new shift weights with the same identification numbers and letters (see illustration). These indicate the weight and the profile of the weights, both of which are specifically tailored to the vehicle.

12 If the shift weights were worn enough to need replacement, replace the rollers as well.

13 Check the bushings in the cover and moving pulley half (see illustrations). Replace them if they're worn or damaged. The bushing in the moving pulley half is Teflon-coated brass. If you can see more exposed brass than Teflon, replace the bushing. Drive or press the bushings out using a bearing driver or socket, then drive or press new ones in using the same tool. On 1985 and 1986 models, the bushing in the moving pulley half has a steel retainer.

Reassembly

14 Assembly is the reverse of the disassembly steps, with the following additions:

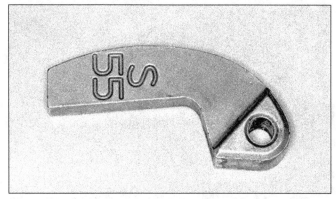

8.11b Replace all three shift weights with weights that have the same markings

Chapter 2 Part C Polaris Variable Transmission (PVT)

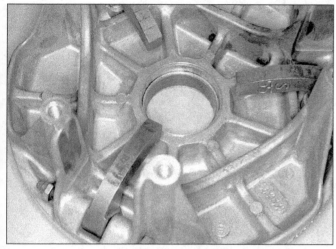

8.13a Inspect the bushing in the moving pulley half (1987-on shown) . . .

8.13b . . . and the bushing in the cover (1987-on design shown)

8.15 The cam holes are numbered 1 and 2; look for the end of the spring in one of them (arrow)

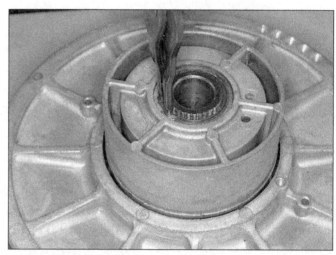

8.16a Push down on the cam and remove the snap-ring . . .

a) Don't use any lubricant on any of the parts during assembly. It may fly off and contaminate the belt when the machine is operated.
b) Align the felt pen marks or factory X marks on the pulley halves, spider and cover with each other when assembling the unit **(see illustration 8.3)**.
c) Be sure the spacer shims are fully seated in their bore **(see illustration 8.7)**.
d) If the roller pins have a flat rubber washer and an O-ring, be sure to position the O-ring on the driving side of the spider pivot and the flat washer on the trailing side.
e) If you're using the late design rubber-backed buttons, be sure to use solid roller pins, not hollow ones.
f) Tighten the spider to 200 ft-lbs, using a holding fixture and a torque wrench. Don't guess at this critical torque setting. If you don't have the necessary tools, have a Polaris dealer or qualified ATV repair shop tighten the spider for you.

Driven clutch

Disassembly
Refer to illustrations 8.15, 8.16a, 8.16b, 8.16c, 8.17 and 8.18

15 Note the location of the spring end in the no. 1 or no. 2 hole of the cam **(see illustration)**. Write this number down so the spring can be positioned in the same hole on assembly.
16 Remove the snap-ring from its groove in the pulley shaft, then lift off the cam **(see illustrations)**. The moving pulley half has three holes for the spring. Note which of the holes the spring is in, then lift the spring out. Again, write down the spring position.

8.16b . . . and the washer beneath it (arrow)

8.16c Lift up the cam and note which hole the other end of the spring is in (arrow)

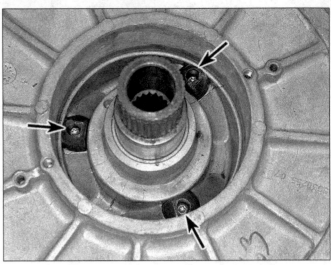

8.17 Inspect the cam contact buttons (arrows)

8.18 These spacer washers adjust drive belt tension

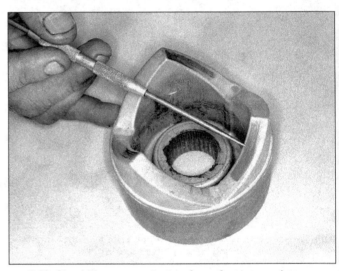

8.20 Check the cam contact surfaces for wear or damage

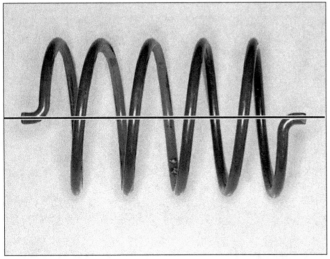

8.21 The ends of the spring should be in a straight line

17   Inspect the cam contact buttons in the moving pulley half **(see illustration)**. If they're worn on the visible side, undo their Torx screws and remove them. If their undersides are not worn, the buttons can be reused; just reinstall them with the good sides showing. Discard them if both sides are worn.

18   Pull the moving pulley half off the fixed half and remove the shims **(see illustration)**. **Note:** *These are the shims used to adjust belt tension (Section 5). At least one shim must be used when the clutch is reassembled.*

Inspection

Refer to illustrations 8.20, 8.21, 8.22a and 8.22b

19   Check the cam contact buttons as described in Step 17 if you haven't already done so. Turn them over and reinstall them if they're worn on one side or install new ones of they're worn on both sides.

20   Check the button contact surfaces on the cam **(see illustration)**. Replace the cam if they're worn or damaged.

21   Lay a straightedge across the spring ends, parallel to the centerline of the spring **(see illustration)**. If the ends don't both line up with the straightedge, replace the spring.

22   Check the bushings in the moving pulley half for wear or damage **(see illustrations)**. Replace the Teflon-coated brass bushing if there's

Chapter 2 Part C Polaris Variable Transmission (PVT)

8.22a Check the bushing in the inner side of the moving pulley half . . .

8.22b . . . and the bushing in the outer side

8.23a Position the spring end in the correct cam hole

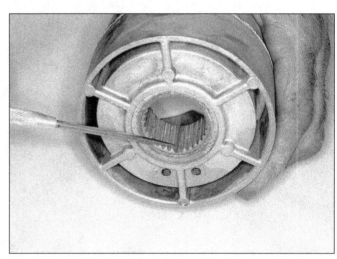

8.23b Align this wide groove in the cam . . .

8.23c . . . with this wide spline on the pulley shaft

more brass than Teflon visible on the surface. Press or drive the bushings out and press new ones in, using a bearing driver or socket.

Reassembly

Refer to illustrations 8.23a, 8.23b and 8.23c

23 Reassembly is the reverse of the disassembly procedure, with the following additions:

a) *Don't use any lubricant during assembly. It may fly off and contaminate the drive belt.*
b) *When you install the spring, position it in the same holes it was removed from* **(see illustration)**. *The standard factory setting is the no. 2 hole at both ends of the spring. Changing the hole at either end will affect the shift characteristics.*
c) *When you install the cam, align its wide spline with the wide spline on the fixed pulley half* **(see illustrations)**. *Push the cam onto the shaft about 1/2 to 3/4-inch, then, while pressing the cam down, turn it counterclockwise 1/3 turn to seat the cam. Install the washer, then the snap-ring, making sure the snap-ring is seated securely in its groove.*

Notes

Chapter 2 Part D
Transmission

Contents

	Section
EZ Shift shifter – disassembly, inspection and reassembly	26
EZ Shift shifter – removal and installation	25
General information	1
Shift linkage - adjustment	14
Transmission – disassembly, inspection and reassembly	13
Transmission – removal and installation	2

Specifications

Transmission types and applications

Type I	
1985	All
1986	All
Type II	
1987 through 1992	All
1993	All except Trail Boss, Trail Blazer, Sportsman, Big Boss 6X6 with EZ shift
Type III	
1993	Trail Boss, Trail Blazer, Sportsman, Big Boss 6X6 with EZ shift
1994-on	All chain drive
Type IV	All shaft drive

Tightening torques

Transmission mounting bolts	25 ft-lbs
Output shaft bearing nuts/bolts	12 ft-lbs
Transmission case cover bolts	
Type II	17 ft-lbs
Type I, Type III and Type IV	12 ft-lbs
Shifter cover screws	12 ft-lbs

Chapter 2 Part D Transmission

1 General information

The vehicles covered in this manual use a chain-drive or gear-drive transmission in addition to the belt-drive transmission (Polaris Variable Transmission, or PVT) covered in Part C of this Chapter. The transmission on these models is a separate unit from the PVT. It allows the operator to select forward range (high or low on some models) or reverse range.

The transmission used on 1985 through some 1990 models is a chain drive unit, referred to as Type I by Polaris. The design used on 1987 through some 1993 models is a gear unit designated Type II. Some 1993 models, and all 1994 and later chain-drive models, use the Type III gear/chain transmission. Shaft drive models use the Type IV transmission.

Transmission designs and model applications are listed in this Chapter's Specifications.

2 Transmission – removal and installation

Type I transmission

1 Remove the rear cargo rack and fenders (see Chapter 8).
2 Remove the outer cover and driven clutch from the PVT (see Part C of this Chapter).
3 Loosen the swingarm pivot bolts just enough to provide clearance to remove the transmission. You'll need to pull the PVT inner cover out slightly for access to the left swingarm bolt.
4 Remove two transmission mounting bolts from the left side and one through-bolt from the bottom.
5 Remove the rear brake caliper from the transmission shaft (see Chapter 7).
6 Remove the air cleaner housing (see Chapter 4).
7 Lift the transmission out of the vehicle.
8 Installation is the reverse of the removal steps, with the following additions:
  a) Tighten the swingarm transmission bolt to the torque listed in the Chapter 6 Specifications, then bend the locktabs to secure it.
  b) Adjust the torque stopper (optional on 1985 and 1986; standard on 1987 and later). Tighten the torque stopper until it touches the transmission. On 1985 and 1986 models, tighten it another 1-1/2 turns. On 1987 and later models, tighten it another 1/2 turn. Tighten the locknut.
  c) Check the transmission oil level and top it up if necessary (see Chapter 1).

Type II transmission

9 Remove the seat, rear rack and rear fenders (see Chapter 8).
10 Remove the air cleaner housing, the exhaust heat shield and the rear of the exhaust system (see Chapter 4).
11 Remove the rear brake caliper and disc (see Chapter 7).
12 Remove the locknut from under the shifter rod where it attaches to the transmission. Separate the shifter rod from the lever on the transmission.
13 Detach the vent line from the transmission and disconnect the wires from the neutral switch.
14 Remove the chain guard that's installed next to the transmission (see Chapter 6). Remove the drive chain and the sprocket that's installed on the transmission output shaft.
15 If there's a locking collar over the transmission output shaft, remove its mounting screw (it's on the side of the collar nearest the transmission) and rotate the collar away from the transmission.
16 Clean all dirt and corrosion from the transmission output shaft. Remove the bearing support bolts from the frame, then slide the bearing off the output shaft. **Note:** *If the bearing won't come off the shaft, heat its inner race carefully.*
17 Remove the transmission mounting bolts, one from the side and one from the bottom of the transmission.
18 Remove the right swingarm bolt just far enough to allow removal

2.27a Remove the locknuts from the undersides of the shifter arms

2.27b Hold the stud with a backup wrench and unscrew the nut with a second wrench

clearance for the transmission (see Chapter 6).
19 Unbolt the rear sprocket and the rear chain guard, then disengage the rear drive chain from the transmission sprocket (see Chapter 6).
20 Remove the outer cover and driven clutch from the PVT (see Part C of this Chapter). Bend the rear portion of the inner cover outward to expose the three transmission mounting bolts and remove them.
21 Remove the left swingarm bolt just far enough to allow removal clearance for the transmission (see Chapter 6).
22 Pull the transmission out of its mounting brackets and move it to the right for access to the torque stop bolt. Remove the torque stop bolt, then lift the transmission out of the frame.
23 Installation is the reverse of the removal steps, with the following additions:
  a) Tighten the swingarm bolts to the torque listed in the Chapter 6 Specifications, then bend the locktabs to secure them.
  b) Adjust the torque stopper. Tighten the torque stopper until it touches the transmission, then tighten it another 1/2 turn and tighten the jam nuts.
  c) Check the transmission oil level and top it up if necessary (see Chapter 1).

Type III transmission

Refer to illustrations 2.27a, 2.27b, 2.28, 2.30a, 2.30b, 2.30c, 2.35, 2.36 and 2.38

24 Remove the seat, rear rack, rear fenders and rear skid plate (see Chapter 8).
25 Remove the carburetor, air cleaner housing and exhaust system (see Chapter 4).
26 Remove the PVT and inner cover (see Chapter 2C).
27 Remove the locknuts from under the shifter rods where they

2.28 The Type III transmission vent line is routed like this

2.30a Right swingarm bolt (left), bearing bracket nuts (center) and transmission mounting bolt (upper right)

2.30b Pull off the bearing plate and note the number of shims used

2.30c Slide the bearing unit off the output shaft

attach to the transmission (see illustrations). Separate the shifter rods from the levers on the transmission.

28 Detach the vent line from the transmission (see illustration). Disconnect the wires from the neutral switch (see Chapter 5).

29 If the vehicle has a center drive chain, remove the chain and its sprocket (see Chapter 6). On all models, remove the rear drive chain. Remove the jam nut, chain guide and washer.

30 If there's a bearing on the transmission output shaft, remove it (see illustrations). Note the number of alignment shims, if any are used, and be sure to reinstall the same number.

31 Remove the right swingarm bolt just far enough to allow removal clearance for the transmission (see Chapter 6).

32 Remove two transmission mounting bolts; a 3/8-16X _ bolt from the top left and another 3/8 bolt from the top right.

33 Loosen the left swingarm bolt just far enough to allow removal clearance for the transmission, then remove the nut and lockwasher form the bolt (see Chapter 6).

34 Drain the transmission oil (see Chapter 1).

35 Remove the transmission lower mounting bolt (see illustration).

36 Loosen the torque stop locknut, then turn the torque stop bolt all the way in to provide removal clearance for the transmission (see illustration).

2.35 Hold the bottom bolt with a backup wrench and unscrew the nut

2.36 Loosen the locknut to adjust the torque stop bolt

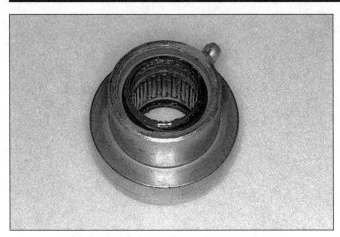

2.38 The bearing unit has a seal on each side

2.42a Shaft drive transmission linkage; high/reverse bellcrank (left) and low bellcrank (right)

37 From the rear of the vehicle, rotate the transmission counterclockwise. Remove the transmission up and to the right through the frame opening.
38 Inspect the output shaft bearing (if equipped) **(see illustrations)**. If the seal on either side is worn or damaged, or if the bearing shows signs of wear, replace the unit.
39 Installation is the reverse of the removal steps, with the following additions:
  a) Tighten the swingarm bolts to the torque listed in the Chapter 6 Specifications, then bend the locktabs to secure them.
  b) Adjust the torque stopper. Tighten the torque stopper until it touches the transmission, then tighten it another 1/2 turn and tighten the locknut.
  c) Adjust the transmission shift linkage (Section 14).
  d) Refill the transmission oil (see Chapter 1).

Type IV transmission

Refer to illustrations 2.42a and 2.42b
40 Remove the air cleaner housing (see Chapter 4).
41 Remove the PVT (see Chapter 2C).
42 Disconnect the shift linkage rods and vent line from the transmission **(see illustrations)**.
43 Jack up the rear end of the vehicle. Securely support the vehicle so it can't fall, then remove the rear wheels. Place an additional jack beneath the transmission as a safety precaution.
44 Remove the caliper and brake disc from the transmission output shaft (see Chapter 7).
45 Detach both ends of the stabilizer bar from the vehicle. Remove the upper right control arm from the rear suspension, then detach the inner end of the upper left control arm and pivot it out of the way (see Chapter 6).
46 Unbolt the lower control arm bracket from each side of the transmission (see Chapter 6).
7 Unbolt the stabilizer bar support bracket and lift it out (see Chapter 6).
8 Unbolt the transmission front mount.
9 Disconnect the inner ends of the driveaxles from the transmission (see Chapter 6). Lift the right driveaxle away from the transmission and tie it out of the way.
10 Remove the mounting bolts from the underside of the transmission.
11 Have an assistant help you lift the transmission out through the right side of the frame.
12 Installation is the reverse of the removal steps, with the following additions:
  a) Lubricate the splines of the front output shaft with anti-seize compound.
  b) Install all of the transmission bolts loosely, then tighten them to the torque listed in this Chapter's Specifications in the following sequence: Upper and lower front support brackets; upper and lower stabilizer bar brackets; bottom bolts; control arm bracket bolts.
  c) If the spacer was removed from the right driveaxle bolt, apply Loctite 272 or equivalent to the threads and tighten the bolt to the torque listed in the Chapter 6 Specifications. Use Loctite 272 on the threads of the left driveaxle bolt. Start the left driveaxle bolt before pushing the driveaxle all the way into the transmission.
  d) Check the transmission oil level and top up if necessary (see Chapter 1).

3 Transmission – disassembly, inspection and reassembly

Type I transmission

Disassembly
1 Drain the transmission oil (if you haven't already done so).
2 Look for a setscrew in the outside of the case. On some models, it secures a detent ball and spring. Remove these if equipped).
3 Remove the sprocket guard and sprocket.
4 Loosen the case bolts evenly, in a criss-cross pattern. Remove the bolts, then carefully tap the case halves apart. **Caution:** *Don't pry against gasket surfaces or they will leak.*
5 Remove the output gear, shafts and chain as an assembly.

Inspection
6 Separate the components and clean them in solvent.
7 Spin the bearings and check them for roughness, looseness or noise. Check the gear teeth for wear or damage and the chain for stretching.
8 If the output shaft bearings need to be replaced, tap the shaft out. If the bearings won't come easily, use a puller or press.
9 If the input shaft bearings need to be replaced, press them off. Remove the snap-ring from the shaft and take off the collar, thrust washer and gear.

Reassembly
10 If the input and output shafts were disassembled, reassemble them, using new components as necessary. If you replace the chain, be sure to get a new one of the same length (11-inch or 15-inch) as the old one.
11 The remainder of disassembly is the reverse of the assembly steps, with the following additions:
  a) Mesh the shafts with each other and install them as a unit.
  b) Lightly coat one of the case mating surfaces (not both) with Loctite 515 Gasket Eliminator or equivalent before assembling the case halves.

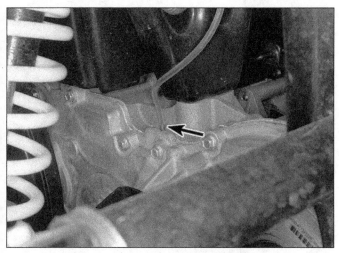

2.42b Transmission vent line (shaft drive models)

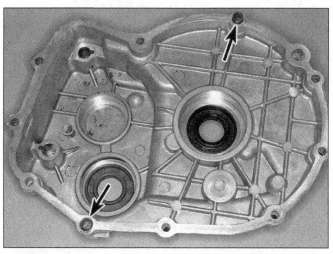

3.14 Remove the case cover and locate the dowels (arrows)

c) *Grease the inner lips of the seals before install the shafts to prevent seal damage.*
d) *Tighten the case bolts evenly to the torque listed in this Chapter's Specifications.*

Type II and Type III transmissions

Refer to illustrations 3.14, 3.15, 3.16, 3.17a, 3.17b, 3.18a, 3.18b, 3.19, 3.20a and 3.20b

12 Type II and Type III transmissions may be equipped with forward and reverse gears or with high and low forward gears as well as reverse. Both types are generally similar in design and service procedures. The Type III transmission with high, low and reverse gears is more complicated (it has two shift forks instead of one) and is illustrated in the following procedure. If you're working on a Type II transmission, use this procedure as a general guide.
13 If you haven't already done so, remove the speedometer drive unit and sprocket from the outside of the case (see Chapters 5 and 6).
14 Loosen the case bolts evenly in a criss-cross pattern, then carefully tap the case halves apart. Lift the input shaft side of the case off the output shaft (sprocket) half and locate the dowels **(see illustration)**.
15 Slide the ball bearing off the sliding gear shaft **(see illustration)**. If it won't come easily, use a small gear puller. Remove the thrust washer, low gear, its needle roller bearing and second thrust washer. If you can't get at the needle roller bearing, slide low gear back-and-forth on the shaft until the bearing is slightly exposed, then use a pointed tool to hook the bearing and pull it off.

3.15 Remove the ball bearing (left arrow) and low gear (right arrow)

16 Make sure the transmission is in neutral (the shafts can rotate independently of each other). If it isn't, turn the fork levers with a wrench on the nuts **(see illustration)** to move the forks and place the transmission in neutral.
17 Turn the low gear shift fork clockwise and slide its shaft out of the case, together with the low gear shift dog **(see illustrations)**.

3.16 The shift shaft bellcranks are secured by self-locking nuts

3.17a The low gear shift fork shaft (left) and high-reverse shift fork shaft (right)

3.17b Type III transmission gear details

A) Low gear engagement dog
B) Low gear shift fork shaft
C) Output gear and sliding gear shaft engagement
D) Oil deflector

3.18a Output gear shaft (arrow) and gear

18 Lift the output gear and shaft out of the case **(see illustration)**. If you haven't already removed the speedometer drive piece from the end of the output shaft, do it now **(see illustration)**.
19 Lift out the sliding gear shaft, input shaft and high/reverse shift fork shaft as a unit **(see illustration)**. Turn the high/reverse shift fork shaft counterclockwise as you remove the shafts. Work the input shaft gear past the oil deflector, but don't remove the oil deflector from the case **(see illustration 3.17b)**.
20 Remove the detent pin cover, detent pins and spring **(see illustrations)**.

3.18b The speedometer drive piece (arrow)

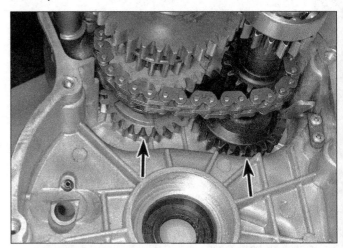

3.19 The sliding gear shaft (left) and input shaft (right)

3.20a Remove the screws (arrows) and the detent cover . . .

3.20b . . . then remove the detent spring and pins

Chapter 2 Part D Transmission 2D-7

3.21 The transmission input shaft (top), sliding gear shaft (center) and output shaft (bottom)

3.22a Spread the snap-ring with snap-ring pliers and lift it with a small screwdriver

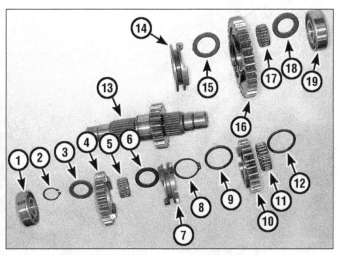

3.22b Sliding gear shaft details

1	Ball bearing	11	Needle roller bearing
2	Snap-ring	12	Thrust washer
3	Thrust washer	13	Sliding gear shaft
4	High gear	14	Low gear engagement dog
5	Needle roller bearing		
6	Thrust washer	15	Thrust washer
7	High-reverse engagement dog	16	Low gear
		17	Needle roller bearing
8	Snap-ring	18	Thrust washer
9	Thrust washer	19	Ball bearing
10	Reverse gear		

3.23 Check the shift dogs for worn corners (arrows)

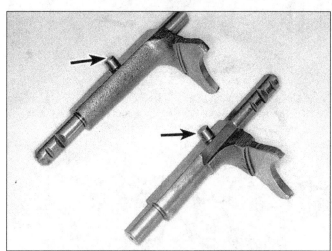

3.26a Check the shift fork shafts for worn pins (arrows) and fork fingers

Inspection

Refer to illustrations 3.21, 3.22a, 3.22b, 3.23, 3.26a, 3.26b, 3.27 and 3.28

21 Lay the shafts on a clean work surface **(see illustration)**. Separate the input shaft and sliding gear shaft from the chain. Spin the ball bearings and check for roughness, looseness or noise. If any of the bearings are in doubtful condition, replace them.

22 Remove the snap-ring from the end of the sliding gear shaft **(see illustration)**. Slide the components off the shaft and lay them in order for inspection **(see illustration)**.

23 Check the gears for worn or chipped teeth and replace any that have defects. Check the corners of the shift dogs on the gears and engagement dogs for wear **(see illustration)**. Rounded corners will cause the transmission to jump out of gear.

24 Inspect the needle roller bearings closely. Since wear on this type of bearing is hard to see and can't be heard or felt when inspecting them, replace them if there's any doubt about their condition.

25 Check the bearing races on the transmission shafts for wear and scoring. Replace the shafts if any problems are found.

26 Check the shift fork shafts and their forks for wear and damage **(see illustration)**, especially at the pins and at the fork fingers. An updated fork design has been made available by Polaris to prevent problems with jumping out of gear **(see illustration)**.

2D-8 Chapter 2 Part D Transmission

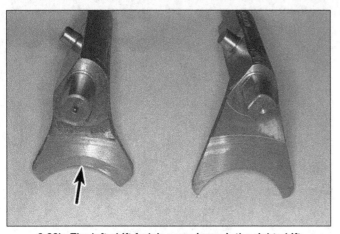

3.26b The left shift fork is worn (arrow); the right shift fork is an updated design

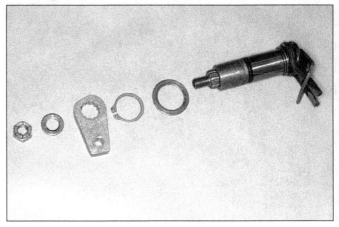

3.27 Shift shaft details

27 Remove the self-locking nuts from the shift shaft bellcranks **(see illustration 3.16)**. Remove the washer and bellcrank from each shaft, then remove the snap-rings and take the shafts out of the case **(see illustration)**.

28 Clean all old sealant from the case mating surfaces **(see illustration)**. Make sure the oil deflector is securely mounted (don't remove it). Tap the old seals out of the case with a seal driver or socket, or carefully pry them out, taking care not to scratch the seal bores.

Assembly

Refer to illustrations 3.31, 3.32 and 3.33

29 Lubricate the shift shafts and install them in the case, using new snap-rings. Install the bellcranks, using new self-locking nuts.

30 Install the detent pins, spring and cover **(see illustrations 3.20b and 3.20a)**.

31 Reassemble the sliding gear shaft, using new snap-rings **(see illustration 3.22b and the accompanying illustration)**. Don't install

3.28 Remove the seals (upper arrow) and make sure the oil deflector (lower arrow) is tight

3.31 The assembled sliding gear shaft looks like this

3.32 Here's how the gears, shafts and shift fork shafts fit together when they're installed

3.33 Here's how the low gear shift fork fits in the engagement dog

Chapter 2 Part D Transmission

4.1a Disconnect the linkage rod (arrow) from the bellcrank . . .

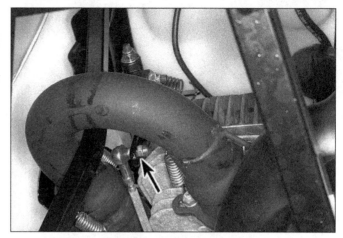

4.1b . . . and disconnect the other end of the rod from the shifter (arrow)

low gear or the low engagement dog yet; they'll be installed after the shafts are in the case. Look closely at the snap-rings; they should have a rounded side and a sharp side. If so, the rounded side faces the direction of thrust (toward the part being retained on the shaft). Make sure the snap-rings are seated securely in their grooves.

32 Mesh the sliding gear shaft and the input shaft and position the chain on their sprockets **(see illustration)**. Position the high-reverse shift fork in the groove of its engagement dog, then install the input

4.2 Move the shift lever (shown here in High) to Neutral

4.5 Make sure the shift block nuts are tight and the lever spring is in position

shaft, sliding gear shaft and high-reverse shift fork shaft in the case **(see illustration 3.19)**. Make sure the pin on the shift fork shaft is positioned between the shift shaft spring fingers.

33 Install the low shift fork shaft and engagement dog, engaging the fork with the groove on the engagement dog **(see illustration)**.

34 Install the thrust washer (with the larger inside diameter) and low gear on the sliding gear shaft. Oil the needle roller bearing and slip it onto the shaft, inside the bore of low gear. Install the thrust washer with the smaller inside diameter, then install the ball bearing so it fits flush with the end of the shaft.

35 Coat the mating surface of one case half with Loctite 515 Gasket Eliminator or equivalent. Install the case cover and tighten the bolts evenly, in a criss-cross pattern, to the torque listed in this Chapter's Specifications.

36 Fill the transmission with the specified lubricant (see Chapter 1).

Type IV transmission

37 Overhaul of the Type IV transmission is a complicated job, requiring special tools not normally available even to well-equipped shops. Overhaul should be done by a Polaris dealer or other shop with the proper equipment and knowledge.

4 Shift linkage - adjustment

Mechanical shifter

Refer to illustrations 4.1a, 4.1b, 4.2 and 4.5

1 If you're working on a 1987 or 1988 model, place the transmission in Neutral, then remove the linkage rod from the transmission bellcrank and shifter **(see illustrations)**. Check the bellcrank and shifter lever to make sure they're still in the neutral positions. Move them slightly if necessary. If the linkage rod now doesn't fit in its holes, loosen the locknuts and rotate the linkage rod ends to change its length. Tighten the locknuts and reinstall the linkage rod.

2 If you're working on a 1989 or later model, place the shift lever in the neutral position **(see illustration)**. Start the engine and let it idle.

3 Slowly move the shift lever to the forward position, listening for the sound of gears engaging. When you hear it, note how far the lever has traveled, then slowly move it back toward reverse and listen again for the sound of gears engaging.

4 The lever should travel an equal distance toward the forward and reverse positions before the gears clash. If not, remove the linkage rod (see Step 1) and adjust its length. Once you've got the rod length adjusted so there's an equal distance between forward and reverse gear clash, tighten the locknuts and reinstall the rod.

5 On all models, check to make sure the nuts on the shift block are tight and the lever spring is in position **(see illustration)**.

2D-10 Chapter 2 Part D Transmission

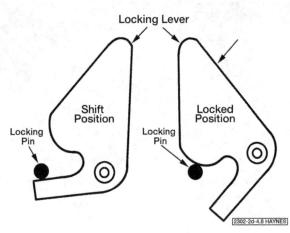

4.8 The locking lever should touch the pin in the shift position and locked position

Locking shifter

Refer to illustration 4.8

6 Correct adjustment of the locking shifter is necessary to keep the transmission from popping out of gear. The transmission may be damaged if the procedure is done incorrectly. This design is used on 1991 through 1993 350L 2X4 models and early 1993 6X6 models.

7 Check the adjustment by squeezing the shift button on the shifter handle and checking the position of the locking lever and pin on the transmission.

8 The tab on the locking lever should touch the locking pin when the shift button is squeezed to the shift position **(see illustration)**. If it doesn't, adjust the length of its cable.

9 When the shift button is released, the main part of the locking lever should touch the locking pin **(see illustration 4.8)**. Again, adjust the length of the cable if it doesn't.

EZ Shift

Refer to illustrations 4.12 and 4.13

10 Start by adjusting the torque stop (see Installation in Section 3).

11 Place the shifter in the Neutral position and disconnect the linkage rod(s) from the transmission **(see illustration 2.27a or 2.42a)**.

12 If you're working on a shaft drive model, disconnect the low gear linkage rod from the center bracket **(see illustration)**.

13 Disconnect the other end of the linkage rod(s) from the shifter **(see illustration)**. **Note:** *The linkage rods may connect to the shifter with both studs facing downward, or with one stud facing upward, depending on model.*

14 Check the bellcrank (high-reverse transmission) or both bellcranks (high-low-reverse transmission) to make sure they're in the neutral position.

 a) *On chain drive models, the transmission is in Neutral when the centerline of the bellcranks is at right angles to the transmission case seam.*
 b) *On shaft drive models, the high-reverse bellcrank is in neutral when its centerline is at right angles to the crankcase seam; the low bellcrank is in neutral when it's offset about 1/8 turn from the neutral position* **(see illustration 2.42a)**.

15 If the bellcranks aren't in the neutral position, place them there. Then try to fit the linkage rods back into their holes. The stud on the transmission end of all linkage rods is installed pointing downward (nut on underside). Be sure to install the studs at the shifter ends of the rods correctly (pointing upward or downward) as follows:

 a) *All 400s except Sport and Scrambler: inner rod stud pointing down, outer rod stud pointing up.*
 b) *All 300s, Magnum 4X4, 1996-on Magnum 2X4: inner rod stud pointing up, outer rod stud pointing down.*
 c) *1995 Magnum 2X4: both studs pointing down.*
 d) *Trail Blazer, Trail Boss, Sport 400, Scrambler: Stud pointing downward, horizontal nut toward engine.*
 e) *All high-reverse transmissions: Stud pointing downward.*
 f) *Sportsman 500 shaft drive: Low rod clevis pins head-up (cotter pins on underside), center bracket stud pointing downward, inner stud at shifter pointing upward, outer stud at shifter pointing downward.*

16 If the rods don't fit easily, loosen the locknuts at the linkage rod ends. Turn the ends to change the linkage rod length so the linkage rod studs align as closely as possible with their holes, then install the linkage rod studs in the bellcranks and shifter slides. Install and tighten the stud nuts, but don't tighten the linkage rod locknuts yet.

17 In this step, you'll make a precise adjustment of the linkage rod length. With the linkage rod studs securely installed, rotate each linkage rod clockwise (viewed from the rear) until you feel resistance. Make a felt pen mark along the length of the rod so you can see how many times you rotate it, then rotate it counterclockwise, counting the number of turns, again until you feel resistance. Now rotate it clockwise again, half the number of turns that you wrote down. This will center the rod ends in the freeplay of the bellcranks and shifter slides.

18 If you're working on a chain drive model, tighten the locknuts on the linkage rods.

19 If you're working on a shaft drive model, there's one more step before you tighten the locknuts. Turn the low gear rod counterclockwise, viewed from the rear of the vehicle, just until you feel slight resistance. This will ease shifting into and out of low gear. Once you've done this, tighten the linkage rod locknuts.

4.12 Here's the linkage center bracket used on shaft drive models

4.13 Disconnect the linkage rods from the shifter

Chapter 2 Part D Transmission 2D-11

5.3a Remove these shifter mounting bolts . . . .

5.3b . . . and one from the other side

20 As a final check, try to rotate each linkage rod. It should rotate a total of 1/4 turn, 1/8 turn in each direction. If it doesn't rotate evenly in both directions, remove and reinstall the linkage rod.

5 EZ Shift shifter – removal and installation

Refer to illustrations 5.3a and 5.3b

1 Remove the seat and the right body panel (see Chapter 8).
2 Remove the heat shield from the exhaust system (see Chapter 4). If the exhaust pipe is in the way of shifter removal, remove it as well.
3 Follow the wiring harness from the shifter to the circuit board and disconnect its connectors. Note how the harness is routed, then cut the tie wraps that secure it.
4 Disconnect the linkage rods for the shifter (Section 4).
5 Remove the shifter mounting bolts **(see illustrations)** and take the shifter out.
6 Installation is the reverse of the removal steps.

6 EZ Shift shifter – disassembly, inspection and reassembly

Refer to illustrations 6.1, 6.2, 6.3, 6.4, 6.5, 6.6, 6.7, 6.8, 6.9, 6.10a, 6.10b, 6.10c, 6.11a, 6.11b and 6.12

1 Loosen the Torx screws that secure the shift lever and cover to the housing in several stages, in a criss-cross pattern, until the cover

6.1 Loosen the cover screws evenly and lift off the cover

lifts up enough to relieve the spring tension **(see illustration)**.
2 If it's convenient, grasp the detent springs with needle-nosed pliers and remove them **(see illustration)**. You can also do this after the shifter is removed.
3 Carefully lift the switch out of the shifter **(see illustration)**. Don't force it – it's easily damaged.

6.2 Lift out the low spring and the high-reverse spring (arrow)

6.3 Lift the switch and carefully pull its pin out of the lever base (arrow)

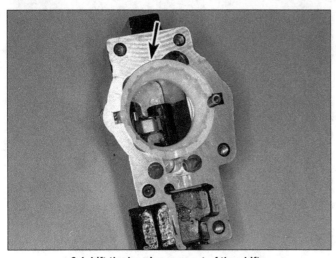

6.4 Lift the bearing cup out of the shifter

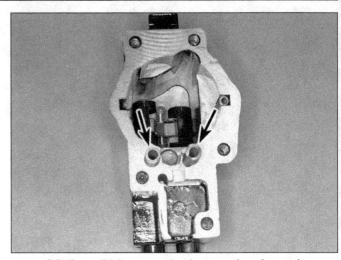

6.5 If you didn't remove the detent springs (arrows), remove them now

4　Lift the bearing cup out of the shifter **(see illustration)**.
5　If you didn't remove the detent springs earlier, do it now **(see illustration)**.
6　Remove the shifter pin spring from its groove in the side of the shifter **(see illustration)**.

7　Lift the detent pins out of the shifter, using a magnet if necessary **(see illustration)**.
8　Pull the stop pin out of its bore **(see illustration)**.
9　Slide the selector slides out of their bores **(see illustration)**. Remove the O-rings from the slides.

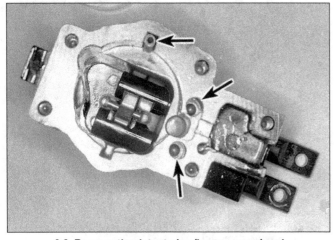

6.6 Remove the detent pins (lower arrows) and pin spring (upper arrow)

6.7 If the detent pins don't come out easily, use a magnet

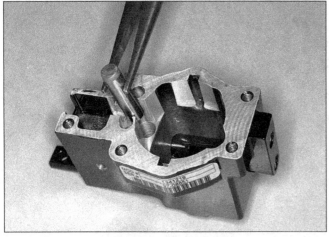

6.8 Lift the stop pin out

6.9 Pull the selector slides out of their bores

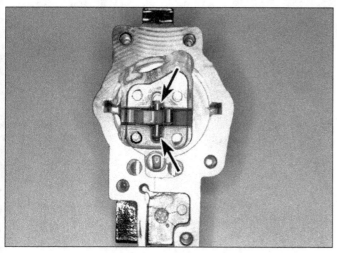

6.10a The interlock rocker pivots on this pin (arrows)

6.10b The ends of the interlock rocker (arrows) . . .

10 Note how the ends of the interlock rocker pin fit in their grooves and remove the rocker **(see illustrations)**.

11 Pry off the shifter cap, then remove the Torx bolt and knob from the top of the shifter **(see illustration)**. Slide the shifter out of the rubber boot **(see illustration)**.

12 Pry open the clamps and remove the shifter boot **(see illustration)**.

Inspection

13 Clean all parts thoroughly in solvent and dry them completely. Check all parts for wear or damage and replace any that have problems.

14 Check the springs for bending or wear at the ends.

15 Check the notches of the selector slides for wear at the corners. Check the detent pins for wear at the ends.

6.10c . . . engage the flat notches in the selector slides (arrow)

6.11a Remove the Torx screw (arrow) from the top of the shift lever

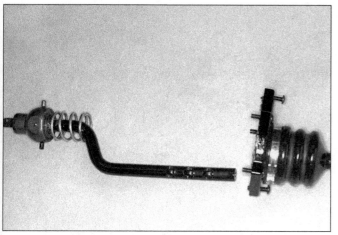

6.11b Slide the shifter lever out of the rubber boot

6.12 Spread the metal band and remove the boot

Assembly

Refer to illustration 6.16

16 Assembly is the reverse of the disassembly steps, with the following additions:

a) Coat new O-rings with Polaris 0W-40 synthetic motor oil or 10W non-detergent motor oil, then slide them onto the selector slides.
b) Install the slides in their bores. Looking at the shifter body from the end with the bore openings, the slide on the right has three notches and the slide on the left has two.
c) Engage the interlock rocker with the flat notches in the selector slides **(see illustrations 6.10a and 6.10b)** and position its pin in the grooves.
d) Once the detent pins and stop pins are installed, fill the shifter body halfway up the slides (but no farther) with Polaris 0W-40 synthetic motor oil or 10W non-detergent motor oil. Don't overfill the shifter or it may lock up in use.
e) Install the switch, engaging its pin with the shifter hole **(see illustration 6.3)**. Be especially careful not to damage the switch.
f) Clean the mating surfaces of the shifter and cover with Loctite Primer T or equivalent, then apply a bead of Loctite 515 Gasket Eliminator or equivalent to the mating surface of the shifter. Don't skip any spots.
g) Tighten the cover screws evenly to the torque listed in this Chapter's Specifications.

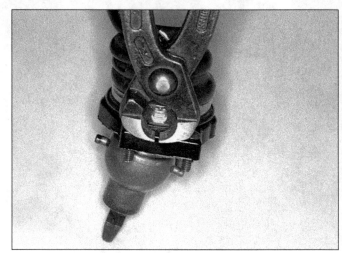

6.16 Crimp the boot bands with a tool like this

h) Clamp the boot bands securely and seal the upper one with silicone sealant **(see illustration)**.

Chapter 3
Cooling system

Contents

	Section		Section
Coolant hoses – replacement	8	General information	1
Coolant level check	See Chapter 1	Radiator – removal and installation	6
Coolant reservoir – removal and installation	3	Radiator cap - check	2
Cooling fan and thermostat switch – check and replacement	4	Thermostat (4-stroke models) – removal, check and installation	5
Cooling system check	See Chapter 1	Water pump – removal, inspection and installation	7
Cooling system draining, flushing and refilling	See Chapter 1		

Specifications

General

Coolant type	See Chapter 1
Mixture ratio	See Chapter 1
Cooling system capacity	See Chapter 1
Radiator cap pressure rating	13 psi
Thermostat	
Opening temperature	176-degrees F
Fully open at	205-degrees F
Opening at 205-degrees F	5/16 inch
Fan switch ratings (1985 through 1995)	
250 and 300	
Fan off	210-degrees F
Fan on	235-degrees F
350 and 400	
Fan off	154-degrees F
Fan on	174-degrees F
425 Magnum	
Fan off	175-degrees F
Fan on	190-degrees F
Fan switch rating (1996-on)	
250 and 300	
Fan off	210-degrees F
Fan on	235-degrees F
Scrambler and Sport models	
Fan off	154-degrees F
Fan on	174-degrees F
All except Scrambler and Sport models	
Fan off	175-degrees F
Fan on	190-degrees F

Tightening torques

Impeller nut (2-stroke)	Not specified*
Impeller nut (4-stroke)	62 to 78 inch-lbs
Thermostat housing bolts (4-stroke)	62 to 78 inch-lbs

*Use Loctite 242 or equivalent on the threads.

Chapter 3 Cooling system

1 General information

The vehicles covered in this manual that are equipped with the 350L or 400L 2-stroke engine or any 4-stroke engine use a liquid cooling system. The system utilizes a water/antifreeze mixture to carry away excess heat produced during the combustion process. The cylinder is surrounded by a water jacket, through which the coolant is circulated by the water pump. The coolant passes through the cylinder head to the radiator.

On 4-stroke models, there's a thermostat in the coolant passage from the cylinder head. When the engine is warm, the thermostat opens and allows coolant to flow to the radiator.

In the radiator, the coolant is cooled by passing air, then routed through another hose back to the water pump, where the cycle is repeated. The water pump is mounted in the crankcase casting on the left side of the engine.

An electric fan, mounted behind the engine and automatically controlled by a thermostatic switch, provides a flow of cooling air through the radiator when the vehicle is standing still or moving slowly. **Note:** *Some air-cooled models have an auxiliary cooling fan. Except for the fact that it's not mounted behind a radiator, it's serviced in the same way as the fan on liquid-cooled models.*

The entire system is sealed and pressurized. The pressure is controlled by a valve which is part of the radiator cap. By pressurizing the coolant, the boiling point is raised, which prevents premature boiling of the coolant. An overflow hose, connected between the radiator and reservoir tank, directs coolant to the tank when the radiator cap valve is opened by excessive pressure. The coolant is automatically siphoned back to the radiator as the engine cools.

Warning 1: *Do not allow antifreeze to come in contact with your skin or painted surfaces of the motorcycle. Rinse off spills immediately with plenty of water. Antifreeze is highly toxic if ingested. Never leave antifreeze lying around in an open container or in puddles on the floor; children and pets are attracted by its sweet smell and may drink it. Check with local authorities about disposing of used antifreeze. Many communities have collection centers which will see that antifreeze is disposed of safely.*

Warning 2: *Do not remove the pressure cap from the radiator when the engine and radiator are hot. Scalding hot coolant and steam may blow out under pressure, which could cause serious injury. To open the pressure cap, let the engine cool. When the engine has cooled, place a thick rag, such as a towel, over the radiator cap; slowly rotate the cap counterclockwise to the first stop. This procedure allows any residual pressure to escape. When the steam has stopped escaping, press down on the on cap while turning counterclockwise and remove it.*

2 Radiator cap - check

If problems such as overheating and loss of coolant occur, check the entire system as described in Chapter 1. The radiator cap opening pressure should be checked by a dealer service department or service station equipped with the special tester required to do the job. If the cap is defective, replace it with a new one.

3 Coolant reservoir – removal and installation

Note: *On some models, the coolant reservoir is molded in one piece with one of the PVT air ducts. Refer to Chapter 2C for removal procedures.*

Removal
Refer to illustration 3.1

1 Remove the reservoir mounting screw and lift it out **(see illustration)**.
2 Disconnect the hose from the reservoir and catch any escaped coolant. Plug the end of the reservoir hose so it doesn't siphon coolant from the system.

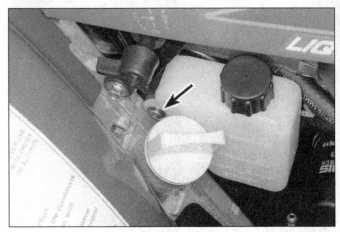

3.1 The reservoir tank on models with a separate reservoir is secured by this screw

4.4 Coolant temperature switch (center arrow) and radiator hose clamps

3 Check the reservoir for cracks, especially if it has required repeated topping up. If cracks are found, replace the reservoir.
4 Installation is the reverse of the removal steps.

4 Cooling fan and thermostat switch – check and replacement

Check
Refer to illustration 4.4

1 If the engine is overheating and the cooling fan doesn't come on, first check the main fuse (see Chapter 5). If the fuse is blown, check the fan circuit for shorts to ground (see the *Wiring diagrams* at the end of this book). Check that the battery is fully charged.
2 If the fuse and battery are good, follow the wiring harness from the fan motor to the electrical connector and disconnect the connector. Using two jumper wires, apply battery voltage to the terminals in the fan motor side of the connector (negative battery terminal to brown wire; positive battery terminal to red/white or orange/black wire). If the fan doesn't work, replace the motor.
3 If the fan does come on, the problem lies in the fan switch or the wiring that connects the components.
4 Disconnect the electrical connector from the switch **(see illustration)**. On some models, the switch is in the top of the radiator; on others, it's above the carburetor. Bypass the switch by connecting a short jumper wire between the terminals in the harness side of the switch connector. The fan should now run. If it does, the fan wiring is good and the switch is defective.

Chapter 3 Cooling system

4.6 Unbolt the fan bracket on each side of the radiator

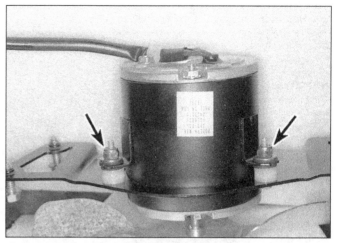

4.7 Remove the self-locking nuts (arrows) and separate the motor from the bracket

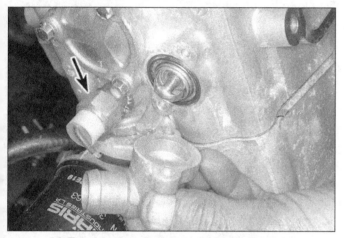

5.3 The thermostat housing is located next to the overheat lamp switch (arrow)

5   If the fan still doesn't run when the switch is bypassed, check the ground side of the circuit. Leave the short jumper wire in place to bypass the switch. Disconnect the electrical connector from the fan motor. Connect a jumper wire between the power wire (red/white or orange/black) terminals in the harness and fan motor sides of the connector. Connect another jumper wire between the brown wire terminal in the fan motor side of the connector and a good ground (bare metal on the engine or frame). If the fan now runs, look for a break or bad connection in the ground wire side of the circuit (brown wire). If it still doesn't run, look for a break or bad connection in the power side of the circuit (red/white or orange/black wire).

Replacement

Fan motor
Refer to illustrations 4.6 and 4.7

Warning: *The engine must be completely cool before beginning this procedure.*

6   Unbolt the bracket on each side of the fan and lift out the fan and bracket **(see illustration)**.
7   Remove the mounting nuts and separate the motor from the bracket **(see illustration)**.
8   Installation is the reverse of the removal steps.

Fan switch

Warning: *The engine must be completely cool before beginning this procedure.*

9   Prepare the new switch by wrapping its threads with Teflon tape or by coating the threads with RTV sealant.

10   Disconnect the electrical connector from the switch and unscrew the switch **(see illustration 4.4)**.
11   Quickly install the new switch, tightening it securely.
12   Connect the electrical connector to the switch. Check the coolant level and top up if necessary (see Chapter 1).

5   Thermostat (4-stroke models) – removal, check and installation

Warning: *The engine must be completely cool before beginning this procedure.*

Removal
Refer to illustration 5.3

1   Remove body panels as necessary for access to the thermostat (see Chapter 8).
2   Detach the cooling system hose from the thermostat housing on the cylinder head near the oil filter and hold the end of the hose over a clean drain pan to catch the coolant.
3   Unbolt the thermostat housing from the cylinder head and take out the thermostat **(see illustration)**.

Inspection

4   Remove any coolant deposits, then visually check the thermostat for corrosion, cracks and other damage. If it was open when it was removed, the thermostat is defective.
5   To check the thermostat operation, submerge it in a pan of water along with a thermometer. The thermostat should be suspended so it does not touch the sides of the container. **Warning:** *Antifreeze is poisonous. DO NOT use a cooking pan to test the thermostat!*
6   Gradually heat the water in the pan with a hot plate or stove and check the temperature when the thermostat just starts to open. Compare the opening temperature to the value listed in this Chapter's Specifications.
7   When the water reaches the fully open temperature listed in this Chapter's Specifications, the thermostat should be open about 5/16 inch.
8   If these specifications are not met, or if the thermostat doesn't open when the water is heated, install a new one.

Installation

9   Install the thermostat in the housing. Place one of its air bleed holes next to the upper bolt hole in the cylinder head.
10   Install the thermostat housing, using a new seal if necessary. Tighten its bolts securely.
11   Connect the hose to the housing.
12   Refill the cooling system (see Chapter 1).

Chapter 3 Cooling system

6 Radiator – removal and installation

Warning: *The engine must be completely cool before beginning this procedure.*

Removal

Refer to illustrations 6.3a, 6.3b and 6.3c

1 Drain the cooling system (see Chapter 1.
2 Remove the fan (Section 4).
3 Remove the radiator mounting bolts and disconnect the siphon hose from the radiator **(see illustration)**. If you're going to remove the siphon hose, follow it and note how it's routed **(see illustrations)**.
4 Disconnect the coolant hoses from the radiator **(see illustration 4.4)**.

Installation

5 Installation is the reverse of the removal steps, with the following additions:
   a) *Don't forget to connect the fan switch wires.*
   b) *Fill the cooling system with the recommended coolant (see Chapter 1).*

7 Water pump – removal, inspection and installation

Warning: *The engine must be completely cool before beginning this procedure.*

2-stroke engines

Removal

Refer to illustrations 7.2 and 7.3

1 Remove the crankcase cover from the engine (see Section 5 in Chapter 2A).
2 Unscrew the impeller nut **(see illustration)**. **Note:** *The nut is secured with Loctite 242. Use a 10mm socket and make sure it fits securely on the nut.*
3 Slide the impeller and water pump housing off the counterbalancer shaft **(see illustration)**. Collect any shims that may be on the counterbalancer shaft behind the water pump housing and be sure to note how many there are.
4 Clean all of the old gasket from the water pump housing and its mating surface on the cover.

6.3a Remove the mounting bolts (arrows) and disconnect the siphon hose . . .

6.3b . . . follow the siphon hose down past the radiator . . .

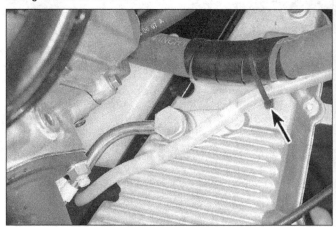

6.3c . . . and remove any tie-wraps (arrow)

7.2 Remove the impeller nut and . . .

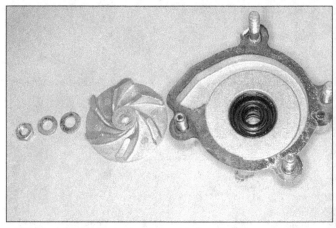

7.3 . . . take off the lockwasher, washer, impeller and housing

Chapter 3 Cooling system

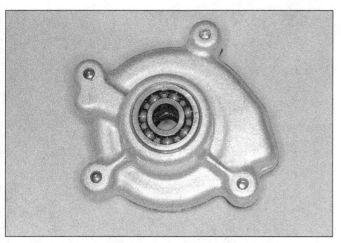

7.6 Turn the housing over and check its bearing

7.12 Disconnect the inlet hose (on the right) and remove the bolts (arrows)

Inspection
Refer to illustration 7.6

5 If there was coolant inside the crankcase cover, the water pump seal is probably leaking. In any case, check it for wear **(see illustration 7.3)**. If there's any doubt about the condition of the seal, pry it out of the housing.

6 Turn the housing over and check the bearing **(see illustration)**. Spin it with a finger and check for roughness, looseness or noise. If there are any problems, replace it.

Installation

7 Installation is the reverse of the removal steps, with the following additions:

  a) *Use Loctite 242 or equivalent on the threads of the impeller shaft.*
  b) *Use a new gasket, coated with a thin layer of sealant on both sides.*

4-stroke engines
Removal
Refer to illustrations 7.12, 7.13, 7.14a and 7.14b

8 Remove the fuel tank, carburetor and exhaust system (see Chapter 4).
9 Remove the PVT (see Chapter 2C).
10 Remove the upper right engine mount (including the bracket). Remove the nut from the left front engine mount.
11 Push the top of the engine to the left far enough to provide access to the water pump housing, then block it in that position.

7.13 Take the cover off the engine

12 Place a pan under the engine to catch dripping coolant, then disconnect the inlet hose from the water pump housing **(see illustration)**.
13 Unbolt the housing from the engine and take it off **(see illustration 7.12 and the accompanying illustration)**.
14 Unscrew the impeller nut, then remove the washer and impeller **(see illustrations)**. Note the number of shims behind the impeller and be sure to install the same number.

7.14a Unscrew the impeller nut . . .

7.14b . . . slide off the impeller and note how many shims are behind it

8.1a The 2-stroke coolant outlet hose is next to the overheat lamp switch (arrow)

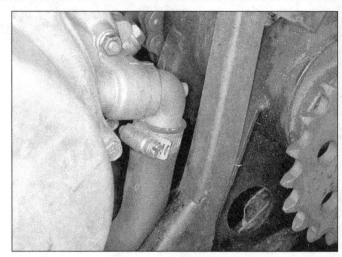

8.1b The 2-stroke water pomp outlet hose attaches to this fitting

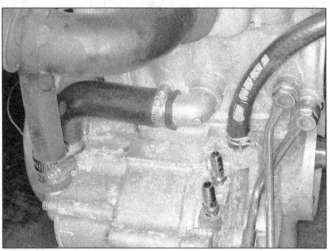

8.2 The 4-stroke water pump outlet hose attaches to this fitting

8.3 Look for clamps and tie-wraps (arrow)

Inspection

15 Clean away any residue from the impeller blades and check them for damage. Replace the impeller if problems are found.

Installation

16 Installation is the reverse of the removal steps.

8 Coolant hoses – replacement

Warning: *The engine must be completely cool before beginning this procedure.*
Refer to illustrations 8.1a, 8.1b, 8.2 and 8.3

1 The coolant hoses are attached to 2-stroke engines at the cylinder head and water pump inlet **(see illustrations)**.

2 On 4-stroke engines, the water pump outlet hose runs to a fitting on the cylinder **(see illustration)**.

3 To remove and install hoses, first drain the cooling system (see Chapter 1). Loosen the clamps and work the hoses free of the fittings. Note the positions of any retainers **(see illustration)**.

4 Installation is the reverse of the removal steps. Replace any cut tie-wraps with new ones.

Chapter 4
Fuel and exhaust systems

Contents

	Section
Air cleaner housing - removal and installation	9
Air filter element - servicing	See Chapter 1
Carburetor overhaul - general information	5
Carburetor - removal and installation	6
Carburetors - disassembly, cleaning and inspection	7
Carburetors - reassembly and float height check	8
Choke cable - removal and installation	11
Exhaust system - check	See Chapter 1
Exhaust system - removal and installation	12

	Section
Fuel pump – disassembly, inspection and reassembly	14
Fuel pump – removal and installation	13
Fuel tank - cleaning and repair	3
Fuel tank - removal and installation	2
General information	1
Idle fuel/air mixture adjustment	4
Spark arrester cleaning	See Chapter 1
Throttle cable - removal, installation and adjustment	10

Specifications

General
Fuel type .... Unleaded or low lead gasoline subject to local regulations; minimum octane 87 pump octane (91 RON)

Carburetor
Jet sizes and settings
  1985 Scrambler and Trail Boss
    Main jet 155
    Pilot jet 30
    Air screw setting .... 1/2 turn out
    Jet needle and clip position 5DP7-3
    Needle jet O-6
  1986 Scrambler and Trail Boss
    Main jet 155
    Pilot jet 50
    Air screw setting .... 1 turn out
    Jet needle and clip position 5DP7-3
    Needle jet O-0
  1987 Trail Boss 4X4
    Main jet 145
    Pilot jet 50
    Air screw setting .... 1 turn out
    Jet needle and clip position 5DP7-3
    Needle jet O-0

Carburetor (continued)

Jet sizes and settings (continued)

1987 Cyclone
- Main jet 200
- Pilot jet 40
- Air screw setting .... 1 turn out
- Jet needle and clip position 6DH5-3
- Needle jet O-4

1988 Trail Boss, Trail Boss 4X4
- Main jet 145
- Pilot jet 35
- Air screw setting .... 1 turn out
- Jet needle and clip position 5DP7-3
- Needle jet O-4

1988 Trail Boss 250R/ES
- Main jet 230
- Pilot jet 45
- Air screw setting .... 1.5 turn out
- Jet needle and clip position 6DH5-3
- Needle jet O-4

1989 (all models)
- Main jet 145
- Pilot jet 40
- Air screw setting .... 1 turn out
- Jet needle and clip position 5DP7-3
- Needle jet O-4

1990 (all 250 models)
- Main jet 145
- Pilot jet 40
- Air screw setting .... 1 turn out
- Jet needle and clip position 5DP7-3
- Needle jet O-4 (169)

1990 (all 350 models)
- Main jet 220
- Pilot jet 30
- Air screw setting .... 1.5 turn out
- Jet needle and clip position 6DH29-3
- Needle jet O-6 (480)

1991 (all 250 models)
- Main jet 145
- Pilot jet 40
- Air screw setting .... 1 turn out
- Jet needle and clip position 5DP7-3
- Needle jet O-4 (169)

1991 through 1993 (all 350 models)
- Main jet 200
- Pilot jet 30
- Air screw setting .... 3/4 turn out
- Jet needle and clip position 6DH29-3
- Needle jet O-6 (480)

1992-on (all 250 models)
- Main jet 145
- Pilot jet 40
- Air screw setting .... 1 turn out
- Jet needle and clip position 5DP7-3
- Needle jet O-4 (169)

1994 and 1995 (all 300 models)
- Main jet 155
- Pilot jet 40
- Air screw setting .... 1.5 turn out
- Jet needle and clip position 5DP7-3
- Needle jet O-4 (169)

1994 and 1995 (all 400 models)
- Main jet 200
- Pilot jet 30
- Air screw setting .... 1.5 turn out
- Jet needle and clip position 6DH29-3
- Needle jet O-6 (480)

Chapter 4 Fuel and exhaust systems

Carburetor (continued)

Jet sizes and settings (continued)

    1995 Sport models
        Main jet 200
        Pilot jet 30
        Air screw setting .... 1.5 turn out
        Jet needle and clip position 6DH29-3
        Needle jet O-6 (480)
    1995 Scrambler models
        Main jet 240
        Pilot jet 30
        Air screw setting .... 1.5 turn out
        Jet needle and clip position 6DH29-2
        Needle jet O-6 (480)
    1995 (all 4-stroke models)
        Main jet 140
        Pilot jet 42.5
        Air screw setting .... 1-3/8 turn out
        Jet needle and clip position 5F81-3
        Needle jet P-8
    1996 Trail Blazer
        Main jet 130
        Pilot jet 40
        Air screw setting .... 1 turn out
        Jet needle and clip position 5DP7-3
        Needle jet O-4 (169)
    1996 Trail Boss
        Main jet 145
        Pilot jet 40
        Air screw setting .... 1 turn out
        Jet needle and clip position 5DP7-3
        Needle jet O-4 (169)
    1996 Xpress 300, Xplorer 300
        Main jet 155
        Pilot jet 40
        Air screw setting .... 1.5 turn out
        Jet needle and clip position 5DP7-3
        Needle jet O-4 (169)
    1996 Scrambler, Sport
        Main jet 250
        Pilot jet 35
        Air screw setting .... 1 turn out
        Jet needle and clip position 6F9-3
        Needle jet O-6 (480)
    1996 Xpress 400
        Main jet 210
        Pilot jet 30
        Air screw setting .... 1 turn out
        Jet needle and clip position 6DH29-3
        Needle jet O-6 (480)
    1996 Sportsman 4X4, Xplorer 400, 400 6X6
        Main jet 210
        Pilot jet 30
        Air screw setting .... 1 turn out
        Jet needle and clip position 6DH2-3
        Needle jet O-6 (480)
    1996 Magnum
        Main jet 140
        Pilot jet 40
        Air screw setting .... 2.5 turns out
        Jet needle and clip position 5F81-3
        Needle jet P-8
    1996 Sportsman 500
        Main jet 142.5
        Pilot jet 42.5
        Air screw setting .... 1.5 turns out
        Jet needle and clip position 5D78-3
        Needle jet P-1

Carburetor (continued)

Jet sizes and settings (continued)

- 1997 Trail Blazer
  - Main jet 130
  - Pilot jet 40
  - Air screw setting .... 1 turn out
  - Jet needle and clip position 5DP7-3
  - Needle jet O-4 (169)
- 1997 Trail Boss
  - Main jet 145
  - Pilot jet 40
  - Air screw setting .... 1 turn out
  - Jet needle and clip position 5DP7-3
  - Needle jet O-4 (169)
- 1996 Xpress 300, Xplorer 300
  - Main jet 155
  - Pilot jet 40
  - Air screw setting .... 1.5 turn out
  - Jet needle and clip position 5DP7-3
  - Needle jet O-4 (169)
- 1997 Scrambler 400, Sport
  - Main jet 230
  - Pilot jet 35
  - Air screw setting .... 1.5 turn out
  - Jet needle and clip position 6CEY6-3
  - Needle jet O-6 (480)
- 1997 Xpress 400, Xplorer 400, Sportsman 400, 400 6X6
  - Main jet 200
  - Pilot jet 30
  - Air screw setting .... 1.5 turn out
  - Jet needle and clip position 6CEY6-3
  - Needle jet O-6 (480)
- 1997 Magnum
  - Main jet 140
  - Pilot jet 40
  - Air screw setting .... 2.5 turns out
  - Jet needle and clip position 5F81-3
  - Needle jet P-8
- 1997 Sportsman 500, Xplorer 500, Scrambler 500
  - Main jet 142.5
  - Pilot jet 42.5
  - Air screw setting .... 2 turn out
  - Jet needle and clip position 5D78-3
  - Needle jet P-1

Float level
- 2-stroke engines .... Parallel with gasket surface
- 4-stroke engines
  - 1985 through 1995 .... 9/16-inch
  - 1996-on 1/2-inch

1 General information

The fuel system consists of the fuel tank, fuel tap, filter screen, carburetor and connecting lines, hoses and control cables.

The carburetor used on 2-stroke models is a Mikuni VM round slide design. There is no separate throttle valve; the throttle slide (piston) is the throttle valve.

The carburetor used on 4-stroke models is a Mikuni CV constant vacuum unit with a butterfly-type throttle valve.

For cold starting on all models, an enrichment circuit is actuated by a cable and the choke lever mounted above the fuel tank.

The exhaust system consists of a pipe and muffler with a spark arrester function.

Many of the fuel system service procedures are considered routine maintenance items and for that reason are included in Chapter 1.

2 Fuel tank - removal and installation

Warning: *Gasoline is extremely flammable, so take extra precautions when you work on any part of the fuel system. Don't smoke or allow open flames or bare light bulbs near the work area, and don't work in a garage where a natural gas-type appliance (such as a water heater or clothes dryer) is present. If you spill any fuel on your skin, rinse it off immediately with soap and water. When you perform any kind of work on the fuel system, wear safety glasses and have a class B type fire extinguisher (flammable liquids) on hand.*

1 The fuel tank is secured to a bracket by a bolt and washer at each rear corner. On 4-stroke models, there's a vertical spacer between each rear corner and the frame crossmember. At the front, the tank is located by tabs; the tank rests on a foam support strip on each side of the frame. On some models, the fuel tap is mounted on the frame and connected to the tank by hoses. On others, it's mounted directly to the tank.

Chapter 4 Fuel and exhaust systems

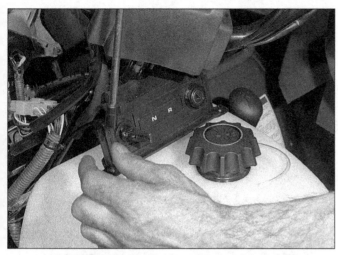

2.3 Unbolt the bracket from the top of the tank

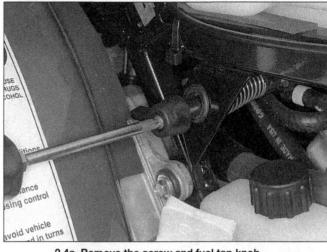

2.4a Remove the screw and fuel tap knob . . .

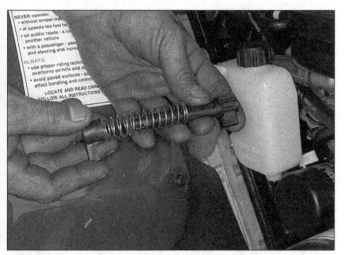

2.4b . . . and the fuel tap extension

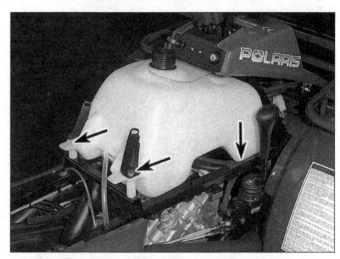

2.5a There's a mounting bolt at each rear corner of the tank and a foam strip on each side (arrows)

Removal

Refer to illustrations 2.3, 2.4a, 2.4b, 2.5a, 2.5b, 2.5c and 2.6

2   Remove the seat, fuel tank cover and headlight housing (see Chapter 8).

3   Disconnect the vent line from the top of the tank. Unbolt the choke cable and warning light bracket from the top of the tank (if equipped) **(see illustration)**.

4   If you're working on a vehicle with a tank-mounted fuel tap, remove the handle extension **(see illustrations)**.

5   Remove the fuel tank mounting bolts, brackets and spacers (if equipped) **(see illustrations)**.

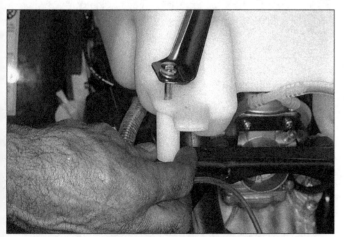

2.5b On 425 and 500 models, the bracket fits on top of the mounting tab and the spacer fits beneath it

2.5c On some models, the tank bracket bolts also secure this foam piece

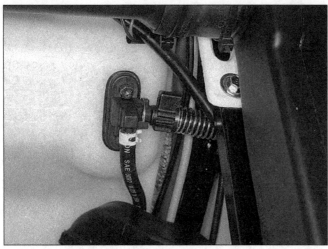

2.6 Disconnect the fuel line from the tap

2.8 Route the vent hose into the steering post without kinking it

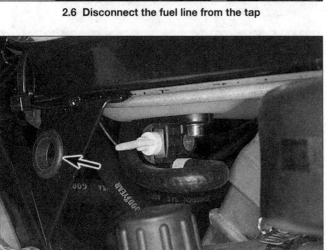

2.9 Pass the extension handle through the bushing (arrow) and engage it with the tap handle

4.3a Drill out the plug, taking care not to damage the screw . . .

6   Lift the tank and disconnect the fuel line from the fuel tap or the fittings on the tank **(see illustration)**.

Installation

Refer to illustrations 2.8 and 2.9

7   Before installing the tank, check the condition of the foam mounting strips - if they're hardened, cracked, or show any other signs of deterioration, replace them **(see illustration 2.5a)**.
8   When installing the tank, reverse the removal procedure. Make sure the vent line is routed correctly into the steering post and do not pinch any control cables or wires **(see illustration)**.
9   If the machine has a tank-mounted fuel tap, pass the extension through the bushing and engage it with the tap handle **(see illustration)**.
10 If the machine has a frame-mounted fuel tap, connect the fuel lines as follows:
   a) *300 and 400 models: Connect the hose from the fuel tank black fitting to the Reserve fitting on the valve.*
   b) *500 models: Connect the silver-colored (short) fitting on the tank to the Reserve fitting on the valve. Connect the hose from the gold-colored (tall) fitting to the On fitting on the valve.*

3   Fuel tank - cleaning and repair

1   All repairs to the fuel tank should be carried out by a professional who has experience in this critical and potentially dangerous work. Even after cleaning and flushing of the fuel system, explosive fumes can remain and ignite during repair of the tank.
2   If the fuel tank is removed from the vehicle, it should not be placed in an area where sparks or open flames could ignite the fumes coming out of the tank. Be especially careful inside garages where a natural gas-type appliance is located, because the pilot light could cause an explosion.

4   Idle fuel/air mixture adjustment

Normal adjustment

4-stroke engines

Refer to illustrations 4.3a and 4.3b

1   Idle fuel/air mixture on 4-stroke engines is preset at the factory and should not need adjustment unless the carburetor is overhauled or the pilot screw, which controls the mixture adjustment, is replaced.
2   The engine must be properly tuned up before making the adjustment (valve clearances set to specifications, spark plug in good condition and properly gapped).
3   To make an initial adjustment, drill out the plug the seals the pilot screw **(see illustration)**. Turn the screw in until it bottoms lightly, then back it out the number of turns listed in this Chapter's Specifications

Chapter 4 Fuel and exhaust systems

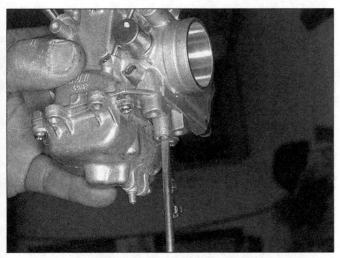

4.3b ... and adjust the pilot screw with a screwdriver

4.6 Here are the 2-stroke pilot screw (left) and idle speed screw (right)

(see illustration). **Caution:** *Turn the screw just far enough to seat it lightly. If it's bottomed hard, the screw or its seat may be damaged, which will make accurate mixture adjustments impossible.*
4   Adjust the idle speed (see Chapter 1).

2-stroke engines

Refer to illustration 4.6
5   The engine must be in good mechanical condition and properly tuned up before making the adjustment (spark plug in good condition and properly gapped).
6   Turn the air screw in until it bottoms lightly, then back it out the number of turns listed in this Chapter's Specifications **(see illustration)**. **Caution:** *Turn the screw just far enough to seat it lightly. If it's bottomed hard, the screw or its seat may be damaged, which will make accurate mixture adjustments impossible.*
7   Adjust the idle speed (see Chapter 1).

High altitude and temperature adjustment

8   The main jet selection and pilot screw setting change, depending both on altitude and the ambient temperature in which the machine is operated.
9   The main jet numbers listed in this Chapter's Specifications are for sea level to 300 feet and 60-degree ambient temperatures. If you plan to operate the machine consistently at higher altitudes or warmer temperatures, you may need a smaller main jet. In colder temperatures, you may need a larger main jet. Consult a Polaris dealer or other qualified shop for the correct jet specification for your operating conditions.
10   For each 30-degrees below 60-degrees F, turn the pilot screw _ turn clockwise from the setting listed in this Chapter's Specifications. For each 30-degrees above 60-degrees F, turn the pilot screw _ turn counterclockwise from the setting listed in this Chapter's Specifications.

5   Carburetor overhaul - general information

1   Poor engine performance, hesitation, hard starting, stalling, flooding and backfiring are all signs that major carburetor maintenance may be required.
2   Keep in mind that many so-called carburetor problems are really not carburetor problems at all, but mechanical problems within the engine or ignition system malfunctions. Try to establish for certain that the carburetor is in need of maintenance before beginning a major overhaul.
3   Check the fuel tap and its strainer screen, the in-tank fuel strainer, the fuel lines, the intake manifold clamps, the O-ring between the intake manifold and cylinder head, the vacuum hoses, the air filter element, the cylinder compression, the spark plug and the ignition timing before assuming that a carburetor overhaul is required. If the vehicle has been unused for more than a month, refer to Chapter 1, drain the float chamber and refill the tank with fresh fuel.
4   Most carburetor problems are caused by dirt particles, varnish and other deposits which build up in and block the fuel and air passages. Also, in time, gaskets and O-rings shrink or deteriorate and cause fuel and air leaks which lead to poor performance.
5   When the carburetor is overhauled, it is generally disassembled completely and the parts are cleaned thoroughly with a carburetor cleaning solvent and dried with filtered, unlubricated compressed air. The fuel and air passages are also blown through with compressed air to force out any dirt that may have been loosened but not removed by the solvent. Once the cleaning process is complete, the carburetor is reassembled using new gaskets, O-rings and, generally, a new inlet needle valve and seat.
6   Before disassembling the carburetors, make sure you have a carburetor rebuild kit (which will include all necessary O-rings and other parts), some carburetor cleaner, a supply of rags, some means of blowing out the carburetor passages and a clean place to work.

6   Carburetor - removal and installation

Warning: *Gasoline is extremely flammable, so take extra precautions when you work on any part of the fuel system. Don't smoke or allow open flames or bare light bulbs near the work area, and don't work in a garage where a natural gas-type appliance (such as a water heater or clothes dryer) is present. If you spill any fuel on your skin, rinse it off immediately with soap and water. When you perform any kind of work on the fuel system, wear safety glasses and have a fire extinguisher suitable for a class B type fire (flammable liquids) on hand.*

Removal

1   Remove the seat and side covers (see Chapter 8). If it's necessary for removal access, remove the fuel tank (see Section 2).
2   If it's necessary for removal access, remove the air cleaner housing (see Section 9).

2-stroke engines

Refer to illustrations 6.3, 6.4, 6.5a, 6.5b and 6.8
3   Loosen the clamp screw that secures the carburetor to the intake

Chapter 4 Fuel and exhaust systems

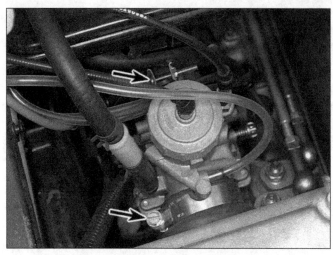

6.3 Loosen the clamping bands on the air cleaner tube and intake tube (arrows)

6.4 Disconnect the fuel and oil lines (arrows)

6.5a Unscrew the carburetor top (left) and starting enrichment valve (right)

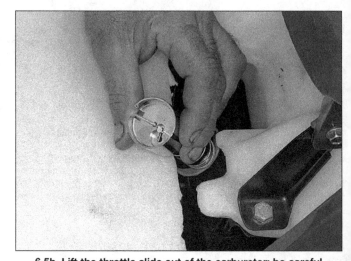

6.5b Lift the throttle slide out of the carburetor; be careful not to damage the jet needle

tube. If you haven't removed the air cleaner housing, also loosen the clamp screw that secures the carburetor to the air cleaner outlet **(see illustration)**.

6.8 Unbolt the intake tube and inspect its O-ring (arrow)

4   Disconnect the fuel line, oil line and vent line **(see illustration)**.
5   Unscrew the carburetor top and lift out the throttle slide **(see illustrations)**. Carefully set the throttle slide out of the way so the jet needle isn't damaged.
6   Unscrew the starting enrichment (choke) valve **(see illustration 6.5a)**.
7   Work the carburetor free of the intake tube (and the air cleaner tube if it's still installed). Lift the carburetor out.
8   Check the intake tube for cracks, deterioration or other damage. If it has visible defects, or if there's reason to suspect its O-ring is leaking, remove it and inspect the O-ring **(see illustration)**.

4-stroke engines

Refer to illustrations 6.9a, 6.9b, 6.10, 6.12 and 6.15

9   Disconnect the upper end of the vent hose from the hole in the frame **(see illustration)**. The hose connects to a Y-fitting, which routes a vent hose to each side of the carburetor **(see illustration)**. The two lower branches of the vent hose are different lengths, so label them before disconnecting them from the carburetor (you don't have to disconnect the vent lines to remove the carburetor from the engine).
10 Disconnect the fuel line from the carburetor **(see illustration)**. Follow the impulse line from the fuel pump to the carburetor and disconnect it.
11 Loosen the clamp that secures the carburetor to the intake tube **(see illustration 6.10)**.

Chapter 4 Fuel and exhaust systems

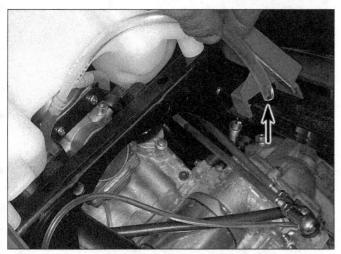

6.9a The upper end of the carburetor vent line on 4-strokes fits through this hole in the frame

6.9b The vent lines run to each side of the carburetor (arrows)

6.10 Loosen the clamp (top) and disconnect the fuel line (bottom)

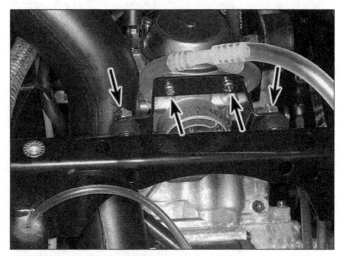

6.12 The mounting bracket is secured to the frame by nuts (outer arrows) and to the carburetor by screws (inner arrows)

12 Remove the nuts that secure the carburetor mounting bracket to the rubber dampers on the frame **(see illustration)**. Work the carburetor free of the intake tube and lift it out.
13 Refer to Sections 10 and 11 and disconnect the throttle and starting enrichment (choke) cables from the carburetor body.
14 The carburetor can now be removed. After the carburetor has been removed, stuff clean rags into the intake tube (or the intake port in the cylinder head, if the tube has been removed) to prevent the entry of dirt or other objects.
15 Check the intake tube for cracks, deterioration or other damage **(see illustration 6.8)**. If it has visible defects, or if there's reason to suspect its O-ring is leaking, remove it and inspect the O-ring **(see illustration)**.

Installation

16 Installation is the reverse of the removal steps, with the following additions: Adjust the throttle and choke freeplay (see Chapter 1).

7 Carburetors - disassembly, cleaning and inspection

Warning: *Gasoline is extremely flammable, so take extra precautions when you work on any part of the fuel system. Don't smoke or allow open flames or bare light bulbs near the work area, and don't work in a garage where a natural gas-type appliance (such as a water heater or clothes dryer) is present. If you spill any fuel on your skin, rinse it off immediately with soap and water. When you perform any kind of work on the fuel system, wear safety glasses and have a fire extinguisher suitable for class B fires (flammable liquids) on hand.*

6.15 The intake adapter is secured to the carburetor by a clamp (center) and to the engine by Allen bolts (outer arrows)

7.2a Check the starter enrichment air passage (left) and idle air passage (right) for clogging

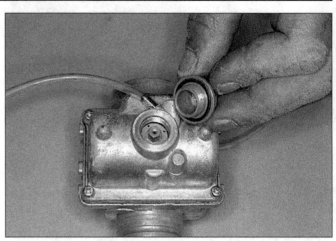

7.2b Unscrew the float chamber drain plug and discard its O-ring

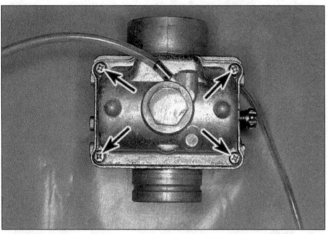

7.2c Remove the screws and take off the float chamber

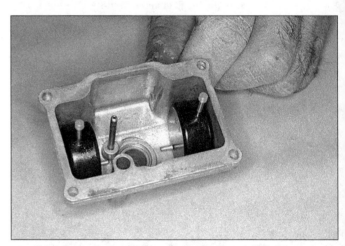

7.2d Remove the retainers from the float guide pins and slide the floats off

Disassembly

1   Remove the carburetor from the machine as described in Section 6. Set it on a clean working surface.

2-stroke engines

Refer to illustrations 7.2a through 7.2h

2   Refer to the accompanying illustrations to disassemble the carburetor **(see illustrations)**.

7.2e Slide out the pin and unscrew the float arm (top); unscrew the main jet (center) and pilot jet (bottom)

7.2f Remove the jet needle retainer

Chapter 4 Fuel and exhaust systems

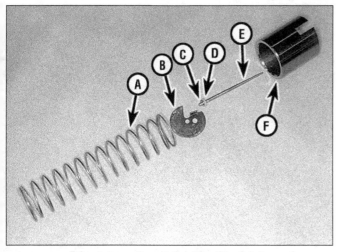

7.2g Jet needle and throttle slide details

A	Throttle spring	D	Washer
B	Retainer	E	Jet needle
C	Clip	F	Throttle slide

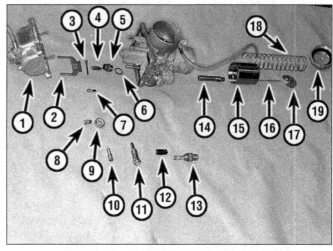

7.2h Carburetor details – 2-stroke models

1	Float chamber and floats	12	Starting enrichment (choke) plunger
2	Float arm	13	Plunger spring and retainer
3	Float pin		
4	Needle valve	14	Needle jet
5	Needle valve seat	15	Throttle slide
6	Needle valve seat gasket	16	Jet needle (with washer and clip)
7	Pilot jet		
8	Main jet	17	Jet needle retainer
9	Baffle	18	Throttle spring
10	Pilot screw	19	Cap and gasket
11	Throttle stop screw		

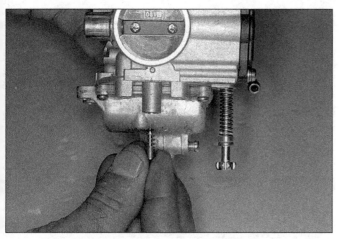

7.3a Unscrew the pilot screw from its passage . . .

4-stroke engines

Refer to illustrations 7.3a through 7.3t

3 Refer to the accompanying illustrations to disassemble the carburetor **(see illustrations)**.

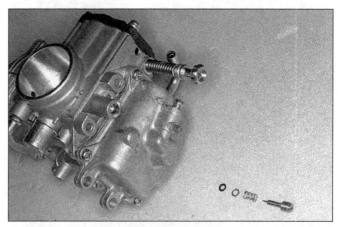

7.3b . . . and remove the screw, spring, washer and O-ring

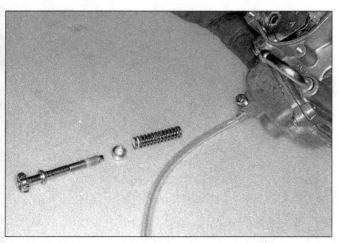

7.3c Remove the throttle stop screw, spring and spacer

7.3d Remove the four screws securing the vacuum chamber cover to the carburetor body (arrows)

4-12 Chapter 4 Fuel and exhaust systems

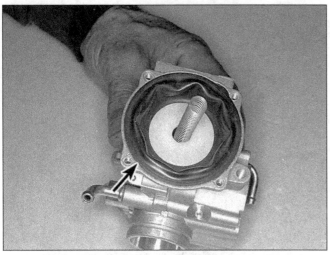

7.3e Lift the cover off and remove the piston spring; note the location of the tab (arrow) which fits into a notch

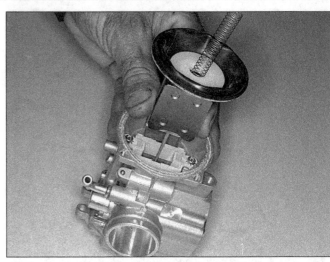

7.3f Carefully peel the diaphragm away from its groove in the carburetor body and lift out the diaphragm/piston assembly

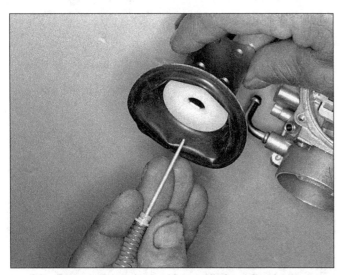

7.3g Remove the spring and jet needle from the piston . . .

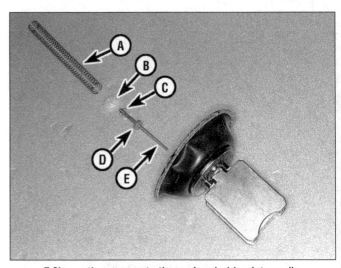

7.3h . . . then separate the spring, holder, jet needle, clip and washer from the piston

- A Spring
- B Holder
- C Clip
- D Washer
- E Jet needle

7.3i Remove the four float chamber screws (arrows) . . .

7.3j . . . then detach the float chamber and remove its O-ring

Chapter 4 Fuel and exhaust systems

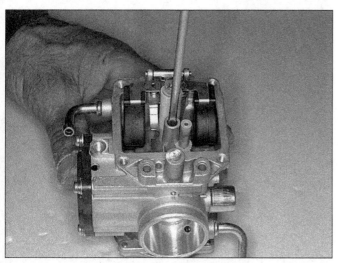

7.3k Unscrew the pilot jet . . .

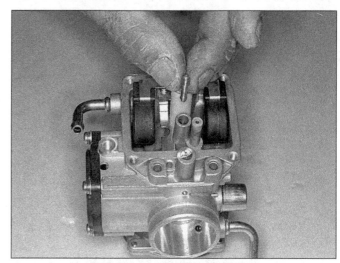

7.3l . . . and lift it out of the bore

7.3m Unscrew the main jet . . .

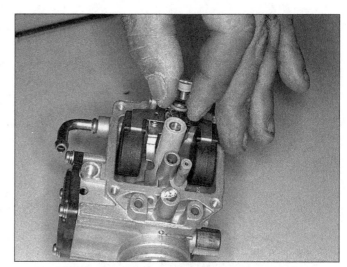

7.3n . . . and lift it out together with its washer

7.3o Remove the screws (arrows) and lift out the needle valve seat retainer, then work the needle valve seat out of its bore with fingers

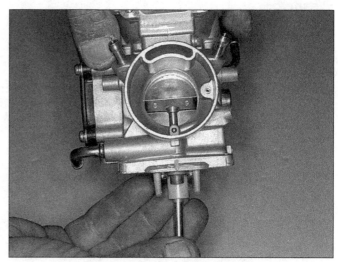

7.3p Insert a screwdriver into the main jet bore and push the needle jet into the venturi . . .

4-14   Chapter 4   Fuel and exhaust systems

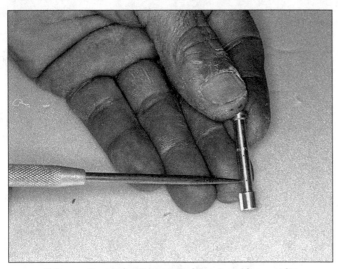

7.3q ... then take the needle jet out and inspect its holes for obstructions

7.3r Unscrew the starter (choke) jet with a screwdriver and the jet bridge screws ...

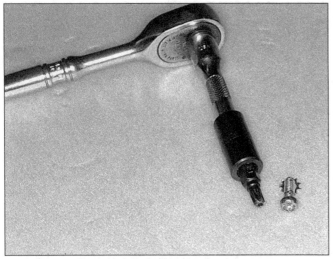

7.3s ... with a Torx driver; be sure to use the correct tool

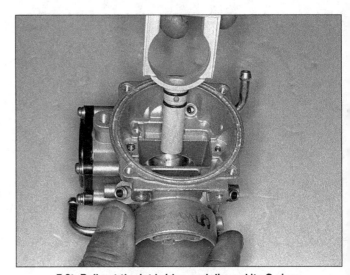

7.3t Pull out the jet bridge and discard its O-rings

Cleaning

Caution: *Use only a carburetor cleaning solution that is safe for use with plastic parts (be sure to read the label on the container).*

4   Submerge the metal components in the carburetor cleaner for approximately thirty minutes (or longer, if the directions recommend it).

5   After the carburetor has soaked long enough for the cleaner to loosen and dissolve most of the varnish and other deposits, use a brush to remove the stubborn deposits. Rinse it again, then dry it with compressed air. Blow out all of the fuel and air passages in the main and upper body. **Caution:** *Never clean the jets or passages with a piece of wire or a drill bit, as they will be enlarged, causing the fuel and air metering rates to be upset.*

Inspection

6   Check the starting enrichment plunger and spring (2-stroke) or valve (4-stroke) for wear or damage and replace it if any defects are found.

7   Check the tapered portion of the pilot screw for wear or damage. Replace the pilot screw if necessary.

8   Check the carburetor body, float chamber and top (2-stroke) or vacuum chamber cover (4-stroke) for cracks, distorted sealing surfaces and other damage. If any defects are found, replace the faulty component, although replacement of the entire carburetor will probably be necessary (check with your parts supplier for the availability of separate components).

9   Check the jet needle for straightness by rolling it on a flat surface (such as a piece of glass). Replace it if it's bent or if the tip is worn.

10   Check the tip of the fuel inlet valve needle. If it has grooves or scratches in it, it must be replaced. Push in on the rod in the other end of the needle, then release it - if it doesn't spring back, replace the valve needle.

11   Check the O-rings on the float chamber jet plug (2-stroke) or float chamber (4-stroke). Replace them if they're damaged.

12   If you're working on a 4-stroke model, operate the throttle shaft to make sure the throttle butterfly valve opens and closes smoothly. If it doesn't, replace the carburetor.

13   Check the floats for damage. This will usually be apparent by the presence of fuel inside one of the floats. If the floats are damaged, they must be replaced.

14   Check the diaphragm (4-stroke) for splits, holes and general deterioration. Holding it up to a light will help to reveal problems of this nature.

15   Insert the throttle slide (2-stroke) or vacuum piston (4-stroke) in the carburetor body and see that it moves up-and-down smoothly. Check the surface of the slide or piston for wear. If it's worn excessively or doesn't move smoothly in the bore, replace the carburetor.

Chapter 4 Fuel and exhaust systems

8.4 The 2-stroke float arm should be parallel to the gasket surface; if not, bend the tang (arrow)

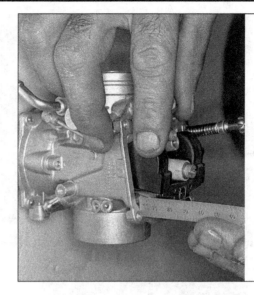

8.5 Measure 4-stroke float height with the float hanging down; bend the float tang to adjust it

8 Carburetors - reassembly and float height check

Caution: *When installing the jets, be careful not to over-tighten them - they're made of soft material and can strip or shear easily.*
Note: *When reassembling the carburetor, be sure to use new O-rings and gaskets.*

Reassembly

1   Reassembly is the reverse of the disassembly steps, with the following additions:
  a) Install the clip on the jet needle if it was removed. Place it in the needle groove listed in this Chapter's Specifications.
  b) Install the pilot screw (if removed) along with its spring, washer and O-ring, turning it in until it seats lightly. Now, turn the screw out the number of turns listed in this Chapter's Specifications.
  c) If you're working on a 4-stroke model, install the diaphragm/vacuum piston assembly into the carburetor body. Lower the spring into the piston. Seat the bead of the diaphragm into the groove in the top of the carburetor body, making sure the diaphragm isn't distorted or kinked **(see illustration 7.3e)**. This is not always an easy task. If the diaphragm seems too large in diameter and doesn't want to seat in the groove, place the vacuum chamber cover over the carburetor diaphragm, insert your finger into the throat of the carburetor and push up on the piston, holding it almost all the way up. Push down gently on the vacuum chamber cover - it should drop into place, indicating the diaphragm has seated in its groove. Once this occurs, install at least two of the vacuum chamber cover screws before you let go of the piston.

Float height check

2-stroke models

Refer to illustration 8.4

2   The carburetor used on 2-stroke models has an unconventional float design. The floats don't pivot on the float pin, as is the common design, but slide on a pair of posts in the float chamber. Each float has a pin that rests on the float arm, which pivots on the float pin.
3   Remove the float chamber (if you haven't already done so) and turn the carburetor upside down.
4   The float arm should be parallel with the float chamber's gasket surface **(see illustration)**. If it isn't, bend the tang that contacts the needle valve. Don't bend the float arm.

4-stroke models

Refer to illustration 8.5

5   Remove the float chamber (if you haven't already done so). To check the float height, hold the carburetor so the float hangs down, then tilt it back until the valve needle is just seated **(see illustration)**. Make sure the spring on the needle valve is no compressed by the weight of the float.
6   Measure the distance from the float chamber gasket surface to the top of the float and compare your measurement to the float height listed in this Chapter's Specifications. If it isn't as specified, bend the float tang that contacts the needle valve.

9 Air cleaner housing - removal and installation

Removal

1   Remove the seat and any other body panels necessary for access (see Chapter 8).

2-stroke models

Refer to illustrations 9.3 and 9.4

2   Loosen the clamp that secures the air cleaner housing to the carburetor (see Section 6).
3   Remove the wingnuts and take off the air cleaner duct **(see illustration)**.

9.3 Remove the wing nuts (one nut hidden) and lift off the air cleaner cover and duct

4-16   Chapter 4   Fuel and exhaust systems

9.4 Remove the housing mounting bolts

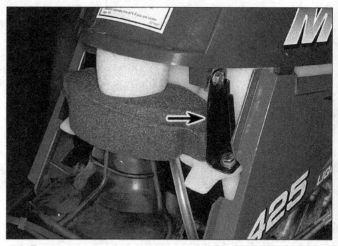

9.5 Remove the bracket on each side (arrow), the duct and foam

9.7 Reach in with a screwdriver and loosen the clamp (arrow)

9.8 Remove the housing mounting bolts (arrows)

4   Remove the mounting bolts and lift the air cleaner housing out of the vehicle **(see illustration)**.

4-stroke models

Refer to illustrations 9.5, 9.7, 9.8, 9.9a, 9.9b, 9.9c and 9.9d

5   Remove the fuel tank cover brackets, the air cleaner duct foam and the forward section of the air cleaner element **(see illustration)**.

6   Remove the air cleaner element (see Chapter 1).
7   Reach beneath the side panel with a long screwdriver and loosen the clamp that secures the air cleaner to the carburetor **(see illustration)**.
8   Remove the air cleaner housing bolts **(see illustration)**.
9   Note the locations of the vent hoses, then lift the air cleaner housing out of the vehicle **(see illustrations)**. Slip a pointed tool between the vent hoses and their fittings to work them free before pulling the

9.9a There's a vent line on the top of the housing . . .

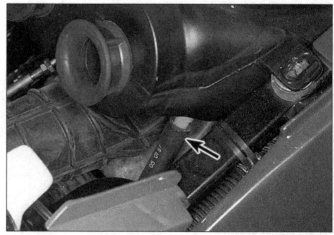

9.9b . . . and one underneath

Chapter 4 Fuel and exhaust systems

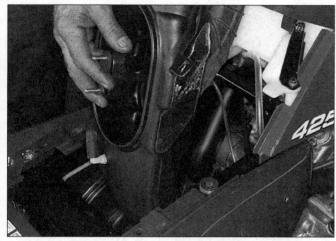

9.9c Lift the air cleaner housing out of the vehicle . . .

9.9d . . . and free the vent lines from the fittings with a pointed tool

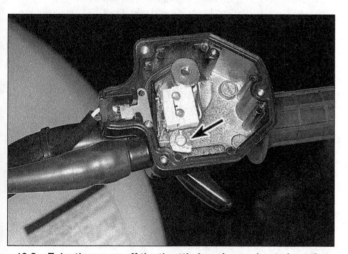

10.2a Take the cover off the throttle housing and note how the cable end fits in the lever (arrow)

10.2b Loosen the locknut and cable adjuster (arrows), then lift the cable end out of the lever

hoses off **(see illustration)**.
10 Installation is the reverse of the removal steps.

10 Throttle cable - removal, installation and adjustment

1 The throttle on these vehicles is operated by a thumb lever on the right handlebar.

Removal

Refer to illustrations 10.2a and 10.2b
2 Remove the cover from the throttle housing on the right handlebar **(see illustration)**. Loosen the throttle cable adjuster at the handlebar to create slack in the cable, then unhook the throttle cable end slug from the lever **(see illustration)**.

2-stroke models

3 Remove the seat (see Chapter 8) and unscrew the carburetor top (see Section 6). Disconnect the cable from the inside of the throttle piston, unscrew the cable fitting from the carburetor top and take the cable out.

4-stroke models

Refer to illustrations 10.5, 10.6a and 10.6b
4 Remove the carburetor partway from the vehicle (see Section 6).

5 Remove the throttle housing cover from the carburetor **(see illustration)**.
6 Loosen the locknut and adjusting nut to create slack in the cable

10.5 Remove the throttle pulley cover from the side of the carburetor

4-18 Chapter 4 Fuel and exhaust systems

10.6a Loosen the locknut and adjusting nut (arrows) . . .

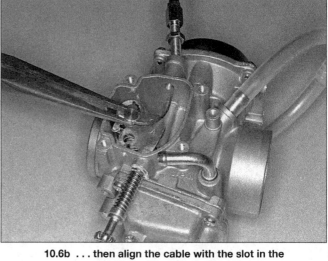

10.6b . . . then align the cable with the slot in the throttle pulley and slip it out

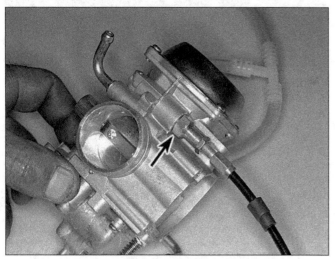

11.3a Unscrew the fitting (arrow) . . .

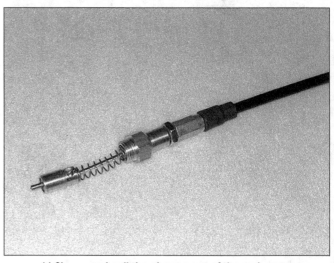

11.3b . . . and pull the plunger out of the carburetor

(see illustration), then rotate the throttle pulley, align the cable with the pulley slot and slip the cable end slug out of the pulley **(see illustration)**.

7 If the cable end slug is worn or damaged, replace it; otherwise, tape it to the end of the cable so it won't be lost.

8 Remove the cable, noting how it's routed.

9 If necessary, remove the throttle housing clamp screws and detach the throttle housing from the handlebar **(see illustration 12.15 in Chapter 1)**.

Installation

10 If the throttle housing was removed, install it (see Chapter 1).

11 Route the cable into place. Make sure it doesn't interfere with any other components and isn't kinked or bent sharply.

12 Lubricate the end of the cable with multi-purpose grease and connect it to the throttle pulley at the carburetor.

13 Reverse the disconnection steps to connect the throttle cable to the handlebar lever. Operate the lever and make sure it returns to the idle position by itself under spring pressure. **Warning:** *If the lever doesn't return by itself, find and solve the problem before continuing with installation. A stuck lever can lead to loss of control of the vehicle.*

Adjustment

14 Follow the procedure outlined in Chapter 1, Throttle operation/grip freeplay - check and adjustment, to adjust the cable.

15 Turn the handlebars back and forth to make sure the cables don't cause the steering to bind.

16 Once you're sure the cable operates properly, install the covers on the throttle housing and carburetor.

17 Install the fuel tank bracket and the tank.

18 With the engine idling, turn the handlebars through their full travel (full left lock to full right lock) and note whether idle speed increases. If it does, the cable is routed incorrectly. Correct this dangerous condition before riding the vehicle.

11 Choke cable - removal and installation

Removal

Refer to illustrations 11.3a, 11.3b and 11.4

1 Remove the seat (see Chapter 8).

2 If you're working on a 2-stroke, unscrew the cable, together with the starting enrichment valve, from the carburetor **(see illustrations 6.5a and 7.2h)**.

3 If you're working on a 4-stroke, remove the carburetor partway (Section 6). Unscrew the cable and the starting enrichment valve **(see illustrations)**.

Chapter 4 Fuel and exhaust systems

11.4 Unscrew the plastic nut and take the lever out of the bracket

12.2 Remove the heat shield bolts if necessary

4 At the choke lever bracket, remove the plastic nut that secures the choke cable and lever to the switch housing **(see illustration)**. Slip the cable out of the bracket.
5 Detach the cable from any clips and remove it, noting how it's routed.
6 If you plan to replace the choke plunger or cable, separate the plunger from the cable and remove the spring.

Installation

7 Installation is the reverse of the removal steps. Make sure the cable is securely fastened in its clips.
8 Check the choke lever operation (see Chapter 1).
9 Install any other components that were previously removed.

12 Exhaust system - removal and installation

Refer to illustrations 12.2, 12.3a, 12.3b, 12.3c, 12.4a, 12.4b, 12.4c, 12.5, 12.6a and 12.6b

1 Remove body panels as needed for access (see Chapter 8).
2 If necessary, unbolt the heat shield and remove it from the exhaust pipe **(see illustration)**.
3 Unhook the muffler springs and remove its mounting nut **(see illustrations)**.

12.3a A spring hook is the safest and easiest way to remove and install the springs

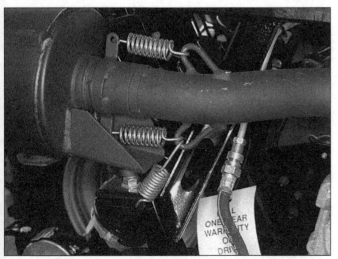

12.3b Unhook the muffler springs (Magnum shown) . . .

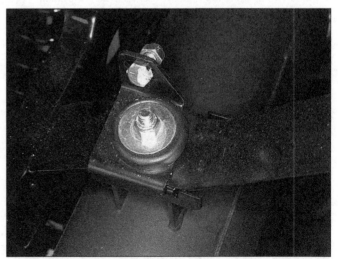

12.3c . . . and remove the mounting nut and insulator

12.4a Unhook the exhaust pipe support spring ...

12.4b ... and the springs that attach it to the engine fitting (this is a 2-stroke) ...

12.4c ... and this is a 4-stroke

12.5 Use a new packing ring at the joint of the exhaust pipe and engine fitting

4   Remove the exhaust pipe support springs and the springs that attach it to the engine fitting **(see illustrations)**.
5   Remove the packing ring that seals the exhaust pipe to the engine fitting **(see illustration)**.
6   Detach the fitting from the cylinder (2-stroke) or cylinder head (4-stroke) **(see illustrations)**.
7   Installation is the reverse of removal, with the following additions:
   a) *Be sure to install a new gasket at the cylinder or cylinder head.*
   b) *Use a new packing ring at the joint of the exhaust pipe and engine fitting.*

12.6a Remove the engine fitting (2-stroke shown) ...

12.6b ... and its gasket (4-stroke shown)

Chapter 4 Fuel and exhaust systems

13.3 Label and disconnect the pump hoses and drill out the bracket rivets

13.4 Lift the pump and remove the mounting nuts from the underside

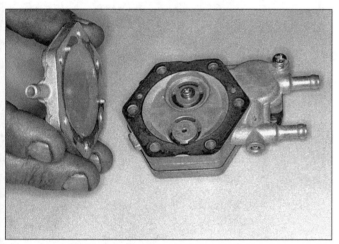

14.2 Remove the screws and lift off the cover

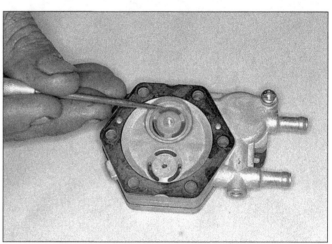

14.3 Carefully pry out the valves with a pointed tool

13 Fuel pump – removal and installation

Refer to illustrations 13.3 and 13.4

1 A fuel pump is used on 4-stroke models. It uses vacuum pulses from the carburetor to operate the pump diaphragm.
2 Remove body panels as necessary for access to the pump (it's mounted on a frame crossmember between the front fenders) (see Chapter 8).
3 Label and disconnect the impulse line, fuel inlet line and fuel outlet line **(see illustration)**. **Note:** *If there's any fuel in the impulse line, the diaphragm is leaking.*
4 Drill out the rivets that secure the pump bracket **(see illustration 13.3)**. Lift up the bracket, remove the pump mounting nuts and detach it from the crossmember **(see illustration)**.
5 Installation is the reverse of the removal steps.

14 Fuel pump – disassembly, inspection and reassembly

Note: *Check with your Polaris dealer to make sure you can get new gaskets and valves before disassembling the pump.*
Refer to illustrations 14.2, 14.3 and 14.4

1 Remove the fuel pump (Section 13).
2 Remove the cover screws and lift off the top cover **(see illustration)**. Peel the gasket off the pump.
3 Carefully pry the valves out of the pump with a pointed tool **(see illustration)**.
4 Turn the pump over and remove the bottom cover **(see illustration)**.
5 Assembly is the reverse of the disassembly steps. Use new gaskets.

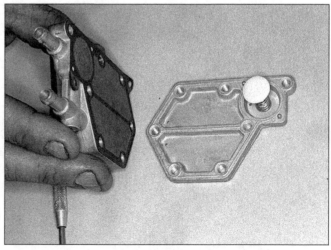

14.4 Remove the bottom cover and gasket

Notes

Chapter 5
Electrical and ignition system

Contents

	Section
Alternator - check and replacement	22
Battery - charging	4
Battery - check	See Chapter 1
Battery - inspection and maintenance	3
Brake switch - check and replacement	10
CDI magneto - check, removal and installation	27
CDI unit - check, removal and installation	28
Charging system - output test	21
Charging system testing - general information and precautions	20
Circuit breaker - check	5
Electrical troubleshooting	2
General information	1
Handlebar switches - check	13
Handlebar switches - removal and installation	14
Headlight aim - check and adjustment	8

	Section
Headlight bulb - replacement	7
Ignition coil - check, removal and installation	26
Ignition main (key) switch - check and replacement	12
Ignition system - check	25
Ignition timing - general information and check	29
Indicator bulbs - replacement	11
Lighting system - check	6
Neutral switch - check and replacement	15
Regulator/rectifier - check and replacement	23
Speedometer cable and drive - removal and installation	24
Starter circuit - check and component replacement	16
Starter motor - disassembly, inspection and reassembly	18
Starter motor - removal and installation	17
Starter pinion - removal, inspection and installation	19
Taillight/brake light bulb - replacement	9
Wiring diagrams	30

Specifications

Battery
Type.. 12V 14 Ah

Bulbs
Headlights
- 1985 through 1988 (except Trail Boss R/ES)........................ 45/45 watts
- 1988 Trail Boss R/ES... 60/60 watts
- 1989 through 1995.. 60 watts (Stanley BF120-59880)
- 1996
    - Trail Blazer ES .... 55 watts
    - Trail Boss 250 .... 60/60 watts
    - Xpress 300, Xplorer 300 .... 30/30 watts
    - Sport 400L .... 55 watts
    - Scrambler 4X4 .... 60 watts
    - Xpress 400L .... 60/55 watts
    - Xplorer 400L .... 60 watts/37.5 watts (2)
    - Sportsman 4X4 .... 60 watts/37.5 watts (2)
    - Big Boss 6X6 400L .... 60 watts/37.5 watts (2)
    - Magnum 4X4, 6X6 .... 60 watts/37.5 watts (2)
    - Sportsman 500 .... 60 watts/37.5 watts (2)
    - Magnum 2X4 .... 60/60 watts
- 1997
    - Trail Blazer .... 55 watts
    - Trail Boss .... 60/60 watts
    - Xpress 300, Xplorer 300 .... 30/30 watts
    - Sport .... 55 watts
    - Scrambler 4X4 .... 35/35 watts
    - Xpress 400 .... 60/55 watts
    - Xplorer 400 .... 60 watts
    - Sportsman 4X4 .... 60 watts
    - 400 6X6 .... 60 watts
    - Magnum 2X4 .... 60/60 watts
    - Magnum 4X4, 6X6 .... 60 watts
    - Sportsman 500 .... 60 watts
    - Xplorer 500 .... 35 watts (2)
    - Scrambler 500 .... 35/35 watts
Taillight (without brake light) .... 5 watts (GE168)
Tail/brake light .... 8.26/26.9 watts

Charging system

Charging output voltage	13.0 to 14.6 volts at 3000-4000 rpm
Charging output amperage	5 amps or more at idle

Stator coil resistance (within +/- 10 per cent)
- 1985 through 1987 (all)
  - Exciter coil (brown/white to white) .... 120 ohms
  - Lighting coil (yellow/red to yellow) .... 0.47 ohms
  - Pulser coil (black/red to brown/white) .... 23 ohms
- 1988 (all)
  - Lighting coil (yellow/red to yellow) .... 0.47 ohms
  - Exciter coil (black/red to brown/white) .... 120 ohms
- All 1989; 1990-on 2-stroke except 400 with 200-watt alternator
  - Lighting coil (yellow/red to yellow) .... 0.3 ohms
  - Exciter coil (black/red to brown/white) .... 120 ohms
- 1990-on 400 with 200-watt alternator
  - Lighting coil (yellow/red to yellow) .... 0.34 ohms
  - Lighting coil (yellow/brown to yellow) .... 0.17 ohms
  - Exciter coil (black/red to red) .... 226 ohms
  - Pulser coil (white/red to white) .... 97 ohms
- 1990 through 1996 4-stroke
  - Lighting coil (yellow/red to yellow)
    - Through 1995 .... 0.34 ohms
    - 1996 .... 0.37 ohms
  - Lighting coil (yellow/brown to yellow) .... 0.17 ohms
  - Exciter coil (red to green) .... 3.2 ohms
  - Exciter coil (red to black/red) .... 450 ohms
  - Exciter coil (green to black/red) .... 446 ohms
  - Pulser coil (white/red to white) .... 97 ohms
- 1997 4-stroke
  - Lighting coil (yellow/red to yellow)
    - 425 and 500 models (200 watts) .... 0.37 ohms
    - 500 models (250 watts) .... 0.13 watts
  - Lighting coil (yellow/brown to yellow)
    - 425 and 500 models (200 watts) .... 0.17 ohms
    - 500 models (250 watts) .... Not applicable
  - Exciter coil (red to green)
    - 425 and 500 models (200 watts) .... 3.2 ohms
    - 500 models (250 watts) .... 1.6 ohms
  - Exciter coil (red to black/red) .... 450 ohms
  - Exciter coil (green to black/red) .... 446 ohms
  - Pulser coil (white/red to white) .... 97 ohms

Starting system

Brush length limit .... 5/16 inch

Ignition System

Ignition coil resistance
- Primary .... 0.3 ohms
- Secondary .... 6300 ohms

Spark plug wire cap resistance (removed from wire)
- Two-stroke engines .... 3700 to 6300 ohms
- Four-stroke engines .... 5000 ohms

Torque specifications

Alternator rotor nut
- 250 and 300 2-stroke engines .... 44 to 62 ft-lbs
- 350 and 400 2-stroke engines .... 29 to 44 ft-lbs
- 4-stroke engines .... 58 to 72 ft-lbs

Starter motor mounting bolts .... 60 to 78 inch-lbs
Stator plate screws/bolts .... 60 to 78 inch-lbs

Chapter 5 Electrical and ignition system

1 General information

All of the machines covered by this manual are equipped with a 12-volt electrical system. The components include a three-phase permanent magnet alternator and a regulator/rectifier unit (except for the 1987 Cyclone and 1990 through 1995 Trail Blazer, which don't have a battery). The regulator/rectifier unit maintains the charging system output within the specified range to prevent overcharging and converts the AC (alternating current) output of the alternator to DC (direct current) to power the lights and other components and to charge the battery. The 1987 Cyclone and 1990 through 1995 Trail Blazer have a regulator, but power the lights and ignition with alternating current which is not rectified to DC.

An electric starter mounted to the engine case is used on most models. A recoil (pull rope) starter is also used. The starting system includes the motor, the battery, the starter relay and the various wires and switches. **Note:** *Keep in mind that electrical parts, once purchased, can't be returned. To avoid unnecessary expense, make very sure the faulty component has been positively identified before buying a replacement part.*

These vehicles are equipped with a fully transistorized, breakerless ignition system. The system consists of the following components:

CDI magneto (exciter coil, pulser coil (some models) and rotor)
CDI unit
Battery and circuit breaker (except 1987 Cyclone and 1990 through 1995 Trail Blazer)
Ignition coil
Spark plug
Engine kill (stop) and main (key) switches
Primary and secondary circuit wiring

The transistorized ignition system functions on the same principle as a breaker point ignition system with the CDI magneto and CDI unit performing the tasks previously associated with the breaker points and mechanical advance system. As a result, adjustment and maintenance of ignition components is eliminated (with the exception of spark plug replacement).

Because of their nature, the individual ignition system components can be checked but not repaired. If ignition system troubles occur, and the faulty component can be isolated, the only cure for the problem is to replace the part with a new one. Keep in mind that most electrical parts, once purchased, can't be returned. To avoid unnecessary expense, make very sure the faulty component has been positively identified before buying a replacement part.

2 Electrical troubleshooting

A typical electrical circuit consists of an electrical component, the switches, relays, etc. related to that component and the wiring and connectors that hook the component to both the battery and the frame. To aid in locating a problem in any electrical circuit, wiring diagrams are included at the end of this Chapter.

Before tackling any troublesome electrical circuit, first study the appropriate diagrams thoroughly to get a complete picture of what makes up that individual circuit. Trouble spots, for instance, can often be narrowed down by noting if other components related to that circuit are operating properly or not. If several components or circuits fail at one time, chances are the fault lies in the fuse or ground connection, as several circuits often are routed through the same fuse and ground connections.

Electrical problems often stem from simple causes, such as loose or corroded connections or a blown fuse. Prior to any electrical troubleshooting, always visually check the condition of the fuse, wires and connections in the problem circuit.

If testing instruments are going to be utilized, use the diagrams to plan where you will make the necessary connections in order to accurately pinpoint the trouble spot.

The basic tools needed for electrical troubleshooting include a test light or voltmeter, a continuity tester (which includes a bulb, battery and set of test leads) and a jumper wire, preferably with a circuit breaker incorporated, which can be used to bypass electrical components. Specific checks described later in this Chapter may also require an ammeter or ohmmeter.

Voltage checks should be performed if a circuit is not functioning properly. Connect one lead of a test light or voltmeter to either the negative battery terminal or a known good ground. Connect the other lead to a connector in the circuit being tested, preferably nearest to the battery or fuse. If the bulb lights, voltage is reaching that point, which means the part of the circuit between that connector and the battery is problem-free. Continue checking the remainder of the circuit in the same manner. When you reach a point where no voltage is present, the problem lies between there and the last good test point. Most of the time the problem is due to a loose connection. Since these vehicles are designed for off-road use, the problem may also be water or corrosion in a connector. Keep in mind that some circuits only receive voltage when the ignition key is in the On position.

One method of finding short circuits is to remove the fuse and connect a test light or voltmeter in its place to the fuse terminals. There should be no load in the circuit. Move the wiring harness from side-to-side while watching the test light. If the bulb lights, there is a short to ground somewhere in that area, probably where insulation has rubbed off a wire. The same test can be performed on other components in the circuit, including the switch.

A ground check should be done to see if a component is grounded properly. Disconnect the battery and connect one lead of a self-powered test light (such as a continuity tester) to a known good ground. Connect the other lead to the wire or ground connection being tested. If the bulb lights, the ground is good. If the bulb does not light, the ground is not good.

A continuity check is performed to see if a circuit, section of circuit or individual component is capable of passing electricity through it. Disconnect the battery and connect one lead of a self-powered test light (such as a continuity tester) to one end of the circuit being tested and the other lead to the other end of the circuit. If the bulb lights, there is continuity, which means the circuit is passing electricity through it properly. Switches can be checked in the same way.

Remember that all electrical circuits are designed to conduct electricity from the battery (the magneto on 1987 Cyclones), through the wires, switches, relays, etc. to the electrical component (light bulb, motor, etc.). From there it is directed to the frame (ground) where it is passed back to the battery. Electrical problems are basically an interruption in the flow of electricity from the battery or back to it.

3 Battery - inspection and maintenance

1 Most battery damage is caused by heat, vibration, and/or low electrolyte levels, so keep the battery securely mounted, check the electrolyte level frequently and make sure the charging system is functioning properly. **Warning:** *Always disconnect the negative cable first and connect it last to prevent sparks which could the battery to explode.*

2 Refer to Chapter 1 for electrolyte level and specific gravity checking procedures.

3 Check around the base inside of the battery for sediment, which is the result of sulfation caused by low electrolyte levels. These deposits will cause internal short circuits, which can quickly discharge the battery. Look for cracks in the case and replace the battery if either of these conditions is found.

4 Check the battery terminals and cable ends for tightness and corrosion. If corrosion is evident, disconnect the cables from the battery, disconnecting the negative (-) terminal first, and clean the terminals and cable ends with a wire brush or knife and emery paper. Reconnect the cables, connecting the negative cable last, and apply a thin coat of petroleum jelly to the cables to slow further corrosion.

5 The battery case should be kept clean to prevent current leakage, which can discharge the battery over a period of time (especially when

4 Battery - charging

1 If the machine sits idle for extended periods or if the charging system malfunctions, the battery can be charged from an external source.
2 To properly charge the battery, you will need a charger of the correct rating, a hydrometer, a clean rag and a syringe for adding distilled water to the battery cells.
3 The maximum charging rate for any battery is 1/10th of the rated amp-hour capacity. As an example, the maximum charge rate for a 14 amp/hour battery would be 1.4 amps. If the battery is charged at a higher rate, it could overheat, causing the plates inside the battery to buckle.
4 Do not allow the battery to be subjected to a so-called quick charge (high charge rate over a short period of time) unless you are prepared to buy a new battery.
5 When charging the battery, always remove it from the machine and be sure to check the electrolyte level before hooking up the charger. Add distilled water to any cells that are low.
6 Loosen the cell caps, hook up the battery charger leads (positive lead to battery positive terminal, negative lead to battery negative terminal), cover the top of the battery with a clean rag, then, and only then, plug in the battery charger. **Warning:** *The hydrogen gas escaping from a charging battery is explosive, so keep open flames and sparks well away from the area. Also, the electrolyte is extremely corrosive and will damage anything it comes in contact with.*
7 Allow the battery to charge until the specific gravity is as specified (refer to Chapter 1 for the specific gravity checking procedure). The charger must be disconnected and disconnected from the battery when making specific gravity checks. If the battery overheats or gases excessively, the charging rate is too high. Either disconnect the charger or lower the charging rate to prevent damage to the battery.
8 If one or more of the cells do not show an increase in specific gravity after a long slow charge, or if the battery as a whole does not seem to want to take a charge, it's time for a new battery.
9 When the battery is fully charged, disconnect the charger first, then disconnect the leads from the battery. Install the cell caps and wipe any electrolyte off the outside of the battery case.
10 If the recharged battery discharges rapidly when left disconnected, it's likely that an internal short caused by physical damage or sulfation has occurred. A new battery will be required. A sound battery will tend to lose its charge at about 1% per day.

5 Circuit breaker - check

1 Models equipped with a battery have a 10-amp or 20-amp circuit breaker (or one of each on some models), mounted in the wire between the starter solenoid and regulator near the battery.
2 If the circuit breaker trips repeatedly, check the wiring for a short.
3 Never, under any circumstances, use a higher rated breaker or bridge the breaker terminals, as damage to the electrical system - or even a fire - could result.
4 Occasionally a circuit breaker will trip for no obvious reason. Corrosion of the breaker terminals may occur and cause poor contact. If this happens, remove the corrosion with a wire brush or emery paper, then spray the terminals with electrical contact cleaner.

it sits unused). Wash the outside of the case with a solution of baking soda and water. Do not get any baking soda solution in the battery cells. Rinse the battery thoroughly, then dry it.
6 If acid has been spilled on the frame or battery box, neutralize it with a baking soda and water solution, then touch up any damaged paint. Make sure the battery vent tube is directed away from the frame and is not kinked or pinched.
7 If the vehicle sits unused for long periods of time, disconnect the cables from the battery terminals. Refer to Section 4 and charge the battery approximately once every month.

6 Lighting system - check

Battery powered models

1 The battery provides power for operation of the headlights, taillight, brake light (if equipped) and indicator lights. If none of the lights operate, always check battery voltage before proceeding. Low battery voltage indicates either a faulty battery, low battery electrolyte level or a defective charging system. Refer to Chapter 1 and Section 3 of this Chapter for battery checks and Sections 20 through 22 for charging system tests. Also, check the condition of the fuse and replace it with a new one if it's blown.

Headlights

2 If both of the headlight bulbs are out with the headlight switch in Lo or Hi and the main key switch On, check the circuit breaker(s) (see Section 5).
3 If only one headlight is out, try installing the bulb from the working headlight. If this solves the problem, replace the defective bulb. If not, test further as described below.
4 Disconnect the electrical connector from the bulb that doesn't light. Connect the negative lead of a voltmeter to the brown wire terminal in the wiring harness. Turn the main key switch On and connect the positive lead to the green wire terminal (low beam) and the yellow wire terminal (high beam). The voltmeter should indicate 12 volts or more.
 a) *If there's voltage at the terminals, the bulb is burned out or the bulb socket is corroded.*
 b) *If there's no voltage, the problem lies in the wiring or one of the switches in the circuit. Refer to Sections 12 and 13 for the switch testing procedures, and also the wiring diagrams at the end of this book.*

Taillight

5 If the taillight fails to work, check the bulb and the bulb terminals first.
6 If the bulb and terminals are good, disconnect the taillight electrical connector. Connect a voltmeter negative lead to the brown wire in the wiring harness and the positive lead to the red/white wire. With the main key switch and lighting switch On, the voltmeter should indicate 12 volts or more.
 a) *If there's voltage at the terminals, the bulb is burned out or the bulb socket is corroded.*
 b) *If there's no voltage, the problem lies in the wiring or one of the switches in the circuit. Refer to Sections 12 and 13 for the switch testing procedures, and also the wiring diagrams at the end of this Chapter.*
7 If no voltage is indicated, check the ground wire and the wiring between the taillight and the lighting switch, then check the switch.

Brake light

8 If the machine is equipped with a brake light, see Section 10 for the brake light switch checking procedure.

Neutral indicator light

9 If the neutral light fails to operate when the transmission is in Neutral, check the fuse and the bulb (see Section 11 for bulb removal procedures). If the bulb and fuse are in good condition, check for battery voltage at the wire attached to the neutral switch on the left side of the engine. If battery voltage is present, refer to Section 15 for the neutral switch check and replacement procedures.
10 If no voltage is indicated, check the wiring to the bulb, to the switch and between the switch and the bulb for open circuits and poor connections.

Non-battery powered models

11 You'll need an ohmmeter and voltmeter that can measure AC voltage (not DC voltage) for this test.
12 Check the resistance of the alternator lighting coil (yellow and yellow/red wires (Section 22). If it's not within the range listed in this

Chapter 5 Electrical and ignition system

7.2a Disconnect the connector from the headlight socket . . .

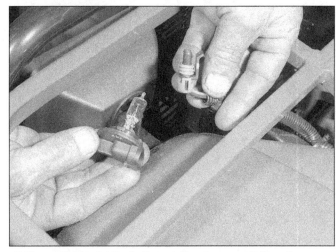

7.2b . . . then twist the bulb socket and remove it from the case

Chapter's Specifications, inspect the rotor and stator.

13 If the ohmmeter reading is correct, connect an AC voltmeter between the lighting coil wires. Start the engine, run it at 3000 rpm and check the voltage reading at the yellow wire. It should be 25 volts AC or more. If it's low, inspect the rotor and stator (Section 22).

14 Reconnect the yellow and yellow/red wires and repeat the test, this time measuring voltage between the yellow wire and ground. It should be 11 to 15 volts. If it's within the specified range, the system is good.

15 If the voltage is high, make sure the regulator is securely mounted to the frame. This provides regulator ground. If it isn't securely mounted, the bad ground will cause excessive voltage. If it is securely mounted, replace the regulator.

16 If the voltage reading is low in Step 14, disconnect the lights and test again. If the voltage is now within the specified range, look for a short circuit in the lighting system. If the voltage stays low with the lights disconnected, replace the regulator.

7 Headlight bulb - replacement

Warning: *If the headlight has just burned out, give it time to cool before changing the bulb to avoid burning your fingers.*
Refer to illustrations 7.2a, 7.2b, 7.3a and 7.3b

1 Remove body panels as necessary for access (see Chapter 8). This may include the seat, headlight cowl and side panels. On some models, you'll need to remove the headlight housing as well.

2 If the bulb has a detachable connector plug, remove it from the headlight socket **(see illustration)**. Pull the socket out of the headlight housing and remove the bulb **(see illustration)**.

3 If the bulb is retained by a wire clip, release it **(see illustration)**. Pull the bulb out of the headlight housing, taking care not to touch the glass **(see illustration)**.

4 Installation is the reverse of the removal procedure, with the following additions:

 a) *Be sure not to touch the bulb with your fingers - oil from your skin will cause the bulb to overheat and fail prematurely. If you do touch the bulb, wipe it off with a clean rag dampened with rubbing alcohol.*
 b) *Align the tab on the metal bulb flange with the slot in the headlight case.*
 c) *Make sure the arrow mark on the headlight cover is upward.*

8 Headlight aim - check and adjustment

Check

1 An improperly adjusted headlight may cause problems for oncoming traffic or provide poor, unsafe illumination of the terrain ahead. Before adjusting the headlight, be sure to consult with local traffic laws and regulations.

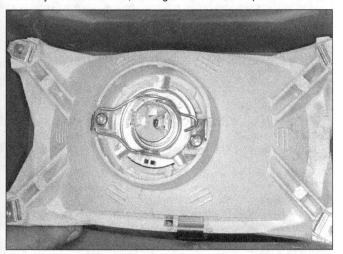

7.3a Pull the cover and wiring connector off the socket and undo the clip . . .

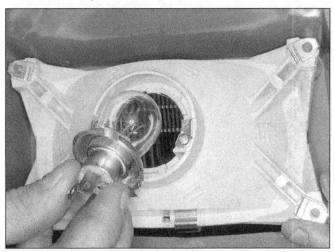

7.3b . . . and pull the bulb out of the socket without touching the glass

Chapter 5 Electrical and ignition system

8.7 There's a headlight adjusting screw at each corner

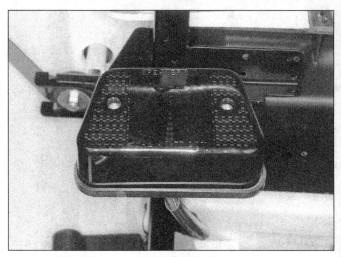

9.1 Remove the lens screws and take off the lens

2   The headlight beam can be adjusted horizontally and vertically. Before performing the adjustment, make sure the fuel tank is at least half full, and have an assistant sit on the seat.
3   Park the on a level surface, vehicle pointing at a blank wall about 25 feet away. Measure from the ground to the center of the headlight beam and write down the distance. At the wall, measure from the ground the same distance you just wrote down and mark the wall at the point.
4   Turn on the engine, let it idle, and turn on the high beam headlights. The brightest spot in the beam should hit the wall about two inches below the mark you made.
5   If the machine has separate low beams, switch them on. The brightest point of the beam should hit the wall about 8 inches below the mark.

Adjustment
Gen II body style
Refer to illustration 8.7

6   This body style includes Trail Boss, Magnum, 6X6, and Sportsman 4X4 models, as well as Xplorer models through 1995.
7   The headlight can be adjusted horizontally or vertically by turning the screws at the four corners of the headlight housing **(see illustration)**.

Gen III body style
8   This body style includes Trail Blazer, Sport and Scrambler models.

10.3 The brake switch is mounted in the hydraulic lines

9   If you're working on a Trail Blazer or Sport model, there's an adjusting screw on each side of the headlight and one in the center just in front of the headlight. The center screw adjusts the headlight vertically and the two side screws move it from side-to-side.
10   If you're working on a Scrambler, the headlight can be adjusted vertically. Loosen the nut and bolt that secure the headlight housing to the handlebars, turn the housing up or down as needed and tighten the bolt.

Gen IV body style
11   This body style includes 1997 and later Xpress 300 and 400, Xplorer 300, 400 and 500 and Sportsman 500 models.
12   Headlight aim on these models is adjusted by turning the adjustment knob on the back of the headlight housing.

9   Taillight/brake light bulb - replacement

Refer to illustration 9.1

1   Remove the lens screws and take off the lens **(see illustration)**.
2   Press the bulb into its socket and turn it counterclockwise to remove.
3   Check the socket terminals for corrosion and clean them if necessary.
4   Press the bulb into its socket, turn clockwise to engage the pins, then release the bulb.
5   Install the lens and tighten the screws securely, but not enough to crack the plastic.

10   Brake switch - check and replacement

Check
Refer to illustration 10.3

1   All models use an electrical switch mounted in the brake lines. The switch is controlled by hydraulic pressure from the brake fluid.
2   Before checking any electrical circuit, check the circuit breaker (see Section 5).
3   Using a test light connected to a good ground, check for voltage to the power wire at the brake light switch **(see illustration)**. If there's no voltage present, check the wire between the switch and the ignition switch (see the Wiring diagrams at the end of the book).
4   If voltage is available, touch the probe of the test light to the other terminal of the switch, then pull the brake lever or depress the brake pedal - if the test light doesn't light up, replace the switch.
5   If the test light does light, check the wiring between the switch and the brake lights (see the Wiring diagrams at the end of the book).

Chapter 5 Electrical and ignition system

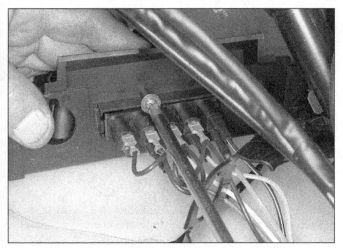

11.2 Remove the screw from the back of the bulb cluster . . .

Switch replacement

6 Disconnect the electrical connector from the switch.
7 Unscrew the switch from the fitting and screw in a new one. Bleed the brakes (see Chapter 7).

11 Indicator bulbs - replacement

1 Two basic types of indicator light display are used. One has the indicator bulbs mounted in the same bracket with the ignition key switch; this design has replaceable bulbs. The other type has the lights mounted in square housings in the headlight housing. This type is replaced as a complete unit.

Replaceable bulb type
Refer to illustrations 11.2 and 11.3

2 Remove the screw that secures the bulb housing to the bracket **(see illustration)**.
3 To replace a bulb, pull it out of the socket **(see illustration)**. **Note:** *A length of rubber hose slipped over the bulb will make a convenient handle.*
4 If the socket contacts are dirty or corroded, they should be scraped clean and sprayed with electrical contact cleaner before new bulbs are installed.
5 Push the new bulb into its socket. Install the bulb housing in the bracket.

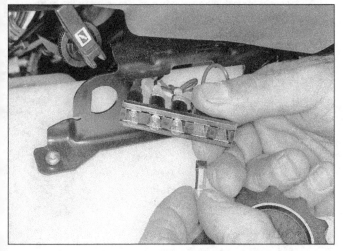

11.3 . . . pull the old bulb out and push a new one in

Replaceable housing type

6 Remove the headlight housing cover (see Chapter 8).
7 Squeeze the locking tab on each side of the indicator light unit and slip it out.
8 Installation is the reverse of the removal steps.

12 Ignition main (key) switch - check and replacement

Check

1 Follow the wiring harness from the ignition switch to the connector and disconnect the connector.
2 Using an ohmmeter, check the continuity of the terminal pairs indicated in the Wiring diagrams at the end of the book. Continuity should exist between the terminals connected by a solid line when the switch is in the indicated position.
3 If the switch fails any of the tests, replace it.

Replacement
Refer to illustration 12.4

4 The ignition switch is secured by a plastic nut **(see illustration)**.
5 If you haven't already done so, disconnect the switch electrical connector. Unscrew the nut and take the switch out.
6 Installation is the reverse of the removal procedure.

13 Handlebar switches - check

1 Generally speaking, the switches are reliable and trouble-free. Most troubles, when they do occur, are caused by dirty or corroded contacts, but wear and breakage of internal parts is a possibility that should not be overlooked. If breakage does occur, the entire switch and related wiring harness will have to be replaced with a new one, since individual parts are not usually available.
2 The switches can be checked for continuity with an ohmmeter or a continuity test light. Always disconnect the battery negative cable, which will prevent the possibility of a short circuit, before making the checks.
3 Trace the wiring harness of the switch in question and disconnect the electrical connectors.
4 Using the ohmmeter or test light, check for continuity between the terminals of the switch harness with the switch in the various positions. Refer to the continuity diagrams contained in the Wiring diagrams at the end of the book. Continuity should exist between the terminals connected by a solid line when the switch is in the indicated position.

12.4 Unscrew this plastic nut and detach the key switch from the bracket

14.1 Remove the two screws (arrows) and separate the switch halves

15.1 The neutral switch is mounted on the transmission

16.2 The starter relay is mounted near the battery

5 If the continuity check indicates a problem exists, refer to Section 14, disassemble the switch and spray the switch contacts with electrical contact cleaner. If they are accessible, the contacts can be scraped clean with a knife or polished with crocus cloth. If switch components are damaged or broken, it will be obvious when the switch is disassembled.

14 Handlebar switches - removal and installation

Refer to illustration 14.1

1 The handlebar switches are composed of two halves that clamp around the bar. They are easily removed for cleaning or inspection by taking out the clamp screws and pulling the switch halves away from the handlebar **(see illustration)**.
2 To completely remove the switches, the electrical connectors in the wiring harness must be disconnected and the harness separated from the tie wraps and retainers.
3 When installing the switches, make sure the wiring harness is properly routed to avoid pinching or stretching the wires.

15 Neutral switch - check and replacement

Refer to illustration 15.1

1 Disconnect the electrical connectors from the switch **(see illustration)**.
2 Connect one lead of an ohmmeter to a good ground and the other lead to the terminal post on the switch being tested.
3 When the transmission is in neutral, the ohmmeter should read 0 ohms between the neutral switch and ground - in any other gear, the meter should read infinite resistance.
4 If the switch doesn't check out as described, replace it.
5 Install the switch in the case with a new sealing washer and tighten it to the torque listed in this Chapter's Specifications.
6 Reconnect the switch wires.

16 Starter circuit - check and component replacement

1 Depending on the procedure, it may be necessary to remove the seat, rear cargo rack or rear fender for access to other parts (see Chapter 7).

Starter relay

Check

Warning: *Make sure the transmission is in Neutral before starting this test.*
Refer to illustration 16.2
2 Locate the starter relay near the battery **(see illustration)**.
3 Follow the wire from the start button on the left handlebar to the relay, then disconnect it.
4 Connect a length of wire from the battery positive terminal to the relay in place of the disconnected start switch wire. The starter should crank the engine. If it doesn't, the problem may be in the starter itself (refer to Sections 17 and 18 for removal and inspection procedures), in the electrical cables between the battery and starter relay, or in the starter relay. If the starter is good and the cables are in good condition and properly connected, the relay is probably at fault.

Replacement

5 Disconnect the negative cable from the battery.
6 Pull back the rubber covers from the terminal nuts, remove the nuts and disconnect the starter relay cables **(see illustration 16.2)**. Disconnect the remaining electrical connector from the starter relay.
7 Remove the relay mounting nuts and take the relay out.
8 Installation is the reverse of removal. Reconnect the negative battery cable after all the other electrical connections are made.

Starter switch

14 The starter switch is part of the switch assembly on the left handlebar. Refer to Section 13 for checking and replacement procedures.

17 Starter motor - removal and installation

Removal

Refer to illustrations 17.2a, 17.2b, 17.3a, 17.3b, 17.4, 17.6a and 17.6b
1 Disconnect the cable from the negative terminal of the battery.
2 If you're working on a small 2-stroke model, pull back the rubber

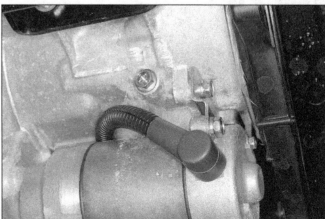

17.2a On small-displacement 2-strokes, disconnect the cable and unbolt the bracket . . .

Chapter 5 Electrical and ignition system

5-9

17.2b . . . then remove the mounting bolts (arrows)

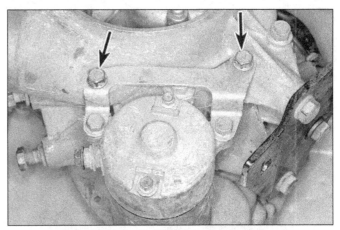

17.3a On large-displacement 2-strokes, unbolt the brace (arrows) . . .

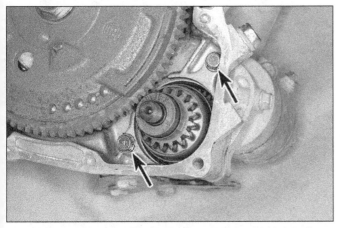

17.3b . . . and remove the starter mounting bolts (arrows)

illustration). Remove the starter mounting bolts; note that one of them secures a ground cable.
5   Lift the outer end of the starter up a little bit and slide the starter out of the engine case. **Caution:** *Don't drop or strike the starter or its magnets may be demagnetized, which will ruin it.*
6   Check the condition of the O-ring or gasket on the end of the starter that fits into the engine and replace it if necessary **(see illustrations)**.

Installation
5   Remove any corrosion or dirt from the mounting lugs on the starter and the mounting points on the crankcase.
6   Apply a little engine oil to the O-ring and install the starter by reversing the removal procedure.

18 Starter motor - disassembly, inspection and reassembly

Note: *Check carefully which components are available as replacements before starting overhaul procedures.*
1   Remove the starter motor (see Section 17). This procedure describes overhaul of the late model two-brush starter. Earlier designs are similar.

Disassembly
Refer to illustrations 18.2, 18.3 and 18.5
2   Look for alignment marks on the starter housing and end cover **(see illustration)**. Make your own marks if they aren't visible.

boot covering the starter cable nut, undo the nut and unbolt the support bracket **(see illustration)**. Remove the starter mounting bolts **(see illustration)**.
3   If you're working on a large 2-stroke model, remove the recoil starter housing from the engine (see Chapter 2A). Disconnect the starter cable, unbolt the starter bracket from the engine, then remove the starter mounting bolts from the crankcase **(see illustrations)**. Remove the starter together with the bracket and starter clutch.
4   If you're working on a 4-stroke model, pull back the rubber boot and remove the nut retaining the starter cable to the starter **(see**

17.4 On 4-strokes, disconnect the cable and remove the mounting bolts (note the ground cable)

17.6a Pull the starter out and inspect the O-ring (arrow)

17.6b On large 2-strokes, inspect the gasket and remove the starter pinion

Chapter 5 Electrical and ignition system

18.2 Make alignment marks between the housing and end covers before disassembly

3 Unscrew the two through-bolts, then remove the brush end cover with its O-ring from the motor (see illustration).
4 Remove the engine end cover with its O-ring from the motor.
5 Slide off the insulating washer and shim(s) from the armature, noting their locations (see illustration). Withdraw the armature from the housing.

Inspection

Refer to illustrations 18.6, 18.7, 18.9a and 18.9b

6 Lift the brush springs and slide the brushes out of their holders (see illustration).
7 The parts of the starter motor that most likely will require attention are the brushes. If one brush must be replaced, replace both of them. The brushes can be replaced separately from the other brush housing components. Brushes must be replaced if they are worn excessively, cracked, chipped, or otherwise damaged. Measure the length of the brushes and compare the results to the brush length listed in this Chapter's Specifications (see illustration). If any of the brushes is worn beyond the specified limits, replace them all.

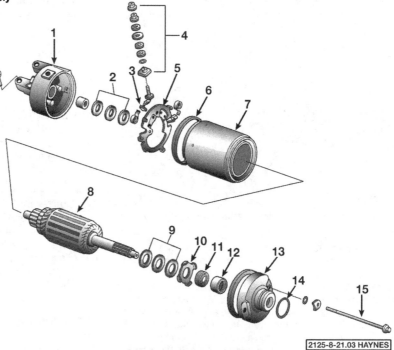

18.3 Starter (late 2-brush shown; others similar)

1 Brush end cover
2 Shims
3 Brush
4 Terminal stud components
5 Brush plate
6 O-ring
7 Starter housing
8 Armature
9 Shims
10 Lockwasher
11 Dust seal
12 Needle roller bearing
13 Engine end cover
14 O-ring
15 Through-bolt

18.5 Unscrew the nut and remove the washers from the terminal stud

18.6 Lift the brush springs and slide the brushes out of their holders

Chapter 5 Electrical and ignition system

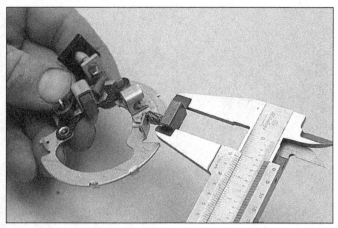

18.7 Measure the brush length and replace them if they're worn

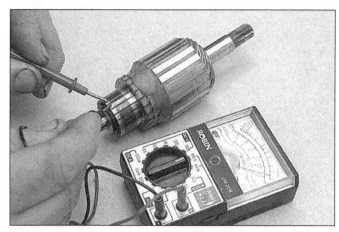

18.9a Check for continuity between the commutator bars . . .

8 Inspect the commutator for scoring, scratches and discoloration. The commutator can be cleaned and polished with 600-grit emery paper, but do not remove copper from the commutator. After cleaning, clean out the grooves and wipe away any residue with a cloth soaked in an electrical system cleaner or denatured alcohol. Measure the commutator diameter and compare it to the diameter listed in this Chapter's Specifications. If it's less than the service limit, the motor must be replaced with a new one.

9 Using an ohmmeter or a continuity test light, check for continuity between the commutator bars **(see illustration)**. Continuity should exist between each bar and all of the others. Also, check for continuity between the commutator bars and the armature shaft **(see illustration)**. There should be no continuity between the commutator and the shaft. If the checks indicate otherwise, the armature is defective.

10 Check the undercut of the mica between the commutator bars. If it isn't deep enough, carefully scrape away mica with a broken-off piece of hacksaw blade until the undercut is as listed in this Chapter's Specifications.

11 Check the seal in the cover for wear or damage. Check the bearing on the armature for roughness, looseness or loss of lubricant. Check with an ATV shop or Polaris dealer to see if the seal and bearing can be replaced separately; if this isn't possible, replace the starter motor.

12 Check the starter pinion for worn, chipped or broken teeth. If the gears are damaged or worn, replace the pinion.

13 Inspect the insulating washers and shims for signs of damage and replace if necessary.

14 Check the magnets inside the armature for damage or loss of magnetism. Replace the starter motor if the magnets are damaged.

Reassembly

Refer to illustrations 18.16, 18.17, 18.19a, 18.19b, 18.20, 18.21, 18.22, 18.23 and 18.24

15 Lift the brush springs and slide the brushes back into position in their holders.

16 Fit the insulator to the housing and install the brush plate. Insert the terminal bolt through the brush plate and housing **(see illustration)**.

17 Slide the rubber ring and small insulating washer onto the bolt, followed by the large insulating washers and the plain washer. Install the nut on the terminal bolt and tighten it securely **(see illustration)**.

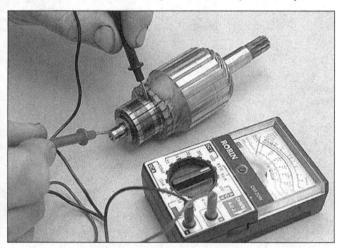

18.9b . . . and for no continuity between each commutator bar and the armature shaft

18.16 Install the brush plate and terminal bolt

18.17 Install the terminal nut and tighten it securely

18.19a Install the shims on the armature, then insert the armature . . .

18.19b . . . and locate the brushes on the commutator

18.20 Fit the lockwasher to the front cover so its teeth engage the ribs

18.21 Install the shims and washer on the armature in the correct order

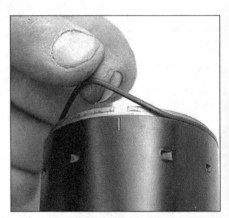

18.22 Install the O-ring . . .

18 Locate the brush assembly in the housing, making sure its tab is correctly located in the housing slot.
19 Install the shims on the armature shaft **(see illustration)** and insert the armature in the housing locating the brushes to the commutator bars. Check that each brush is securely pressed against the commutator by its spring and is free to move easily in its holder **(see illustration)**.
20 Install the lockwasher on the engine end cover so its teeth are correctly located with the cover ribs **(see illustration)**. Apply a smear of grease to the cover dust seal lip.
21 Slide the shim(s) onto the other end of the armature shaft, then install the insulating washer **(see illustration)**. Fit the O-ring to the housing and carefully slide the cover into position, aligning the marks made on removal.
22 Ensure the brush plate inner tab is correctly located in the housing slot and fit the O-ring to the housing **(see illustration)**.
23 Align the cover groove with the brush plate outer tab and install the cover **(see illustration)**.
24 Check that the marks made on disassembly are correctly aligned, then install the through-bolts and tighten them securely **(see illustration 18.3b)**. If the through-bolt washers have straight sides, make sure the straight sides align with the bosses on the bracket **(see illustration)**.
25 Install the starter as described in Section 17.

18.23 . . . and install the rear cover, aligning its groove with the brush plate outer tab

18.24 If the washers have flat sides, face them like this

Chapter 5 Electrical and ignition system

19.2a Pull the starter pinion off the engine . . .

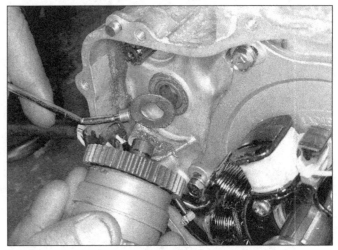

19.2b . . . and remove the wire terminal

19 Starter pinion - removal, inspection and installation

Removal

Refer to illustrations 19.2a and 19.2b

1   If you're working on a 2-stroke, remove the starter and remove the pinion from it (Section 16).
2   If you're working on a 4-stroke, refer to Section 22 and remove the alternator cover and rotor. Remove the starter pinion from the crankcase **(see illustrations)**.

Inspection

Refer to illustration 19.5

3   Check the gears for worn or broken teeth. Check the shafts and the friction surface on the gears for wear or damage and replace the pinion if there are any defects.
4   The garter spring can be replaced separately. Twist the pinion to its extended position, then unhook the spring from its groove with a pointed tool. Install a new spring, making sure it is correctly located between the alignment pins and within the anti-kickout shoes.
5   Check the pinion bushing in the alternator cover for wear or damage and replace as necessary **(see illustration)**.

Installation

6   Installation is the reverse of the removal steps. Lubricate the bushing with a small amount of non-petroleum grease.

19.5 Replace the bushing if it's worn or damaged

20 Charging system testing - general information and precautions

1   If the performance of the charging system is suspect the system as a whole should be checked first, followed by testing of the individual components (the alternator and the regulator/rectifier). **Note:** *Before beginning the checks, make sure the battery is fully charged and that all system connections are clean and tight.*
2   Checking the output of the charging system and the performance of the various components within the charging system requires the use of an ohmmeter; voltmeter or ammeter (depending on model); or the equivalent multimeter.
3   When making the checks, follow the procedures carefully to prevent incorrect connections or short circuits, as irreparable damage to electrical system components may result if short circuits occur.
4   If the necessary test equipment is not available, it is recommended that charging system tests be left to a dealer service department or a reputable motorcycle repair shop.

21 Charging system - output test

1   If a charging system problem is suspected, perform the following checks. Start by checking the circuit breaker (Section 5) and battery (Section 3 and Chapter 1). If necessary, charge the battery (Section 4).
2   Start the engine and let it warm up to normal operating temperature. Connect an inductive tachometer to the spark plug wire.
3   Touch the positive probe of a 0-20 volt voltmeter to the positive (+) battery terminal and the negative probe to the battery negative (-) terminal.
4   Slowly increase the engine speed to 3000-400 rpm and compare the voltmeter reading to the value listed in this Chapter's Specifications.
5   If the output is as specified, the alternator is functioning properly.
6   Low voltage may be the result of damaged windings in the alternator stator coils or wiring problems. Make sure all electrical connections are clean and tight, then refer to the following Sections to check the alternator stator coils and the regulator/rectifier.
7   Output above the specified range indicates a defective voltage regulator/rectifier. Refer to Section 23 for regulator testing and replacement procedures.
8   Locate the yellow and yellow/red wires at the alternator connector. Disconnect them and connect an ammeter between them. With the engine idling (not revved above idle), the alternator should produce at least 5 amps.
9   Disconnect the test equipment.

22.4a Locate the cover dowels (this is a large-displacement 2-stroke) . . .

22.4b . . . and this is a 4-stroke

22 Alternator - check and replacement

Stator coil check

1 Locate and disconnect the coil connector on the left side of the vehicle frame. The connector can be identified by its wire colors. **Note:** *Some of the wires go to the ignition pulse coil.*

2 Connect an ohmmeter between the terminals in the side of the connector that runs back to the engine (connect the positive lead to one of the terminals and the negative lead to each of the two remaining terminals in turn). Refer to Chapter's Specifications for wire colors. If the readings are outside the range listed in this Chapter's Specifications, replace the stator coils as described below.

3 Connect the ohmmeter between a good ground on the vehicle and each of the connector terminals in turn. The meter should indicate infinite resistance (no continuity). If not, remove the rotor and replace the stator coils.

Rotor removal

Refer to illustrations 22.4a, 22.4b, 22.5, 22.6 and 22.8

4 Remove the alternator cover from the engine and locate the cover dowels **(see illustrations)**.

5 **Note:** *To remove the alternator rotor, a special puller will be required* **(see illustration)**. *Don't try to remove the rotor without the proper puller, as it's almost sure to be damaged. Pullers are readily available from ATV dealers and aftermarket tool suppliers.* Remove the rotor nut and washer. Thread the puller shaft and bolts into the rotor. Hold the rotor with a strap wrench and turn the shaft with another wrench to separate the rotor from the crankshaft.

6 Pull the rotor off **(see illustration)**. You may need to pull firmly to overcome the resistance of the rotor magnets, but don't use excessive force. If the rotor seems to be stuck, check to make sure all fasteners have been removed.

7 Check the rotor Woodruff key in the end of the crankshaft; if it's not secure in its slot, pull it out and set it aside for safekeeping. A convenient method is to stick the Woodruff key to the magnets inside the rotor, but be certain not to forget it's there, as serious damage to the rotor and stator coils will occur if the engine is run with anything stuck to the magnets.

8 Check the magnets inside the rotor for mechanical damage and to make sure they haven't come unglued from the inside of the rotor **(see illustration)**.

Rotor installation

9 Degrease the center of the rotor and the end of the crankshaft.

10 Make sure the Woodruff key is positioned securely in its slot. Take a look to make sure there isn't anything stuck to the inside of the rotor.

11 Align the rotor slot with the Woodruff key. Place the rotor on the crankshaft.

12 The remainder of installation is the reverse of the removal steps.

22.5 Install the puller and use its center bolt to free the rotor . . .

22.6 . . . then pull off the puller and rotor together

Chapter 5 Electrical and ignition system 5-15

22.8 Check the inside of the rotor for damaged magnets and make sure nothing is stuck to them

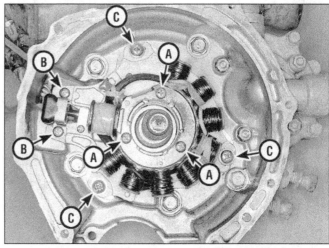

22.14a Stator coil screws (A), pulse coil screws (B) and stator plate screws (C) on a large-displacement 2-stroke

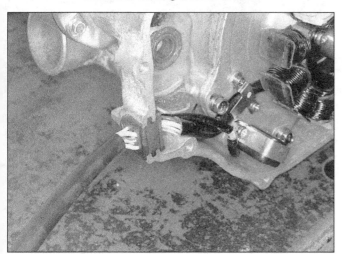

22.14b The harness grommet fits like this on 4-strokes

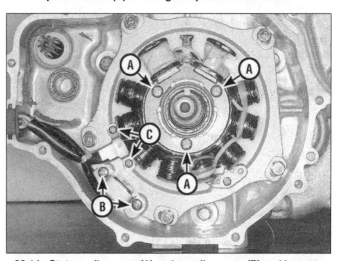

22.14c Stator coil screws (A), pulse coil screws (B) and harness retainer screws (C) on a 4-stroke

Stator coil and ignition pulse coil replacement

Refer to illustrations 22.14a through 22.14e

13 Remove the alternator rotor as described above.

14 Remove the stator coil screws and the pulse coil screws **(see illustrations)**, then remove the stator coils and pulse coil together (they're replaced as a unit).

15 Installation is the reverse of the removal steps.

22.14d The pulse coil is held down by this metal retainer

22.14e Remove the ground wire screw (arrow)

5-16   Chapter 5   Electrical and ignition system

22.17a One of the timing marks on this large-displacement 2-stroke is hidden by a stator plate screw

22.17b Here are the 4-stroke timing marks

Stator plate replacement

Refer to illustrations 22.17a through 22.17e

16   Remove the stator coil and ignition pulse coil as described above.

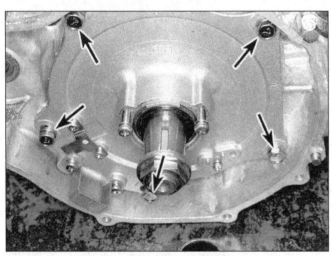

22.17c Remove the rotor nut and washer and the stator plate bolts (arrows)

17   Note the alignment of the ignition timing marks, then remove the stator plate screws or bolts **(see illustration 22.14a and the accompanying illustrations)**.

18   Check the seal and bushing in the stator plate for wear or damage. If the seal needs to be replaced, pry it out and press in a new one with a socket or bearing driver. If the bushing needs to be replaced, press it out and press a new one in.

23   Regulator/rectifier - check and replacement

Check

1   The regulator/rectifier is tested by process of elimination (when all other possible causes of charging system failure have been checked and eliminated, the rectifier/regulator is defective). Since it's easy to miss a problem, it's a good idea to have the charging system tested by a Polaris dealer or substitute a known good unit before buying a new one.

Replacement

Refer to illustration 23.2

2   Disconnect the rectifier/regulator electrical connector **(see illustration)**. Remove the mounting screws and lift it off the frame.

3   Installation is the reverse of the removal steps.

22.17d Remove the stator plate and its O-ring; DO NOT forget to reinstall the oil passage O-ring (arrow)

22.17e Check the seal for leaks and the bushing for wear

Chapter 5 Electrical and ignition system

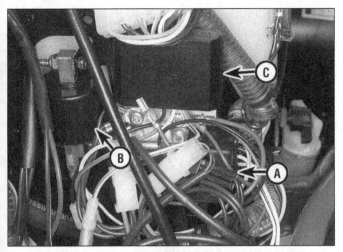

23.2 Magnum rectifier/regulator (A), speed limiter (B) and ignition coil (A)

24.1a Unscrew the speedometer cable nut from the speedometer . . .

24 Speedometer cable and drive - removal and installation

Refer to illustrations 24.1a, 24.1b, 24.2a and 24.2b

1   To disconnect the speedometer cable, unscrew its fittings from the speedometer and the drive unit on the transmission **(see illustrations)**.
2   To remove the drive unit, unscrew its mounting bolts and nut and take it off the transmission **(see illustrations)**.

25 Ignition system - check

Refer to illustrations 25.5 and 25.13
Warning: *Because of the very high voltage generated by the ignition system, extreme care should be taken when these checks are performed.*

1   If the ignition system is the suspected cause of poor engine performance or failure to start, a number of checks can be made to isolate the problem.
2   Make sure the ignition kill (stop) switch is in the Run or On position.

Engine will not start

3   Refer to Chapter 1 and disconnect the spark plug wire. Connect the wire to a spare spark plug and lay the plug on the engine with the threads contacting the engine. If necessary, hold the spark plug with an insulated tool. Crank the engine over and make sure a well-defined, blue spark occurs between the spark plug electrodes. **Warning:** *Don't*

24.1b . . . and from the drive unit on the transmission

remove the spark plug from the engine to perform this check - atomized fuel being pumped out of the open spark plug hole could ignite, causing severe injury!

4   If no spark occurs, the following checks should be made:
5   Unscrew the spark plug cap from the plug wire and check the cap resistance with an ohmmeter **(see illustration)**. If the resistance is not as listed in this Chapter's Specifications, replace it with a new one.

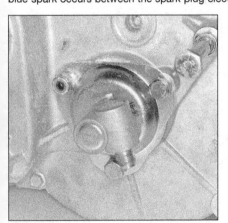

24.2a Remove the nut and bolts . . .

24.2b . . . and take off the drive unit and insulator; don't lose the small drive piece (arrow)

25.5 Unscrew the plug cap from the plug wire and measure its resistance with an ohmmeter

Chapter 5 Electrical and ignition system

25.13 A simple spark gap testing fixture can be made from a block of wood, two nails, a large alligator clip, a screw and a piece of wire

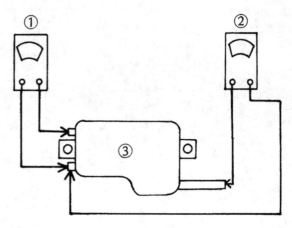

26.4 Ignition coil test

1. Measure primary winding resistance
2. Measure secondary winding resistance
3. Ignition coil

6 Make sure all electrical connectors are clean and tight. Check all wires for shorts, opens and correct installation.
7 Check the battery voltage with a voltmeter (except 1987 Cyclone). If the voltage is less than 12.46 volts, recharge the battery.
8 Check the circuit breaker and its connections (see Section 5).
9 Refer to Section 26 and check the ignition coil primary and secondary resistance.
10 Refer to Section 27 and check the pickup coil and source coil resistance.
11 If the preceding checks produce positive results but there is still no spark at the plug, refer to Section 28 for information on the CDI unit.

Engine starts but misfires

12 If the engine starts but misfires, make the following checks before deciding that the ignition system is at fault.
13 The ignition system must be able to produce a spark across a six millimeter (1/4-inch) gap (minimum). A simple test fixture **(see illustration)** can be constructed to make sure the minimum spark gap can be jumped. Make sure the fixture electrodes are positioned six millimeters apart.
14 Connect one of the spark plug wires to the protruding test fixture electrode, then attach the fixture's alligator clip to a good engine ground.
15 Crank the engine over with the key in the On position and see if well-defined, blue sparks occur between the test fixture electrodes. If the minimum spark gap test is positive, the ignition coil is functioning properly. If the spark will not jump the gap, or if it is weak (orange colored), refer to Steps 5 through 11 of this Section and perform the component checks described.

26 Ignition coil - check, removal and installation

Check

Refer to illustration 26.4

1 In order to determine conclusively that the ignition coil is defective, it should be tested by an authorized Polaris dealer service department which is equipped with the special electrical tester required for this check.
2 However, the coil can be checked visually (for cracks and other damage) and the primary and secondary coil resistances can be measured with an ohmmeter. If the coil is undamaged, and if the resistances are as specified, it is probably capable of proper operation.
3 To check the coil for physical damage, it must be removed (see Step 9). To check the resistance, remove the fuel tank (see Chapter 4), disconnect the primary circuit electrical connectors from the coil and remove the spark plug wire from the spark plug. Mark the locations of all wires before disconnecting them.
4 To check the coil primary resistance, attach one ohmmeter lead to the primary terminals and the other ohmmeter lead to the coil base **(see illustration)**.
5 Place the ohmmeter selector switch in the Rx1 position and compare the measured resistance to the value listed in this Chapter's Specifications.
6 If the coil primary resistance is as specified, check the coil secondary resistance by disconnecting the meter leads and attaching them between the spark plug wire terminal and the primary terminal **(see illustration 26.4)**.
7 Place the ohmmeter selector switch in the Rx100 position and compare the measured resistance to the values listed in this Chapter's Specifications.
8 If the resistances are not as specified, unscrew the spark plug cap from the plug wire and check the resistance between the primary terminal and the end of the spark plug wire. If it is now within specifications, the spark plug cap is bad. If it's still not as specified, the coil is probably defective and should be replaced with a new one.

Removal and installation

Refer to illustration 26.9

9 To remove the coil, refer to Chapter 8 and remove the seat and/or body panels as necessary for access, then disconnect the spark plug wire from the plug. Unplug the coil primary circuit electrical connector **(see illustration 23.2 and the accompanying illustration)**.

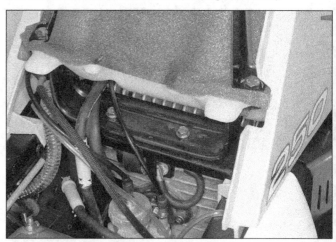

26.9 The Trail Boss 250 ignition coil is under the fuel tank bracket

Chapter 5 Electrical and ignition system

28.3 Here's a typical CDI unit

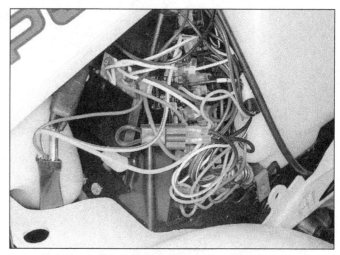

30.3 Here's a terminal board; see the wiring diagrams for terminal identification

10   Support the coil with one hand and remove the coil mounting bolts, then lift the coil out.
11   Installation is the reverse of removal.

27   CDI magneto - check, removal and installation

The pulse coil and rotor are checked as part of the alternator check procedure (see Section 22).

28   CDI unit - check, removal and installation

Check

1   If there's no spark, disconnect the black wire from the CDI module and test again. This will prevent the engine kill switch and reverse speed limiter from affecting the ignition circuit. If there's now spark, test the kill switch (Section 13). If it's good, have the speed limiter tested by a Polaris dealer or other qualified shop.
2   Beyond this test, the CDI unit is checked by process of elimination (when all other possible causes have been checked and eliminated, the CDI unit is at fault). Because the CDI unit is expensive and can't be returned once purchased, consider having a Polaris dealer test the system before you buy a new CDI unit.

Replacement

Refer to illustration 28.3

3   Locate the CDI unit (it's generally beneath the rear fender). Disconnect the electrical connector from the CDI unit **(see illustration)**. Remove its mounting screws with a Torx bit and take it off the frame.
4   Installation is the reverse of the removal steps.

29   Ignition timing - general information and check

General information

1   Ignition timing need be checked only if you're troubleshooting a problem such as loss of power. Since ignition timing isn't routinely adjusted and since none of the ignition system parts is subject to mechanical wear, there's no need for regular checks.

2   The ignition timing is checked with the engine running at the idle speed listed in the Chapter 1 Specifications. Inexpensive neon timing lights should be adequate in theory, but in practice may produce such dim pulses that the timing marks are hard to see. If possible, one of the more precise xenon timing lights should be used, powered by an external source of the appropriate voltage. **Note:** *Don't use the vehicle's own battery as an incorrect reading may result from stray impulses within the electrical system.*

Check

3   Warm the engine to normal operating temperature, make sure the transmission is in Neutral, then shut the engine off.
4   Refer to Valve clearance - check and adjustment in Chapter 1 and remove the timing hole plug.
5   Connect the timing light and a tune-up tachometer to the engine, following manufacturer's instructions.
6   Start the engine. Make sure it idles at the speed listed in the Chapter 1 Specifications. Adjust if necessary.
7   Point the timing light into the timing window. At idle, the timing mark in the timing window should align with the line next to the F mark on the alternator rotor.
8   If the timing is incorrect, check timing mark alignment on the stator plate and crankcase **(see illustration 22.17a or 22.17b)**.
9   When the check is complete, disconnect the test equipment. Refer to *Valve clearance - check and adjustment* in Chapter 1 and install the timing hole plug.

30   Wiring diagrams

Refer to illustration 30.3

Prior to troubleshooting a circuit, check the fuses to make sure they're in good condition. Make sure the battery is fully charged and check the cable connections.

When checking a circuit, make sure all connectors are clean, with no broken or loose terminals or wires. When disconnecting a connector, don't pull on the wires - pull only on the connector housings themselves.

The wires on all models run to a terminal board mounted at the front of the vehicle **(see illustration)**. This is a convenient place to test circuits, since the connectors are easy to reach. Terminal board connectors and wire colors are illustrated in the wiring diagrams.

Notes

Chapter 6
Steering, suspension and final drive

Contents

	Section
Drive chain and sprockets - removal, cleaning, inspection and installation	14
Driveaxles - boot replacement and U-joint overhaul	17
Driveaxles - removal and installation	16
Eccentrics - removal, inspection and installation	15
Front forks (3-wheel models) - disassembly, inspection and reassembly	8
Front forks (3-wheel models) - removal and installation	7
Front gearcase and driveshaft (shaft drive models) - removal and installation	18
Front struts - ball-joint replacement	5
Front struts - removal and installation	4
Front struts - spring and cartridge replacement	6

	Section
General information	1
Handlebar - removal, inspection and installation	2
Rear axle - removal, inspection and installation	19
Rear shock absorbers - removal and installation	9
Rear stabilizer struts (shaft drive models) - removal and installation	11
Steering shaft and tie-rods - removal, inspection and installation	3
Suspension - check	See Chapter 1
Suspension arms - removal, inspection and installation	10
Swingarm - removal and installation	13
Swingarm bearings - check	12

Specifications

Torque specifications (replace all self-locking nuts with new ones)

Handlebar bracket bolts	11 to 12 ft-lbs
Tie-rod ends	
Tie-rod end bolts/nuts	25-28 ft-lbs
Tie-rod end castellated nuts	23 to 24 ft-lbs
Tie-rod locknuts	12 to 14 ft-lbs
Swingarm pivot bolts/nuts	55 ft-lbs
Rear axle and 6X6 center axle nuts	150 ft-lbs (1)
Front suspension arm pivot bolts	30 ft-lbs
Strut top nuts	15 ft-lbs
Strut pinch bolts	15 ft-lbs
Rear shock absorber bolts	
Chain drive (except 6X6)	25 ft-lbs
Lower bolt (6X6)	18 ft-lbs
Upper nut (6X6)	16 ft-lbs
Shaft drive	30 ft-lbs
Ball-joint Allen screws	8 ft-lbs (1)

Chapter 6 Steering, suspension and final drive

Torque specifications (replace all self-locking nuts with new ones)

Center chain drive sprocket bolt (on transmission shaft)	17 ft-lbs
Center chain driven sprockets (on eccentric)	30 ft-lbs
Rear chain sprocket (including 6X6)	35 ft-lbs
Center and front eccentric pinch bolts	50 ft-lbs (2)
Center swingarm eccentric bolts (6X6)	55 ft-lbs (2)
Rear swingarm eccentric bolts (with trailer hitch)	60 ft-lbs (2)
Shaft drive rear suspension	
Upper arm pivot bolts (inner and outer)	35 ft-lbs
Lower arm pivot bolts (inner and outer)	30 ft-lbs
Stabilizer bracket to frame	17 ft-lbs
Stabilizer end link nut and bolt	17 ft-lbs
Stabilizer strut bolts	Not specified
Wheel nuts	
Front	15 ft-lbs
Rear except Sportsman 500	50 ft-lbs
Sportsman 500 rear	15 ft-lbs
Rear hub nuts	
All except Sportsman 500	80 ft-lbs
Sportsman 500	100 ft-lbs

* Use Loctite 242 or equivalent on the threads.
** Specified torque is in addition to the rolling torque of the self-locking nuts.

1 General information

The steering system consists of a handlebar, steering shaft and two tie-rods. When the handlebar turns, it rotates the steering shaft, which pushes or pulls the tie-rods to turn the suspension struts and with them, the wheel hubs.

The front suspension on 4-wheel and 6-wheel models consists of a single lower control arm on each side of the vehicle, connected to the frame by a strut (a combined shock absorber and coil spring that supports the wheel laterally). The lower end of the strut contains the suspension ball-joint. The upper end contains the damper cartridge and coil spring.

The rear suspension on chain drive models consists of a single coil-over shock absorber and a steel swingarm.

The rear suspension on shaft drive models is independent, consisting of upper and lower control arms and a coil-over shock absorber on each side. A torsion-type stabilizer bar connects the two sides of the suspension. Hubs mounted between the upper and lower control arms support the rear wheels.

Final drive on chain drive models is by one or more chains. On all models, a chain from the transmission output shaft turns a sprocket on the rear axle (4-wheel models) or center axle (6-wheel models). On 6-wheel models, a second chain connects the center axle to the rear axle. On chain-drive vehicles with all-wheel drive, a drive chain connects to a shaft between the front wheels. The shaft is connected to the front wheel hubs by a pair of driveaxles.

Final drive on shaft drive models is by driveaxles at front and rear. The rear driveaxles are mounted in the rear of the transmission and are turned by the transmission gears. The front driveaxles are mounted between the wheel hubs and a front drive gearbox. The front drive gearbox is connected to the transmission by a driveshaft located on the right side of the machine.

2 Handlebar - removal, inspection and installation

1 The handlebar is a one-piece tube. The tube fits into a bracket, which is attached to the steering shaft. If the handlebar must be removed for access to other components, such as the steering shaft, simply disconnect the electrical connectors, remove the bolts and lift the

2.3 Remove the screws and lift off the handlebar cover

handlebar off the bracket. It's not necessary to detach the cables or wires from the handlebar, but it is a good idea to support the assembly with a piece of wire or rope, to avoid unnecessary strain on the cables.

2 If the handlebar is to be removed completely, refer to Chapter 4 for the throttle housing removal procedure and Chapter 5 for the switch removal procedure.

Removal

Refer to illustrations 2.3 and 2.5

3 If the vehicle has a handlebar trim cover, remove the screws and lift it off **(see illustration)**. On some models, you'll need to remove the upper half of the headlight housing (see Chapter 8).

4 Disconnect the electrical connectors for the handlebar switches.

5 Remove the handlebar bracket bolts and lift off the upper half of the bracket **(see illustration)**. Separate the upper and lower halves and take the handlebar out of the bracket.

6 If necessary, remove the nuts and take the lower half of the bracket off the steering shaft.

Chapter 6 Steering, suspension and final drive

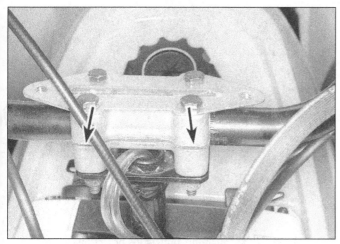

2.5 When you install the handlebar, make sure the gap (arrows) is even at front and rear of the bracket

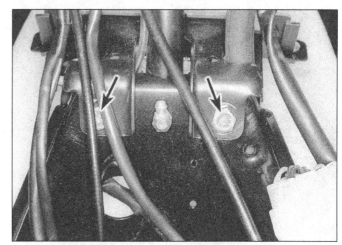

3.3a Remove the bolts (arrows) and bracket . . .

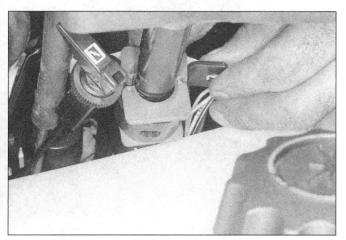

3.3b . . . and separate the halves of the steering shaft upper bushing

3.5 Remove the cotter pin and nut, metal, rubber and metal washers and the O-ring

Inspection

7   Check the handlebar and bracket for cracks and distortion and replace them if any undesirable conditions are found.
8   Clean the knurled area of the handlebar of any corrosion or dirt.

Installation

9   Installation is the reverse of the removal steps, with the following addition: Tighten the bracket bolts to the torques listed in this Chapter's Specifications, leaving a gap between the upper bracket half and lower bracket half. The gaps should be even at front and rear.

3   **Steering shaft and tie-rods - removal, inspection and installation**

Steering shaft

Removal

Refer to illustrations 3.3a, 3.3b, 3.5 and 3.6

1   Remove the handlebar and its brackets (Section 2).
2   Remove body panels as necessary for access (see Chapter 8).
3   Unbolt the steering shaft bracket from the frame to expose the upper bushing and seals **(see illustrations)**.
4   Refer to Section 5 and disconnect the inner ends of the tie-rods from the steering shaft.
5   Remove the cotter pin, nut, metal washer, rubber washer, metal

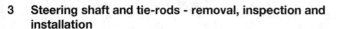

3.6 The steering shaft lower bushing plate is secured by rivets (outer arrows) and has a metal washer and O-ring on top (center arrow)

washer and O-ring from the bottom of the steering shaft **(see illustration)**.
6   Lift the steering shaft from the vehicle and remove the metal washer and O-ring that sit on top of the lower bushing plate **(see illustration)**.

3.12 Remove the cotter pin and nut and separate the tie rod

3.13a At this inner end, both studs point downward

3.13b This design uses bolts and self-locking nuts, with the tie-rod end below the steering arm

3.13c On some models the tie-rod end is between the upper and lower steering arms . . .

Inspection

7 Clean all the parts with solvent and dry them thoroughly, using compressed air, if available.

8 Check the steering shaft bushings and seals or O-rings for wear, deterioration or damage. Replace them if there's any doubt about their condition. If the lower bushing plate needs to be replaced, drill out its rivets **(see illustration 3.6)**.

9 Check the steering shaft and its integral steering arm for bending or other signs of damage. Do not attempt to repair any steering components. Replace them with new parts if defects are found.

Installation

10 Installation is the reverse of removal, with the following additions:
  a) *Lubricate the steering shaft bushings through the grease nipples (see Chapter 1).*
  b) *Use new locknuts and cotter pins and tighten all fasteners to the torques listed in this Chapter's Specifications.*

Steering tie-rods

Removal

Refer to illustrations 3.12, 3.13a, 3.13b, 3.13c and 3.13d

11 Turn the handlebar to the left or right to provide access to the tie rod ends.

12 Remove the cotter pin and nut from the outer end of the tie rod **(see illustration)**. Separate the tie rods ends from the knuckle arm. If they won't come easily, use a small tie-rod puller **(see illustration 4.6b)**.

13 The inner tie rod ends are secured by nuts on studs or by nuts and bolts, depending on model **(see illustrations)**. Remove the cotter pin and nut. If they're secured by nuts and bolts, push the bolts out; if they're secured by nuts on studs, use a small tie-rod puller **(see illustration 4.6b)**.

14 Repeat the above steps to remove the other tie-rod.

Inspection

15 Check the tie rod ends for looseness or damaged dust boots. If any problems are found, mark the threads so you can screw the new tie-rod end to the same position. Loosen the locknut, unscrew the

3.13d . . . and on some models, there's a spacer (arrow) between the upper and lower steering arms

Chapter 6 Steering, suspension and final drive

4.2 Remove the nut, washers and bushing at the upper end of the strut

4.4 Remove the driveaxle CV joint protector

tie-rod end and screw on the new one.
16 Adjust toe-in (see Chapter 1), then tighten the locknuts.

Installation

17 Installation is the reverse of the removal steps. On 1988 and later models, be sure to install the studs or bolts pointing up or down as listed below. Be sure to place the tie-rod ends above or below the steering shaft arm, as follows:
  a) 1988 through 1995 2X4 (except Magnum): inner end studs pointing down (nut on bottom); outer end studs pointing up.
  b) Magnum 2X4: inner and outer studs facing down.
  c) 1993-on 4X4 and 6X6 (except late 1995 Xplorer and Scrambler): inner bolt heads on bottom; left inner end beneath lower steering shaft arm with spacer between upper and lower steering shaft arms; right inner end between upper and lower steering shaft arms; outer studs pointing down.
  d) Late 1995 Xplorer and Scrambler: may be the same as "c" above; alternatively, both inner ends may be between the upper and lower steering shaft arms, with no spacer; outer studs pointing down on all models.
  e) 1996-on 2X4 (except Trail Boss 250): inner bolts pointing down (nut on bottom); inner end between steering shaft arms; outer stud pointing up.
  f) 1996-on Trail Boss 250: inner stud pointing down; outer stud pointing up.

18 Install the nuts and tighten them to the torque listed in this Chapter's Specifications. Secure the nuts with new cotter pins.
19 Turn the steering from full left lock to full right lock and back.

Carefully check the inner and outer tie-rod ends to make sure they're not hitting anything. **Warning:** *If the tie-rod ends hit anything during their travel, you may lose control of the vehicle. Do this step carefully. If you find a problem, fix it before riding the vehicle.*

4 Front struts - removal and installation

Removal

Refer to illustrations 4.2, 4.4, 4.6a and 4.6b

1 Securely block both rear wheels so the vehicle won't roll. Refer to Chapter 7 and remove the front wheels.
2 Support the outer ends of the front suspension arms with jackstands so they won't drop when the struts are removed. Remove the nut, washers and bushing at the upper end of the strut **(see illustration)**.
3 If you're working on a 1985 or 1986 model or a 1987 2wd model, remove the two mounting bolts at the lower end of the strut.
4 If you're working on an all-wheel drive model, remove the CV joint protector from the suspension arm **(see illustration)**. Follow the wiring harness from the hub coil to its electrical connector and disconnect it.
5 Detach the brake caliper from the strut and support it so it doesn't hang by the brake hose (see Chapter 7). On some models, the upper pinch bolt on the strut secures a brake hose clamp. In this case, remove the bolt.
6 Remove the ball-joint nut at the bottom of the strut **(see illustration)**. Carefully separate the ball-joint stud from the suspension arm. You can use a "pickle fork" type separator tool or a small automotive

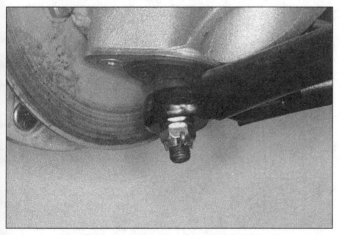

4.6a Remove the cotter pin and ball-joint nut

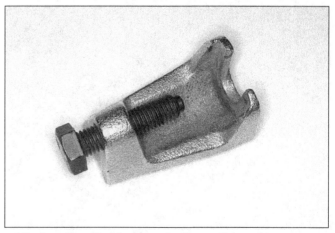

4.6b A small automotive tie-rod puller can be used to separate the ball-joint

5.3a Remove the Allen screws, metal plate and dust boot

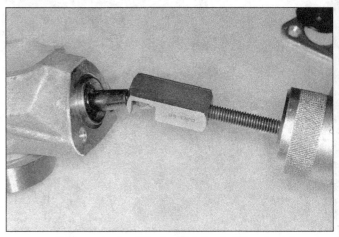

5.3b . . . and pull the ball-joint out of the strut

tie-rod separator **(see illustration)**.
7   If you're working on an all-wheel drive model, carefully separate the outer end of the driveaxle from the bearing in the bottom of the strut.
8   Separate the shock absorber from the steering knuckle and lift it out.
9   Check the shock absorber for signs of wear or damage such as oil leaks, bending, a weak spring and worn bushings. Replace both shock absorbers as a pair if any problems are found.

Installation

10 Installation is the reverse of the removal steps. Tighten the nuts and bolts to the torques listed in this Chapter's Specifications and the Chapter 7 Specifications.

5 Front struts - ball-joint replacement

Refer to illustration 5.3a and 5.3b
1   Remove the ball-joint nut and separate the suspension arm from the strut (Section 4).
2   If you're working on a 1985 or 1986 model or a 1987 2wd model, remove the ball-joint pinch bolt from the groove and take the ball-joint out.
3   If you're working on a 1986 4wd model or any 1988-on model, remove the Allen screws, plate and dust boot from the bottom of the strut **(see illustration)**. Pull the ball-joint out; if it won't come easily, hook it with a slide hammer and pull it out **(see illustration)**.
4   Check the ball-joint socket in the strut for wear or damage. If any problems are found, replace the strut.
5   Installation is the reverse of the removal steps, with the following additions:
   a) Tap the new ball-joint in with a hollow driver.
   b) Use non-permanent thread locking agent on the threads of the Allen screws.
   c) Tighten all fasteners to the torques listed in this Chapter's Specifications.
   d) Use a new cotter pin and position its open end toward the rear of the vehicle.

6 Front struts - spring and cartridge replacement

Warning: *Do not attempt this procedure without a spring compressor designed for the coil springs used in suspension struts. Make sure the spring compressor fits the strut you're working on and is securely installed. The coil spring is under considerable tension and could fly out, causing serious injury. Spring compressors that are safe for this use can be rented from tool outlets and auto parts stores.*
Refer to illustrations 6.3, 6.4, 6.5 and 6.6
1   Remove the strut from the vehicle (See Section 4).
2   Place a coil spring compressor on the spring and compress it enough to relieve the spring tension.
3   Place a wrench on the flats of the stud on the upper end of the strut **(see illustration)**.
4   Lift the pivot bushing away from the flats of the upper nut and unscrew it **(see illustration)**.

6.3 With a spring compressor securely installed, hold the flats on the end of the shaft . . .

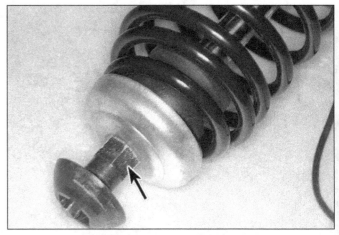

6.4 . . . lift up the pivot bushing and unscrew the nut

Chapter 6 Steering, suspension and final drive

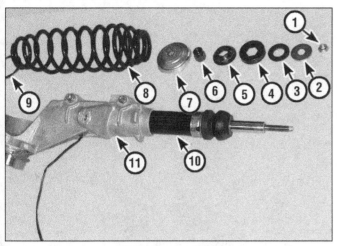

6.5 Front strut details

1	Upper nut	7	Spring retainer
2	Washer	8	Spring
3	Plastic washer	9	Spring seat
4	Upper bushing	10	Cartridge
5	Pivot bushing	11	Strut housing
6	Pivot bushing nut		

5 Slowly release the spring compressor until the spring tension is relieved. Take off the spring retainer, spring and base washer **(see illustration)**.
6 Remove the pinch bolts **(see illustration)**. Try to work the cartridge bolts out of the strut housing. If they won't come, clean the threads of the pinch bolt holes with a 3/8-inch X16 tap. Slip a flat washer into the gap where the upper pinch bolt was removed, then install the upper pinch bolt and tighten it against the washer just enough to open the strut housing slightly so the cartridge can be pulled out.
7 Assembly is the reverse of the disassembly steps, with the following addition: Use new self-locking nuts on the strut pinch bolts and at the top of the strut.

7 Front forks (3-wheel models) - removal and installation

Removal

1 Support the machine securely upright so it can't fall over during this procedure.
2 Refer to Chapter 7 and remove the front wheel.
3 If you plan to disassemble the forks, read the steps in Section 8 that involve loosening loosen the damper rod bolts. This can be done later, but it be easier while the springs and caps are still installed in the forks.
4 Unbolt the brake cable retainers from the fork legs.
5 Loosen the upper and lower triple clamp bolts.
6 Lower the fork leg out of the triple clamps, twisting it if necessary. Guide the boot breather hose out of the steering stem.

Installation

7 Slide each fork leg into the lower triple clamp.
8 Slide the fork legs up, installing the tops of the tubes into the upper triple clamp. Position the forks so the top of each tube is flush with the top surface of the upper triple clamp. Position each fork leg's boot breather tube in the steering stem.
9 Tighten the triple clamp bolts to the torque listed in this Chapter's Specifications.
10 The remainder of installation is the reverse of the removal steps.

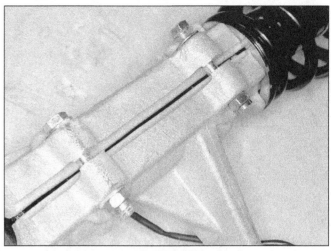

6.6 Remove the bolts, self-locking nuts and washers

8 Front forks (3-wheel models) - disassembly, inspection and reassembly

Disassembly

Refer to illustrations 8.2, 8.5, 8.6, 8.7 8.8 and 8.9

1 Remove the forks following the procedure in Section 3. Work on one fork at a time to prevent mixing up the parts.
2 Loosen the damper rod bolt at the bottom of the fork with an Allen bolt bit. The bolt is threaded into the damper rod inside the fork and secured with Loctite. For this reason, it has a tendency to spin the damper rod, instead of loosening, when you try to unscrew it. An air wrench is the easiest way to loosen the bolt **(see illustration)**. If you don't have one, loosen the bolt now while the cap is still on the fork. The tension of the fork spring pressing against the damper rod may keep it from spinning.
3 If there's a rubber cap on top of the fork, pull it off. Remove the plug from inside the top of the fork. The plug is secured by a wire ring that fits in a groove inside the top of the fork. There's a good chance that you won't be able to see the wire ring because of built-up corrosion or dirt. If you can't see it, clean the recess inside the top of the fork thoroughly.
4 To remove the plug, press it into the fork (against the tension of the fork spring) with a Phillips screwdriver, punch, or similar tool. When it's moved away from the wire ring, pry the wire ring out with a small

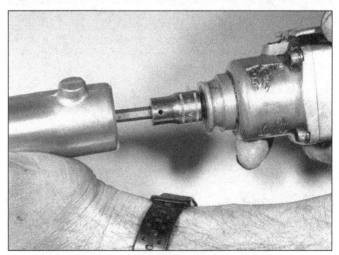

8.2 An air wrench is the easiest way to loosen the damper rod bolt

6-8 Chapter 6 Steering, suspension and final drive

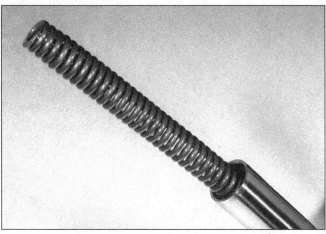

8.5 The tightly wound coils of the spring go on top

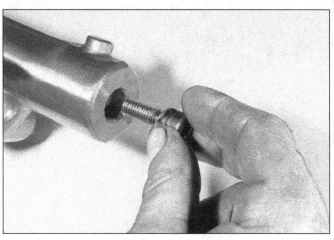

8.6 Unscrew the damper rod bolt; use a new copper washer on reassembly

8.7 Let the oil drain; pump the fork tube up and down to expel the oil

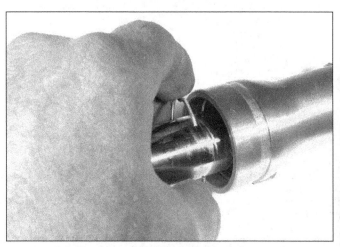

8.8 Pry the retaining ring out of its groove and slide it off the inner fork tube

screwdriver or pointed tool. Release the spring tension and let the plug out of the fork.
5   Pull the spring out of the fork **(see illustration)**. Invert the fork over a drain pan and pour out the oil.
6   Unscrew the damper rod bolt from the bottom of the fork **(see illustration)**.
7   Let the oil drain from the damper rod hole **(see illustration)**. Pump the fork tubes in and out to expel any remaining oil.
8   Remove the fork boot and pry the retaining ring from its groove in the lower fork tube **(see illustration)**. Yank the tubes apart; f they don't come on the first try, slide them together and yank them apart several times until they separate.
9   Pry the oil seal out of the fork tube without scratching its bore in the fork tube **(see illustration)**.

Inspection

10   Clean all parts in solvent and blow them dry with compressed air, if available. Check the inner and outer fork tubes and the damper rod for score marks, scratches, flaking of the chrome and excessive or abnormal wear. Look for dents in the tubes and replace them if any are found. Check the fork seal seat for nicks, gouges and scratches. If damage is evident, leaks will occur around the seal-to-outer tube junction. Replace worn or defective parts with new ones.
11   Have the fork inner tube checked for runout at a dealer service department or other repair shop. **Warning:** *If the tube is bent, it should be replaced with a new one. Don't try to straighten it.*

12   Check the fork spring for cracks or other damage. Measure its free length and compare it to the value listed in this Chapter's Specifications If it's defective or sagged, replace both fork springs with new ones. Never replace only one spring.
13   Check the O-ring on the plug at the top of the fork for wear or damage. If it has been leaking or you aren't sure of its condition, replace it.

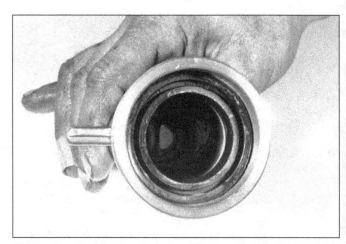

8.9 Pry the oil seal out of its bore without scratching the fork tube

Chapter 6 Steering, suspension and final drive

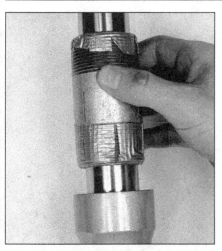

8.14a A taped section of pipe can be used as a seal driver

8.14b Make sure the retaining ring seats in its groove

8.14c Pour the specified amount of oil into the fork

Reassembly

Refer to illustrations 8.14a, 8.14b and 8.14c

14 Assembly is the reverse of the disassembly steps, with the following additions:

a) Insert the inner fork tube/damper rod assembly into the outer fork tube until the Allen-head bolt (with copper washer) can be threaded into the damper rod from the lower end of the outer tube **(see illustration 8.6)**. **Note:** *Apply Loctite 680 or equivalent to the threads of the bolt. Temporarily install the fork spring and cap to place tension on the damper rod so it won't spin inside the fork tube while you tighten the Allen bolt. Tighten the Allen bolt securely, then remove the fork cap and spring.*

b) Lubricate the lips and outer diameter of the fork seal with the recommended fork oil (see this Chapter's Specifications). Slide the seal down the inner tube with the lips facing down. Drive the seal into position with a fork seal driver. If you don't have access to one of these, it is recommended that you take the fork to a Polaris dealer or other repair shop for seal installation. You can also make a substitute tool **(see illustration)**. *If you're very careful, the seal can be driven in with a hammer and drift punch. Work around the circumference of the seal, tapping gently on the outer edge of the seal until it's seated. Be careful - if you distort the seal, you'll have to disassemble the fork and end up taking it to a dealer anyway!*

c) Make sure the retaining ring is completely seated in its groove **(see illustration)**.
d) Compress the fork fully and add the recommended type and quantity of fork oil listed in this Chapter's Specifications **(see illustration)**. *Measure the fork oil level from the top of the fork tube. If necessary, add or remove oil to bring it to the proper level.*
e) Install the fork spring with its closely-wound coils facing up. Install the fork top plug and wire ring.

15 Install the fork, following the procedure outlined in Section 3.

9 Rear shock absorbers - removal and installation

Removal

Refer to illustrations 9.2, 9.3. 9.4 and 9.5

1 Securely block both rear wheels so the vehicle won't roll. Loosen the rear wheel nuts with the tires still on the ground, then jack up the rear end, support it securely on jackstands and remove the rear wheels.
2 Support the swingarm (chain drive models) or rear hub (shaft drive models) with a jack **(see illustration)**.
3 Remove the nut and pivot bolt from the lower end of the shock **(see illustration)**.
4 On all except 6X6 models, remove the nut and pivot bolt from the

9.2 Support the suspension with a jack

9.3 Remove the nut and bolt at the bottom of the shock . . .

6-10 Chapter 6 Steering, suspension and final drive

9.4 ... and at the top

9.5 Remove the nut, washer and bushing at the top of the shock

upper end of the shock **(see illustration)**.
5   If you're working on a 6X6 model, remove the nut, washer and bushing from the shock absorber upper stud **(see illustration)**.

Installation

6   Installation is the reverse of the removal steps. Tighten all fasteners to the torque listed in this Chapter's Specifications.

10 Suspension arms - removal, inspection and installation

Front suspension arms
Removal
Refer to illustrations 10.3a and 10.3b
1   Securely block both rear wheels so the vehicle won't roll. Loosen the front wheel nuts with the tires still on the ground, then jack up the front end, support it securely on jackstands and remove the front wheels.
2   Refer to Section 3 and detach the outer end of the suspension arm from the suspension ball-joint. If you're working on a 4wd or 6wd model, remove the CV joint shield.
3   Remove the pivot bolts and nuts and take the suspension arm out **(see illustrations)**.

Inspection
4   Check the suspension arm(s) for bending, cracks or other damage. Replace damaged parts. Don't attempt to straighten them.
5   Check the rubber bushings at the inner end of the suspension arm for cracks, deterioration or wear of the metal insert. Check the pivot bolts for wear as well. Replace the bushings and pivot bolts if any problems are visible.
6   Check the ball-joint boot for cracks or deterioration. Twist and rotate the threaded stud. It should move easily, without roughness or looseness. If the boot or stud show any problems, refer to Section 3 and replace the ball-joint.

Installation
7   Installation is the reverse of the removal steps, with the following addition: Tighten the nuts and bolts slightly while the vehicle is jacked up, then tighten them to the torque listed in this Chapter's Specifications while the vehicle's weight is resting on the wheels.

Rear suspension arms (shaft drive models)
Removal
Refer to illustrations 10.10 and 10.11
8   Securely block both front wheels so the vehicle won't roll. Loosen the rear wheel nuts with the tires still on the ground, then jack up the rear end, support it securely on jackstands and remove the front wheels.
9   Remove the rear shock absorber (Section 9).

10.3a Remove the bolt and lockwasher (left arrow), bushing and plastic washer (right arrows)

10.3b The arrangement is the same at both ends of the suspension arm (arrows)

Chapter 6 Steering, suspension and final drive

10.10 Rear suspension mounting points (shaft drive models)
A Stabilizer strut lower mounting bolt
B Upper suspension arm-to-hub carrier bolt
C Lower suspension arm-to-hub carrier bolt (rear bolt shown)

10.11 Remove the pivot bolts at the inner ends of the rear suspension arms (shaft drive models)

10   At the outer end of the suspension arms, remove the through-bolt that attaches the upper arm to the hub carrier **(see illustration)**. Unbolt the lower end of the stabilizer bar strut from the upper arm, then remove the two bolts that attach the lower arm to the hub carrier.
11   At the inner ends of the suspension arms, remove the pivot bolts and nuts that attach the upper arm to the stabilizer bar bracket and frame bracket **(see illustration)**. Remove the front and rear pivot bolts from the lower arm, then take the arms out.
12   Inspect the suspension arms and bushing as described in Steps 4 and 5 above.

Installation

13   Installation is the reverse of the removal steps, with the following additions:

  a) The heads of all the lower arm bolts, at inner and outer ends, face the rear of the vehicle.
  b) Tighten the nuts and bolts slightly while the vehicle is jacked up, then tighten them to the torque listed in this Chapter's Specifications while the vehicle's weight is resting on the wheels.

11 Rear stabilizer struts (shaft drive models) - removal, inspection and installation

Removal

Refer to illustrations 11.2 and 11.4

1   Securely block both front wheels so the vehicle won't roll. Loosen the rear wheel nuts with the tires still on the ground, then jack up the rear end, support it securely on jackstands and remove the rear wheels.
2   Note how the notches in the strut bushings are aligned **(see illustration)**.
3   Unbolt the lower end of the strut from the upper suspension arm **(see illustration 10.10)**.
4   Unbolt the upper end of the strut from the torsion link **(see illustration)**. Remove the nut and bolt from the torsion link and slide it off the shaft.

Installation

5   Installation is the reverse of the removal steps, with the following additions: Align the bushing notches facing the center of the vehicle **(see illustration 10.10)**. Tighten all nuts and bolts to the torque listed in this Chapter's Specifications.

11.2 The bushing notches face the center of the vehicle

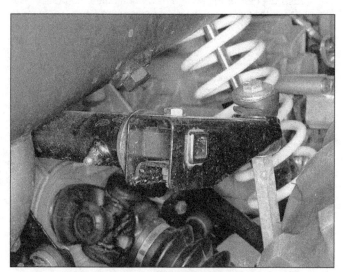

11.4 Remove the bolt from the stabilizer strut and the nut and bolt from the bracket

13.10 Bend the lockwasher tabs away from the bolt head and unscrew it

13.11a To hold the pivot bolt nut that's next to the transmission . . .

13.11b . . . slip an open-end wrench into the gap

13.13a The pivot bolt and bushing on the chain guard side are installed like this . . .

12 Swingarm bearings - check

1 Refer to Chapter 7 and remove the rear wheels, then refer to Section 9 and remove the rear shock absorber.
2 Grasp the rear of the swingarm with one hand and place your other hand at the junction of the swingarm and the frame. Try to move the rear of the swingarm from side-to-side. Any wear (play) in the bearings should be felt as movement between the swingarm and the frame at the front. The swingarm will actually be felt to move forward and backward at the front (not from side-to-side). If any play is noted, the bearings should be replaced with new ones (see Section 14).
3 Next, move the swingarm up and down through its full travel. It should move freely, without any binding or rough spots. If it does not move freely, refer to Section 13 for servicing procedures.

13 Swingarm - removal and installation

1 If the swingarm is being removed just for bearing replacement or driveshaft removal, the rear sprocket, eccentric and rear axle need not be removed from the swingarm.

Removal

Refer to illustrations 13.10, 13.11a, 13.11b, 13.13a, 13.13b and 13.13c

2 Remove the seat, rear cargo rack and rear fenders (see Chapter 8).
3 Remove the PVT, including the inner cover (see Chapter 2C).
4 Raise the rear end of the vehicle off the ground with a jack.

Support the vehicle securely so it can't be knocked over while it's jacked up.
5 If you're planning to remove the rear wheels, do it now (see Chapter 7).
6 Support the swingarm so it won't drop, then refer to Section 9 and detach the lower end of the shock absorber from the swingarm. If you're planning to remove the shock absorber completely, do it now.
7 If you're working on a 4X4 or 6X6 model, remove the auxiliary brake arm (see Chapter 7) and the center chain and sprockets (Section 14).
8 If you're planning to remove the rear axle, do it now (see Section 19).
9 If the vehicle has an output shaft brake caliper and disc, remove them (see Chapter 7).
10 Bend back the tabs from the pivot bolt on each side of the swingarm **(see illustration)**.
11 Hold the pivot bolt nuts with a wrench and unscrew the pivot bolt from each side of the swingarm **(see illustrations)**.
12 Pull the swingarm back and away from the vehicle. Remove the chain slider.
13 Check the pivot bushings in the swingarm for dryness or deterioration **(see illustrations)**. Lubricate or replace them as needed.

Installation

Refer to illustrations 13.16a and 13.16b

14 If the chain slider was removed from the swingarm, install it.
15 Lift the swingarm into position in the frame. Align the pivot bolt holes in the swingarm with those in the frame.

Chapter 6 Steering, suspension and final drive

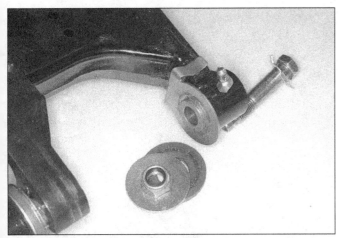

13.13b . . . here are the pivot bolt and bushing on the left side

13.13c Push the bushing out of the swingarm

13.16a Place the swingarm between the pivots and install the bolts loosely

full travel to seat the bearings and pivot bolts. Look at the grease fittings and make sure they're pointed so they'll be easily accessible for lubrication later.

18 The remainder of installation is the reverse of the removal steps.

14 Drive chain and sprockets - removal, cleaning, inspection and installation

1 Several drive chains are used. On 2X4 models (rear-wheel drive only), a single chain connects the transmission output shaft sprocket to the rear axle sprocket. On 4X6 and 6X6 models, a chain connects the transmission output shaft and the center axle double sprocket, and another chain connects the center axle double sprocket to the rear axle sprocket. On 4x4 and 6x6 models, a second sprocket on the transmission output shaft connects to a center eccentric; a second sprocket on the center eccentric connects to a sprocket on the front eccentric, which turns the front axle shafts.

Removal

Refer to illustrations 14.2, 14.4a, 14.4b, 14.5, 14.6a, 14.6b, 14.7a, 14.7b, 14.7c, 14.8, 14.9a and 14.9b

16 Loosely install both pivot bolts and their nuts to hold the swingarm in the frame **(see illustration)**. Don't forget the lockwashers; position one tab on each lockwasher over the locating tab on the frame **(see illustration)**. Make sure the bushing tabs are position correctly, then tighten the bolts and nuts to the torque listed in this Chapter's Specifications.

17 Raise and lower the swingarm several times, moving it through its

2 Original equipment chains are equipped with a master link **(see illustration)**.

3 Refer to the chain adjustment procedure in Chapter 1 to create slack in the chain. Remove chain guards as necessary for access.

4 To remove the chain, slide the clip off the pins, remove the plate

13.16b Bend a lockwasher tab against this tab on the swingarm pivot on the frame

14.2 The master link clip opening faces away from the direction of chain travel

14.4a Remove the clip from the plate . . .

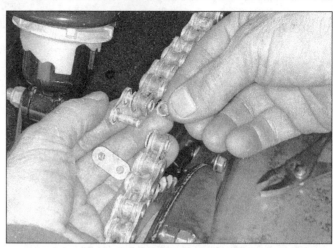

14.4b . . . then remove the plate, master link and O-rings

14.5 The front eccentric shaft sprocket

14.6a The center eccentric shaft sprocket (left side)

14.6b The sprocket protrusion faces the eccentric

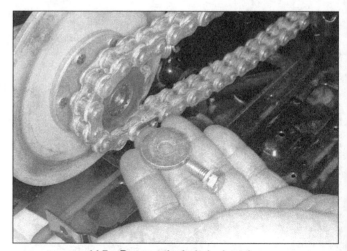

14.7a Remove the bolt, lockwasher and lockwasher

and pull the master link out **(see illustrations)**.
5   To remove the front sprocket, remove the driveaxle (Section 16). Remove the chain from the sprocket and slide the sprocket off the eccentric shaft **(see illustration)**.
6   To remove the front drive sprocket from the center eccentric, remove its bolt and washer **(see illustration)**. Slide the sprocket off the

eccentric shaft **(see illustration)**.
7   To remove the center chain sprockets, unbolt the outermost sprocket from the transmission output shaft **(see illustration)**. Remove the cotter pin, nut and washer from the eccentric shaft, then pull the chain and sprockets off as a unit **(see illustrations)**.
8   To remove the rear drive sprocket from the transmission output

Chapter 6 Steering, suspension and final drive   6-15

14.7b Pull out the cotter pin and remove the nut

14.7c The center chain and sprockets can be removed together

14.8 The rear chain's forward sprocket is mounted on the transmission output shaft

14.9a Remove the nuts and the sprocket cover . . .

14.9b . . . slide the spacer ring off the studs, then remove the nuts and sprocket

shaft, remove the center drive sprocket (Step 7) and the brake disc (Chapter 7). Remove the snap-ring and slide the sprocket off the output shaft **(see illustration)**.

9   To remove the rear axle sprocket, remove the right rear wheel and hub (see Chapter 7). Remove the sprocket cover and spacer ring, then remove the sprocket from the axle **(see illustrations)**.

14.10 Master link details

Inspection

Refer to illustration 14.10

10   Check the master link O-rings for damage or deterioration **(see illustration)**. Replace them if problems are found.

11   Clean the rest of the chain with hot soapy water and a soft brush. Don't use any solvent, as it may damage the O-rings or wash the lubricant out from between the O-rings.

12   Stretch the chain and measure the length of any 20 pitches (the spaces between pins), from the center of the first pin in the run to the center of the last pin. Standard is 12-1/2 inches; if any twenty-pitch length exceeds 12-7/8 inches, replace the chain.

Chapter 6 Steering, suspension and final drive

14.15 The protrusion on the output shaft sprocket faces the transmission

15.3 Remove the eccentric pinch bolt

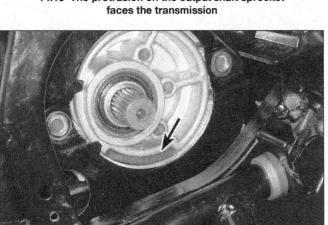

15.4a Turn the eccentric so its recess and plastic spacer (arrow) . . .

15.4b . . . align with the bracket like this, allowing removal

13 Lubricate the chain with Polaris O-ring chain lube or another lubricant recommended for O-ring chains.
14 Check the sprocket teeth for wear. If the sprockets are worn, replace them as a set with their chain. Don't install a worn chain on new sprockets or vice versa.

Installation

Refer to illustration 14.15

15 Installation is the reverse of the removal steps, with the following additions:
   a) *The protrusion on the transmission output shaft sprocket and the center eccentric's front drive sprocket face toward the center* **(see illustration 14.6b and the accompanying illustration)**.
   b) *Use new cotter pins.*
   c) *When installing the front sprocket, don't forget the wave washer between the sprocket and the driveaxle on the right side.*
   d) *Tighten the nuts and bolts to the torque listed in this Chapter's Specifications.*

15 Eccentrics - removal, inspection and installation

Front eccentric

Removal

Refer to illustrations 15.3, 15.4a, 15.4b, 15.5 and 15.6
1 Remove the front chain guard, drive chain and front sprockets

(Section 14).
2 Disconnect the driveaxles from the eccentric shaft (Section 16).
3 Working beneath the vehicle, remove the eccentric pinch bolt **(see illustration)**.
4 Turn the eccentric until its recess aligns with the bracket in such a way that the eccentric can be removed **(see illustrations)**. Remove the eccentric out the left-hand side of the bracket.
5 Locate the black plastic piece that fits in the recess of the eccen-

15.5 Remove the housing bolts . . .

Chapter 6 Steering, suspension and final drive

15.6 ... and any shims, then remove the housing

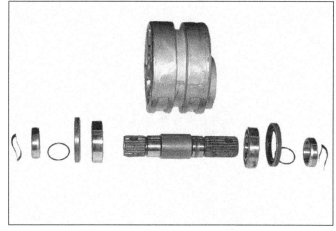

15.7 Front eccentric details

tric and eccentric housing and remove it **(see illustration)**.
6   If necessary, remove the eccentric housing through-bolts and any shims **(see illustration)**, then remove the housing.

Inspection
Refer to illustration 15.7

7   Remove the spacer and O-ring from each side of the eccentric **(see illustration)**.
8   Tap on one end of the shaft with a soft-faced hammer (or a hammer and block of wood) to drive out the shaft and one of the bearings. Take the bearing off the shaft.
9   Now insert the shaft into the other bearing from the far side, then tap on the shaft to drive out the remaining bearing.
10   Clean the bearings in solvent and rotate them with fingers. If they're rough, loose or noisy when rotated, replace them.
11   Replace the shaft if there's any step wear on the spines, if the bearing races are scored, or if the shaft is too corroded to clean up.
12   Place the bearings in a freezer for ten minutes or more to shrink them. Heat the left side of the shaft (the side with the grease nipple) to expand it, and tap in its bearing. Tap only on the outer race so the bearing won't be damaged.
13   Support the bearing inner race with a piece of pipe or a socket, then tap in the shaft from the opposite end, with the shorter splined end going into the bearing.
14   Fit another bearing over the longer splined end and start it into the eccentric. Tap the bearing the rest of the way in, using a socket or piece of pipe that seats on the bearing inner race only.
15   Tap in new seals, using a socket or piece of pipe the same diameter as the seal.
16   Install the O-rings and spacers. Don't lose track of the wave washers that fit over the splined shaft on each side.

Installation
17   Installation is the reverse of the removal steps, with the following additions: Tighten the housing bolts (if removed) securely. Tighten the pinch bolt to the torque listed in this Chapter's Specifications.

Center eccentric

Removal
18   Remove the sprocket from each side of the eccentric (Section 14).
19   Remove both of the pinch bolts completely. Push the eccentric out of the housing.

Inspection
20   Three types of center eccentric have been used; one with two ball bearings, one with four ball bearings and one with two tapered roller bearings. Disassembly, inspection and assembly of the ball bearing types is basically the same as for front eccentrics, described above. The main difference is that there is a spacer between the two inner bearings.

Tapered roller bearing type
Refer to illustrations 15.21a, 15.21b, 15.22a, 15.22b and 15.24

21   To disassemble the eccentric, pull out the shaft and remove the collar **(see illustrations)**.

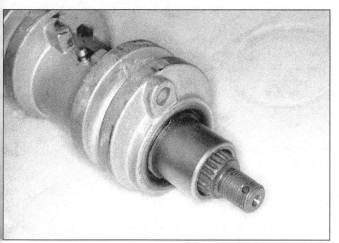

15.21a Tap out the shaft and remove the collar

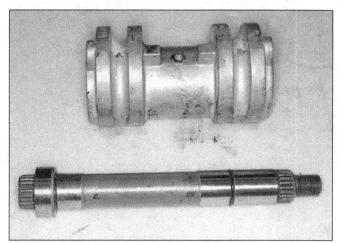

15.21b Center eccentric and shaft (tapered roller bearing design)

Chapter 6 Steering, suspension and final drive

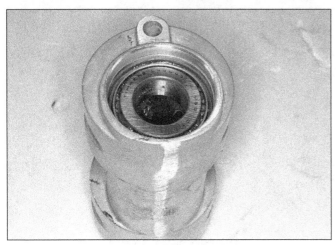
15.22a Pry out the seals to expose the bearings

15.22b Lift the bearing cones out of the eccentric

15.24 Insert a punch from the opposite side into the removal notch, against the bearing outer race

15.32 Remove the locknuts and pinch bolts

22 Pry out the seals and lift out the bearing cones **(see illustrations)**.
23 Clean all parts in solvent. Check the bearings for roughness, looseness or noise when rotated with fingers. Check the outer races of the bearings for scoring, roughness or corrosion. If any problems are found with either a bearing cone or its outer race, replace both of them as a set.
24 Insert a drift (a punch or large screwdriver) through the removal notch and place it against the outer race on one side **(see illustration)**. Tap on the drift to drive the outer race out.
25 Drive out the opposite outer race in the same way.
26 Drive in new outer races, seating them firmly against the shoulders inside the eccentric, with a socket or bearing driver.
27 Pack the bearing cones with the specified grease and lay them in the outer races.
28 Tap in new seals, using a socket or seal driver the same diameter as the seal.
29 Install the shaft and the spacer.

Installation
30 Installation is the reverse of the removal steps, with the following additions: Tighten all fasteners to the torques listed in this Chapter's Specifications. Adjust chain tension (see Chapter 1).

Rear eccentric
Removal
Refer to illustration 15.32
31 Remove the rear axle (Section 16).

32 Remove the axle pinch bolts **(see illustration)**. On some models, these secure the trailer hitch as well.
33 Push the eccentric out of the swingarm.

Inspection
34 Inspection is basically the same as for front eccentrics, described above.

Installation
35 Installation is the reverse of the removal steps, with the following additions: Tighten all fasteners to the torques listed in this Chapter's Specifications. Adjust chain tension (see Chapter 1).

16 Driveaxles - removal and installation

Front driveaxles
Removal
Refer to illustration 16.5
1 Securely block both rear wheels so the vehicle won't roll. Loosen the front wheel nuts with the tires still on the ground, then jack up the front end, support it securely on jackstands and remove the front wheels.
2 Remove the lower balljoint-to-suspension arm nut and separate the lower suspension arm from the strut (see Section 6).
3 Remove the wheel hub ((see Chapter 7).
4 Swing the strut outward and pull the end of the driveaxle out.

Chapter 6 Steering, suspension and final drive

16.5 Tap the roll pin out of the driveaxle and spindle with a punch

16.6 The roll pin fits flush with the driveaxle

16.11 Rear driveaxles are secured to the transmission by a bolt (arrow)

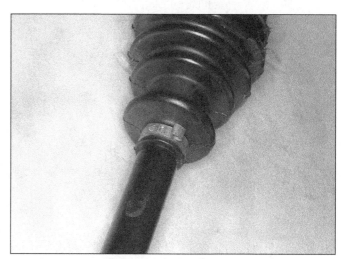

17.4 Release the small clamp . . .

5 The inner end of the driveaxle is secured to the shaft in the front eccentric or gearcase by a roll pin. Tap the pin out with a punch **(see illustration)**. With the outer end of the driveaxle free of the strut, grasp the joint at the inner end and pull the driveaxle off of the shaft.

Installation
Refer to illustration 16.6
6 Installation is the reverse of the removal steps. After installing the driveaxle on the shaft, tap in a new roll pin until it's flush with the driveaxle surface **(see illustration)**.

Rear driveaxles
Removal
Refer to illustration 16.11
7 Securely block both front wheels so the vehicle won't roll. Loosen the rear wheel nuts, but don't remove them. Remove the hub cover and the outer hub nut (see Chapter 7).
8 Unbolt the hub carrier from the upper and lower suspension arms **(see illustration 10.10)**.
9 Jack up the vehicle so the wheel is off the ground. **Warning:** *Support the vehicle securely so the next step doesn't knock it off the supports.*
10 Pull the wheel and hub carrier off the end of the driveaxle.
11 Remove the bolt that secures the inner end of the driveaxle to the transmission **(see illustration)**. Pull the drive axle out of the transmission.

Installation
12 Installation is the reverse of the removal steps, with the following additions: Tighten all fasteners to the torques listed in this Chapter's Specifications. Note that on some models, the driveaxle has a wide inner spline so the driveaxle can only be installed one way.

17 Driveaxles - boot replacement and U-joint overhaul

Boot replacement (front driveaxles)
Disassembly
Refer to illustrations 17.4, 17.5a and 17.5b
1 If you're working on a 1987 or 1988 model, the CV joint can't be taken apart to replace the boot. A split boot kit is available from Polaris dealers, so you can cut the old boot off and install the new one without taking the joint apart. Follow the instructions that come with the kit. The following steps apply to 1989 and later models.
2 Remove the driveaxle from the vehicle (see Section 16).
3 Mount the driveaxle in a vise. The jaws of the vise should be lined with wood or rags to prevent damage to the driveaxle.
4 Open the boot clamps with a small screwdriver and slide the clamps off the boot **(see illustration)**.

6-20 Chapter 6 Steering, suspension and final drive

17.5a ... and the large one, then slide the boot off the CV joint ...

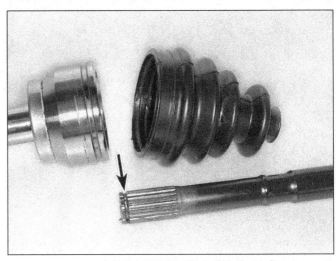

17.5b ... yank the CV joint off the clip (arrow) and remove the boot

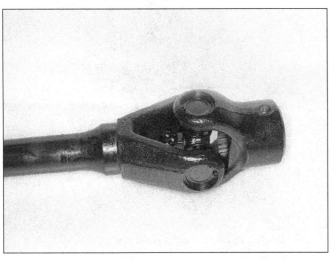

17.12 Remove the snap-rings from the U-joint

17.13a Tap the yoke down to expose a bearing ...

5   Slide the large boot off the axleshaft and pull sharply on the CV joint to separate it from the wire ring retainer **(see illustrations)**.

Inspection
6   Clean the components with solvent to remove all traces of grease. This is especially important if the vehicle has been operated with a damaged boot, as the grease may be contaminated. **Note:** *Because the joint cannot be disassembled, it is difficult to wash away all the old grease and to rid the bearing of solvent once it's clean. But it is imperative that the job be done thoroughly, so take your time and do it right.* Polaris recommends using its own brand of CV joint cleaner. Inspect the cage, balls and races for pitting, score marks, cracks and other signs of wear and damage. Shiny, polished spots are normal and will not adversely affect CV joint performance.
7   If the CV joint is worn or damaged, replace it.

Reassembly
8   Wrap the axleshaft splines with tape to avoid damaging the boot. Slide the small boot clamp and boot onto the axleshaft, then remove the tape **(see illustration 17.4)**.
9   The remainder of reassembly is the reverse of the disassembly steps, with the following addition: Pack the CV joint with the recommended grease listed in this Chapter's Specifications.
10 Install the driveaxle as described in Section 16.

Universal joints
11 Universal joints are used at the inner ends of front driveaxles, and at the inner and outer ends of rear driveaxles.

Disassembly
Refer to illustrations 17.12, 17.13a, 17.13b and 17.14
12 Place the driveaxle in a vise. Make alignment marks on the yoke, spider and driveaxle so the U-joint can be assembled in the same relationship. Remove the snap-rings form the U-joints **(see illustration)**.
13 Tap the end of the driveaxle downward to expose one of the bearings **(see illustration)**. Pull the bearing out, then remove the opposite bearing in the same manner **(see illustration)**.
14 Separate the yoke from the driveaxle and repeat the procedure to remove the spider from the other yoke **(see illustration)**.

Inspection
15 Thoroughly wash the components in clean solvent and blow them dry with compressed air, if available.
16 Inspect the bearings and their journals on the spider for signs of wear. If the bearings are damaged or worn, replace them.
17 If you're working on a rear driveaxle, the opposite bearing bores in each yoke must be in perfect alignment. To check, insert a rod of the same diameter through both bores of each yoke. If it won't go straight through, the yoke is twisted or bent. Replace it.

Chapter 6 Steering, suspension and final drive

17.13b ... and pull it out ...

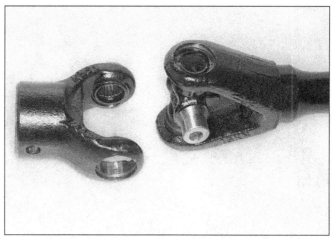

17.14 ... then remove the yoke and the other bearing

Reassembly

18 Assembly is the reverse of the disassembly steps, with the following additions:

a) *Pack the bearings with waterproof grease.*
b) *Align the marks made before disassembly.*

18 Front gearcase and driveshaft (shaft drive models) - removal and installation

Driveshaft

Removal

Refer to illustration 18.2

1 The driveshaft is secured to the front gearcase by a roll pin similar that used for driveaxles. To disconnect it from the gearcase, drive out the roll pin with a punch **(see illustration 3.27 in Chapter 1)**.

2 Pull the driveshaft rearward off the gearcase shaft. Then yank it forward off the shaft in the transmission front output housing **(see illustration)**.

Installation

3 Installation is the reverse of the removal steps, with the following addition: Use a new roll pin where the driveshaft joins the front gearcase.

Gearcase

Removal

4 Refer to Section 16 and detach both front driveaxles from the gearcase.
5 Disconnect the driveshaft from the gearcase as described above.
6 Disconnect the breather hose from the gearcase.
7 Remove the mounting bolts from underneath the gearcase.
8 Lift the gearcase and remove it to the right side of the vehicle.

Inspection

9 Check the oil seals at the driveaxle holes and the driveshaft hole for signs of leakage. Turn the shafts by hand to rotate the gears and listen for obvious signs of damage such as broken gear teeth.
10 Gearcase overhaul is a complicated procedure that requires several special tools, for which there are no readily available substitutes. If there's visible wear or damage, or if the rotation is rough or noisy, take it to a Polaris dealer for disassembly and further inspection.

Installation

11 Installation is the reverse of the removal steps, with the following additions:

18.2 Separate the driveshaft from the output shaft housing (arrow)

a) *Lubricate the splines of the shafts with moly-based multi-purpose grease.*
b) *Tighten the gearcase bolts to the torque listed in this Chapter's Specifications.*
c) *Fill the differential with the recommended type and amount of oil (see Chapter 1).*

19 Rear axle - removal, inspection and installation

Removal

Refer to illustrations 19.3a, 19.3b, 19.4 and 19.5

Note 1: *This procedure requires a pair of 1-3/4 inch open-end wrenches. The axle nuts are secured with Loctite and torqued to 150 ft-lbs, so the wrenches must fit exactly.*

Note 2: *Procedures for the 6x6 center axle are basically the same as for the rear axle.*

1 Securely block the front wheels so the vehicle won't roll. Loosen the rear wheel nuts with the vehicle on the ground. Jack up the rear end and support it securely, positioning the jackstands so they won't obstruct removal of the axle. The supports must be secure enough so the vehicle won't be knocked off of them while the axle is removed. Remove the rear wheels.
2 Remove the rear wheel hubs (see Chapter 7).
3 Loosen the axle locknut (away from the axle nut) with a 1-3/4-inch

6-22 Chapter 6 Steering, suspension and final drive

19.3a Remove the locknut and axle nut with a pair of 1-3/4 inch wrenches

19.3b Remove the washer

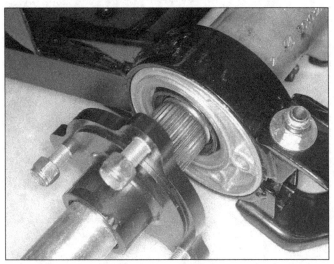

19.4 You can slide off the sprocket hub or leave it installed

19.5 Tap the axle out from right to left

wrench **(see illustration)**. Unscrew the locknut, then the axle nut and take them off the axle and remove the washer **(see illustration)**.
4 The drive chain, sprocket and sprocket hub can be left where they are while the axle is driven out - just be prepared to support them as the axle passes through the sprocket hub **(see illustration)**. If you do plan to remove the sprocket and hub, it will be easier to undo the nuts now while the axle splines are still supporting the hub and keeping it from turning.
5 Clean any foreign material from the right side of the axle so it won't be pulled into the eccentric during removal. Install the right wheel hub and thread the nut on until it's flush with the end of the axle to protect the threads. Tap on the right end of the axle with a soft faced hammer to free it, then remove the hub and pull the axle out of the eccentric **(see illustration)**.

Inspection

Refer to illustrations 19.8a and 19.8b
6 Check the axle for obvious damage, such as step wear of the splines or bending, and replace it as necessary.
7 Place the axle in V-blocks and set up a dial indicator to contact each of the outer ends in turn. Rotate the axle and check for bending. If the axle is bent, replace it.
8 Check the locating collar on the axle for wear or damage **(see illustration)**. If problems are found, tap the collar away from the circlip, then remove the circlip **(see illustration)**. Slide the locating collar off the axle, slide a new one on and install a new circlip. Then tap the collar back over the circlip.

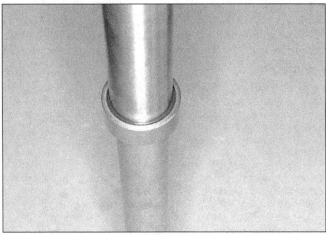

19.8a The locating collar fits over its retainer like this

Chapter 6 Steering, suspension and final drive

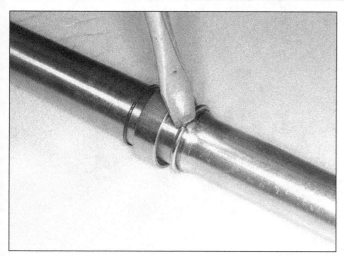

19.8b Slide the collar along the shaft and pry the retainer off

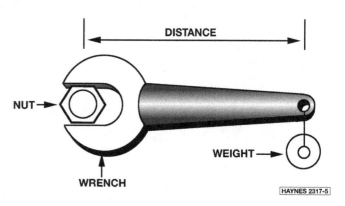

19.11 Length in feet times weight in pounds equals foot-pounds of torque

Installation

Refer to illustration 19.11

9   Lubricate the axle splines and the lip of the eccentric oil seal with multipurpose grease. Install the axle from the left side of the vehicle. Place the left hub on the axle and thread the hub nut on until it's flush with the end of the axle shaft, then tap the axle into position. If the sprocket hub, sprocket and chain are in position, pass the axle through them.
10 Install the rear sprocket hub and sprocket if they were removed.
11 Coat the axle nut and its locknut with Loctite 242 or equivalent and tighten them to 150 ft-lbs **(see illustration)**.
12 The remainder of installation is the reverse of the removal steps.
13 Check rear chain play and adjust as necessary (see Chapter 1).

Notes

Chapter 7
Brakes, wheels and tires

Contents

	Section
Brake check.. See Chapter 1	
Brake discs - inspection, removal and installation	5
Brake drum and shoes - removal, inspection and installation	9
Brake hoses and lines - inspection and replacement	7
Brake pads - replacement	2
Brake system bleeding	8
Front brake master cylinder - removal, overhaul and installation	6
Front hubs (except 3-wheel models) – removal, inspection and installation	11

	Section
General information	1
Hydraulic brake calipers - overhaul	4
Hydraulic brake calipers - removal and installation	3
Mechanical rear caliper – removal, inspection and installation	10
Rear hubs – removal and installation	12
Tires – general information	13
Wheels and tires - general check .... See Chapter 1	
Wheels – inspection, remvoal and installation	14

Specifications

Disc brakes

Brake fluid type..	See Chapter 1
Brake pad minimum thickness	See Chapter 1
Front disc thickness	
Standard...	0.150 to 0.164 inch
Limit..	0.140 inch
Output shaft and 6X6 center axle disc thickness	
Standard...	0.177 to 0.187 inch
Limit..	0.167 inch
Disc runout limit	
Front...	0.020 inch
Output shaft and 6X6 center axle	0.010 inch

Wheels and tires

Tire pressures	See Chapter 1
Tire tread depth	See Chapter 1

Torque specifications

Front axle nuts	
All-wheel drive...	See text
2X4	40 ft-lbs
Front axle holder nuts	
1983 through 1987 models	10 to 14 Nm (7 to 10 ft-lbs)
1988-on models	12 Nm (9 ft-lbs)
Rear axle nut	
1983 through 1987 models	80 to 110 Nm (58 to 72 ft-lbs)
1988-on models	95 Nm (69 ft-lbs)
Caliper mounting bolts	
Front caliper	18 ft-lbs
Output shaft and 6X6 center axle calipers	15 ft-lbs
Brake line banjo bolts	12 ft-lbs
Master cylinder clamp bolts	55 inch-lbs
Master cylinder reservoir cover bolts	45 inch-lbs
Front brake disc mounting bolts..	18 ft-lbs
Center axle brake disc (6X6)..	24 ft-lbs

Chapter 7 Brakes, wheels and tires

2.5 Unscrew the pad adjuster

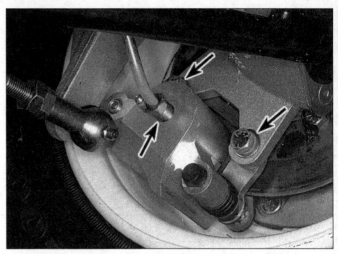

2.6 Front caliper mounting bolts (right arrows) and fluid line flare nut (left arrow)

1 General information

All 1985 and 1986 models, as well as 1987 2X4 models, use a mechanical drum brake at the front and mechanical disc brake at the rear. 1987 4X4 models, and all 1988-on models, use hydraulic disc brakes at front and rear.

The front calipers are mounted at the wheels; the rear caliper is mounted on the transmission output shaft. In addition to these three calipers, 6X6 models have a single caliper or pair of calipers mounted on the center axle. Several different caliper designs have been used, but service procedures are generally the same.

The front calipers and auxiliary shaft caliper are actuated by a single master cylinder mounted of the left handlebar. The middle axle caliper(s) on 6X6 models have their own master cylinder on the right handlebar.

All models are equipped with solid steel wheels. **Caution:** *Disc brake components rarely require disassembly. Do not disassemble components unless absolutely necessary. If any hydraulic brake line connection in the system is loosened, the entire system should be disassembled, drained, cleaned and then properly filled and bled upon reassembly. Do not use solvents on internal brake components. Solvents will cause seals to swell and distort. Use only clean brake fluid or alcohol for cleaning. Use care when working with brake fluid as it can injure your eyes and it will damage painted surfaces and plastic parts.*

2 Brake pads - replacement

Warning: *The dust created by the brake system may contain asbestos, which is harmful to your health. Never blow it out with compressed air and don't inhale any of it. An approved filtering mask should be worn when working on the brakes.*

Front caliper

Refer to illustrations 2.5, 2.6 and 2.9

1   Securely block both rear wheels so the vehicle won't roll. Jack up the front end, support it securely and remove the wheels.
2   The caliper hose doesn't need to be disconnected for pad replacement.
3   If you're working on a 1987 Trail Boss 4X4, unscrew the pad pins and remove the pads.
4   On 1988 through 1995 models, remove the setscrew from the upper pin bore in the caliper.
5   If you're working on a 1996 or later model, unscrew the pad adjuster screw with an Allen wrench **(see illustration)**.
6   Remove the caliper mounting bolts and take the caliper off

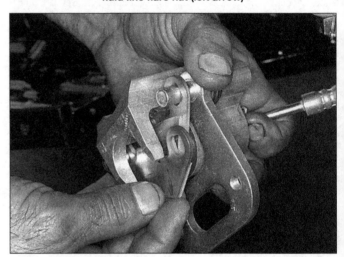

2.9 Slip the pad off the pin, rotate it and slip it off the other pin

(see illustration).
7   Compress the piston into the pad with a C-clamp or similar tool. **Note:** *This will force brake fluid into the master cylinder. Watch the level to make sure it doesn't overflow. Remove fluid if necessary. If the pistons stick, remove the caliper and overhaul it as described in Section 3.*
8   If you're working on a 1996 or later model, compress the pin from which the adjuster screw was removed.
9   Slide the pad loop off the pin, then rotate the pad and slide it off the other pin **(see illustration)**. Remove the inner pad in the same way.
10   If you're working on a 1987 Trail Boss 4X4, inspect the pad spring and replace it if it's rusted or damaged.
11   Refer to Chapter 1 and inspect the pads.
12   Check the condition of the brake disc (see Section 4). If it's in need of machining or replacement, follow the procedure in that Section to remove it. If it's okay, deglaze it with sandpaper or emery cloth, using a swirling motion.
13   Install the spring (if removed), new pads and the retaining pins (if removed). Tighten the retaining pins to the torque listed in this Chapter's Specifications.
14   Operate the brake lever or pedal several times to bring the pads into contact with the disc.
15   If you're working on a 1996 or later model, turn the pad adjuster screw until the fixed pad touches the disc, then loosen it 1/2 turn.
16   Check the operation of the brakes carefully before riding the vehicle.

Chapter 7 Brakes, wheels and tires

2.18a The auxiliary brake shaft on a chain drive model

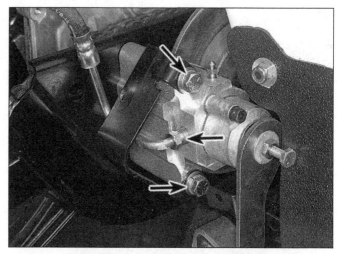

2.18b Caliper mounting bolts and fluid line (chain drive auxiliary shaft caliper)

2.18c Auxiliary brake connections and caliper mounting bolts (shaft drive)

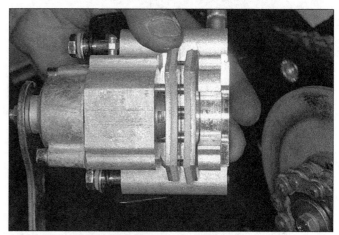

2.19a The pads fit in the caliper like this and are held in by the pins

Output shaft caliper

Refer to illustrations 2.18a, 2.18b, 2.18c, 2.19a, 2.19b, 2.19c, 2.21 and 2.23

17 Remove body panels as necessary for access (see Chapter 8). On chain drive models, remove the center chain guard.
18 If the vehicle has a mechanical auxiliary brake, disconnect the auxiliary brake rod from the caliper lever. Remove the caliper mounting bolts **(see illustrations)**. Lift the caliper off the brake disc, leaving the brake line(s) connected.
19 Turn the caliper over for access to the pads **(see illustration)**. Grasp the pins with pliers and work them partway out, then grasp the exposed ends and pull them the rest of the way **(see illustrations)**.
20 Pull the pads out of the caliper.

2.19b Start the pins out with a pair of needle-nosed pliers . . .

2.19c . . . then grasp the exposed ends and pull them the rest of the way out

Chapter 7 Brakes, wheels and tires

2.21 Replace the pad spring if it's damaged or corroded

2.23 The pad pins fit flush with the caliper body when installed

21 Check the pad spring and replace it if it's corroded or damaged **(see illustration)**.
22 Squeeze the piston into the bore using a C-clamp or similar tool.
23 Install new pads and secure them with the pins. Push the pins in until they're flush with the caliper body **(see illustration)**.

Middle axle calipers (6X6 models)

Refer to illustration 2.24

24 The middle axle on 6X6 models is equipped with either two calipers of the same design as the 1996 and later front calipers **(see illustration)** or with one dual-piston caliper.
25 Pad replacement for vehicles equipped with two single-piston calipers is the same as for front calipers, described above.
26 If you're working on a vehicle with a dual-piston caliper, remove its mounting bolts and lift it off the disc. Unclip the pad pin (don't open it any farther than necessary) and slide it out of the pads. Pull the pads out of the caliper, slip new ones in and reinstall the clip. Install the caliper on the disc and tighten its mounting bolts to the torque listed in this Chapter's Specifications.

3 Hydraulic brake calipers - removal and installation

Warning: *If a caliper indicates the need for an overhaul (usually due to leaking fluid or sticky operation), all old brake fluid flushed from the system. Also, the dust created by the brake system may contain asbestos, which is harmful to your health. Never blow it out with compressed air and don't inhale any of it. An approved filtering mask should be worn when working on the brakes. Do not, under any circumstances, use petroleum-based solvents to clean brake parts. Use brake cleaner or denatured alcohol only!*
Note: *If you are removing the caliper only to remove other components, don't disconnect the hose from the caliper.*
Note: *If you're planning to disassemble the caliper, read through the overhaul procedure, paying particular attention to the steps involved in removing the pistons with compressed air. If you don't have access to an air compressor, you can use the vehicle's hydraulic system to force the pistons out instead. To do this, remove the pads and pump the brake lever. If one piston comes out before the other on a dual-caliper piston, push it back into its bore and hold it in with a C-clamp while pumping the brake lever to remove the remaining piston.*

Front caliper

Removal

1 Securely block both rear wheels so the vehicle won't roll. Jack up the front end, support it securely and remove the wheels.

2.24 Some 6X6 models use two calipers on the center axle; each has a bleed fitting (arrow)

2 **Note:** *If you're just removing the caliper to remove the wheel, ignore this step.* Disconnect the brake line from the caliper. Unscrew the brake line flare nut and separate the line from the caliper **(see illustration 2.6)**. Cap the end of the line or wrap a plastic bag tightly around it to prevent excessive fluid loss and contamination.
3 Unscrew the caliper mounting bolts and lift it off the strut, being careful not to strain or twist the brake hose if it's still connected.

Installation

4 Installation is the reverse of the removal steps, with the following additions: Tighten the caliper mounting bolts to the torque listed in this Chapter's Specifications. Bleed the brakes (Section 00).

Output shaft caliper

Removal

5 Remove chain guards as necessary for access.
6 Detach the auxiliary brake lever from the caliper **(see illustration 2.18a or 2.18c)**.
7 **Note:** *If you're just removing the caliper to replace the pads, ignore this step.* Disconnect the brake line from the caliper. Unscrew the brake line flare nut and separate the line from the caliper **(see illustration 2.18b)**. Cap the end of the line or wrap a plastic bag tightly around it to prevent excessive fluid loss and contamination.
8 Unscrew the caliper mounting bolts and lift it off the disc, being careful not to strain or twist the brake hose if it's still connected.

Chapter 7 Brakes, wheels and tires

4.2a Make alignment marks on the auxiliary brake housing and caliper, then remove the bolts . . .

4.2b . . . and take the housing off . . .

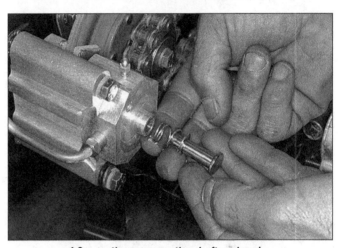

4.2c . . . then remove the shaft and spring

Installation

9 Installation is the reverse of the removal steps, with the following additions: Tighten the caliper mounting bolts to the torque listed in this Chapter's Specifications. Use a new cotter pin to connect the auxiliary brake shaft. Bleed the brakes (Section 8).

Middle axle caliper (6X6)

Removal

10 Remove chain guards as necessary for access.
11 **Note:** *If you're just removing the caliper to replace the pads, ignore this step.* If you're working on a vehicle with two single-piston calipers, label the brake lines and disconnect them **(see illustration 2.24)**. If you're working on a vehicle with one dual-piston caliper, hold the fitting on the caliper with a backup wrench and unscrew the brake line flare nut from the fitting with a flare nut wrench.
12 Unscrew the caliper mounting bolts and lift it off the disc, being careful not to strain or twist the brake hose if it's still connected.

Installation

13 Installation is the reverse of the removal steps, with the following additions: Tighten the caliper mounting bolts to the torque listed in this Chapter's Specifications. Bleed the brakes (Section 8).

4 Hydraulic brake calipers - overhaul

Refer to illustrations 4.2a, 4.2b, 4.2c, 4.3a, 4.3b, 4.3c, 4.4, 4.6 and 4.7
Note: *This procedure shows overhaul of the auxiliary brake mechanism on a transmission-mounted caliper. If you're working on a front caliper or a middle axle caliper, ignore the steps that don't apply.*

1 Remove the brake pads and anti-rattle spring from the caliper (see Section 2, if necessary). Clean the exterior of the caliper with denatured alcohol or brake system cleaner.
2 Remove the Allen bolts, take the auxiliary brake housing off the caliper and remove the shaft and spring **(see illustrations)**. **Note:** *This can be done with the caliper still mounted on the vehicle.*
3 Unbolt the auxiliary brake lever from the cam **(see illustration)**. Remove the cam from the housing and remove its seal from the groove with a pointed tool **(see illustrations)**.

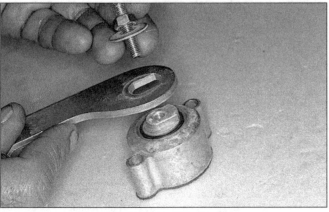

4.3a Unbolt the auxiliary brake arm

4.3b Remove the cam from the housing

7-6   Chapter 7 Brakes, wheels and tires

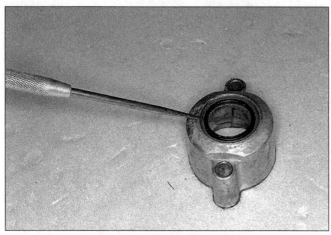

4.3c Remove the seal from the housing

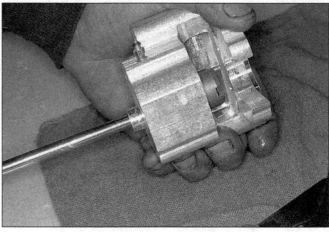

4.4 Push the piston out of the bore

4.6 Remove the dust seal and piston seal from the bore

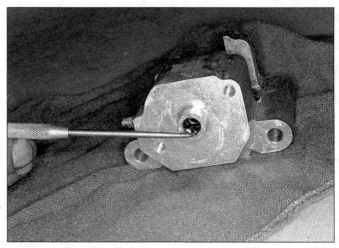

4.7 Remove the auxiliary brake shaft seals from the shaft bore

4   On single-piston calipers, push the piston out of the bore with a screwdriver or punch **(see illustration)**.
5   On dual-piston calipers, pack a shop rag into the space that holds the brake pads. Use compressed air, directed into the caliper fluid inlet, to remove the pistons. Use only enough air pressure to ease the pistons out of the bore. If a piston is blown out forcefully, even with the rag in place, it may be damaged. **Warning:** *Never place your fingers in front of the piston in an attempt to catch or protect it when applying compressed air, as serious injury could occur.*
6   Using a wood or plastic tool, remove the piston seals **(see illustration)**. Metal tools may cause bore damage.
7   If you're working on an auxiliary brake caliper, remove the seals form the shaft bore with a pointed tool **(see illustration)**.
8   Clean the piston(s and the bore(s) with denatured alcohol, clean brake fluid or brake system cleaner and blow dry with filtered, unlubricated compressed air. Inspect the piston and caliper bore surfaces for nicks and burrs and loss of plating. If surface defects are present, the caliper must be replaced. If the caliper is in bad shape, the master cylinder should also be checked.
9   Lubricate the piston seal with clean brake fluid and install it in its groove in the caliper bore **(see illustration 4.6)**. Make sure it seats completely and isn't twisted.
10   Lubricate the dust seal with clean brake fluid and install it in its groove, making sure it seats correctly.
11   If you're working on a dual-piston caliper, install the dust seal and piston seal in the other bore.
12   Lubricate the piston (both pistons on dual-piston calipers) with clean brake fluid and install it into the caliper bore. Using your thumbs,

push the piston all the way in, making sure it doesn't get cocked in the bore.
13   If you're working on an auxiliary brake caliper, coat the shaft seals with clean brake fluid and install them in the bore **(see illustration 4.7)**. Install the spring and shaft, then bolt the housing to the caliper, aligning the marks made during removal.
14   Make sure the pad spring is in position in the caliper brackets **(see illustration 4.6)**.

---

5   Brake discs - inspection, removal and installation

Inspection

1   If you're working on a front disc, place a jack beneath the vehicle and raise the wheel being checked off the ground. Be sure the vehicle is securely supported so it can't fall.
2   Visually inspect the surface of the disc(s) for score marks and other damage. Light scratches are normal after use and won't affect brake operation, but deep grooves and heavy score marks will reduce braking efficiency and accelerate pad wear. If the discs are badly grooved they must be machined or replaced.
3   To check disc runout, mount a dial indicator with the plunger on the indicator touching the surface of the disc about 1/2-inch from the outer edge. Slowly turn the wheel and watch the indicator needle, comparing your reading with the limit listed in this Chapter's Specifications. If the runout is greater than allowed, check the hub bearings for

Chapter 7 Brakes, wheels and tires

5.7 With the hub removed, unscrew the brake disc bolts (two bolts shown)

5.9 Push the circlip out of the shaft groove and slide the disc off the shaft

play (see Chapter 1). If the bearings are worn, replace them and repeat this check. If the disc runout is still excessive, the disc will have to be replaced.

4 The disc must not be machined or allowed to wear down to a thickness less than the minimum allowable thickness, listed in this Chapter's Specifications. The thickness of the disc can be checked with a micrometer. If the thickness of the disc is less than the minimum allowable, it must be replaced.

Removal

5 Remove the brake caliper (see Section 3).

Front brake disc

Refer to illustration 5.7

6 Remove the front hub together with the brake disc (Section 00).
7 Remove the disc mounting bolts and take the disc off the hub **(see illustration)**.

Output shaft disc

Refer to illustration 5.9

8 Remove the chain, sprocket and brake caliper (see Chapter 6 and Section 3).
9 Push the circlip out of its groove with a pair of needle-nosed pliers **(see illustration)**. Slide the disc off the shaft.

Middle axle disc (6X6 models)

10 Remove the rear axle (see Chapter 6).

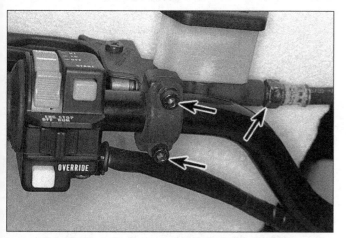

6.4 Master cylinder mounting bolts (left) and brake hose flare nut (right)

11 Drive the spacer, left bearing housing and disc to the right to expose the circlip inside the bearing housing. Remove the circlip, then slide the spacer, O-ring, bearing housing and disc off the left end of the axle.

Installation

12 Installation is the reverse of the removal steps, with the following addition:
Use a new clip on the axle (6X6) or transmission output shaft if the old one is weak or bent.
13 Install the wheel.
14 Operate the brake lever several times to bring the pads into contact with the disc. Check the operation of the brakes carefully before riding the vehicle.

6 Front brake master cylinder - removal, overhaul and installation

1 If the master cylinder is leaking fluid, or if the lever doesn't produce a firm feel when the brake is applied, and bleeding the brakes doesn't help, master cylinder overhaul is recommended. Before disassembling the master cylinder, read through the entire procedure and make sure that you have the correct rebuild kit. Also, you will need some new, clean brake fluid of the recommended type, some clean rags and internal snap-ring pliers. **Note:** *To prevent damage to the paint from spilled brake fluid, always cover the gas tank when working on the master cylinder.*
2 **Caution:** *Disassembly, overhaul and reassembly of the brake master cylinder must be done in a spotlessly clean work area to avoid contamination and possible failure of the brake hydraulic system components.*

Removal

Refer to illustration 6.4

3 Place rags beneath the master cylinder to protect the paint in case of brake fluid spills.
4 Brake fluid will run out of the upper brake hose during this step, so either have a container handy to place the end of the hose in, or have a plastic bag and rubber band handy to cover the end of the hose. Loosen (but don't remove) the flare nut or remove the banjo bolt that connects the brake hose to the master cylinder **(see illustration)**.
5 Remove the master cylinder mounting bolts **(see illustration 6.4)**. Take the master cylinder off the handlebar. If it has a flare nut, unscrew the master cylinder from the brake hose.

7-8 Chapter 7 Brakes, wheels and tires

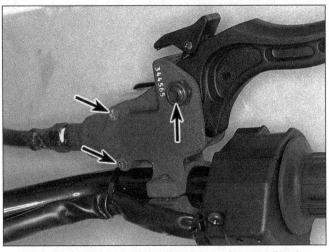

6.7 Reservoir screws (left) and pivot bolt locknut (right)

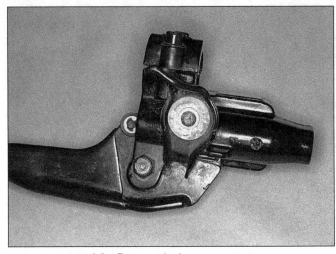

6.9a Remove the lever screw . . .

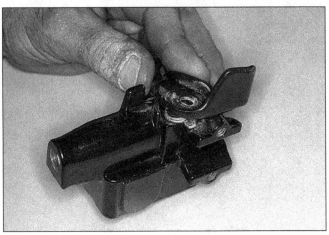

6.9b . . . and the lever

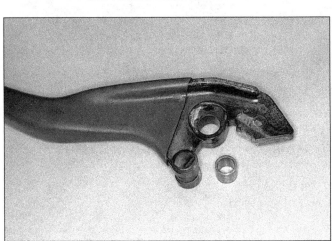

6.9c Remove the lever pivot bushing . . .

Overhaul

Refer to illustrations 6.7, 6.9a through 6.9e, 6.10 and 6.12

6 Remove the master cylinder cover and diaphragm (see Chapter 1).
7 Remove the locknut from the underside of the lever pivot bolt **(see illustration)**.
8 If you're working on a master cylinder with a plastic reservoir, remove the reservoir screws, then lift off the reservoir and its O-ring.

9 Remove the parking brake lever screw or bolt and lever **(see illustrations)**. Remove the bushing from the brake lever **(see illustration)**. Remove the snap-ring and lever rollers **(see illustration)**.
10 Slide out the piston assembly and spring out of the bore **(see illustration)**.
11 Clean all of the parts with brake system cleaner (available at auto parts stores), isopropyl alcohol or clean brake fluid. **Caution:** *Do not, under any circumstances, use a petroleum-based solvent to clean*

6.9d . . . the snap-ring that secures the rollers . . .

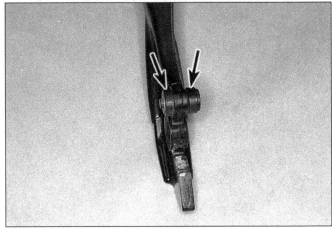

6.9e . . . and the rollers (arrows)

Chapter 7 Brakes, wheels and tires

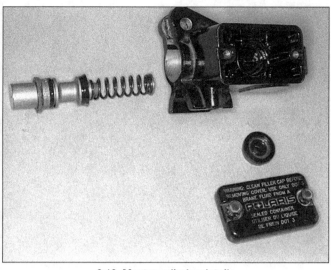

6.10 Master cylinder details

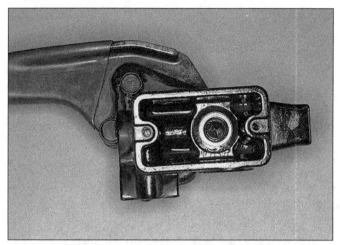

6.12 There's a baffle in the bottom of cast aluminum reservoirs

brake parts. If compressed air is available, use it to dry the parts thoroughly (make sure it's filtered and unlubricated). Check the master cylinder bore and piston for corrosion, scratches, nicks and score marks. If damage or wear can be seen, the master cylinder must be replaced with a new one. If the master cylinder is in poor condition, then the calipers should be checked as well.

12 Check the baffle in the bottom of the reservoir to make sure it's securely seated **(see illustration)**.

13 Polaris supplies a new piston in its rebuild kits. If the cup seals are not installed on the new piston, install them, making sure the lips face away from each other **(see illustration 6.10)**. Use the new piston regardless of the condition of the old one.

14 Before reassembling the master cylinder, soak the piston and the rubber cup seals in clean brake fluid for ten or fifteen minutes. Lubricate the master cylinder bore with clean brake fluid, then carefully insert the piston and related parts in the reverse order of disassembly. Make sure the lips on the cup seals do not turn inside out when they are slipped into the bore.

15 The remainder of assembly is the reverse of the disassembly steps.

Installation

16 Installation is the reverse of the removal steps, with the following additions:

a) If the brake hose has a banjo bolt, use new sealing washers at the connection of the brake hose to the master cylinder. Tighten the hose fitting to the torque listed in this Chapter's Specifications.

b) If the brake hose has a flare nut, screw the master cylinder onto the hose, using only fingers at this time (don't use a wrench). Once the master cylinder is installed, tighten the flare nut an additional 1/2 turn, using a wrench.

c) Attach the master cylinder to the handlebar, then tighten the bolts to the torque listed in this Chapter's Specifications. If there's a gap between the clamp and master cylinder, don't try to close it by tightening the bolts beyond the specified torque; you'll break the clamp.

17 Refer to Section 8 and bleed the air from the system.

7 Brake hoses and lines - inspection and replacement

Inspection

1 Once a week, or if the vehicle is used less frequently, before every ride, check the condition of the brake hoses and metal lines.

2 Twist and flex the rubber hoses while looking for cracks, bulges and seeping fluid. Check extra carefully around the areas where the hoses connect with the metal fittings, as these are common areas for hose failure.

Replacement

Refer to illustration 7.5

3 The pressurized brake hoses on some models have banjo fittings at one or both ends.

4 You'll need a flare nut wrench to undo the flare nuts used on the metal lines. An open-end wrench will probably round the nuts off, especially if they haven't been loosened in some time.

5 Cover the surrounding area with plenty of rags and unscrew the banjo bolt or flare nut on either end of the hose. Detach the hose or line from any clips that may be present and remove the hose **(see illustration)**.

6 Position the new hose or line, making sure it isn't twisted or otherwise strained. On hoses equipped with banjo fittings, make sure the metal tube portion of the banjo fitting is located against the stop on the component it's connected to, if equipped. Install the banjo bolts, using new sealing washers on both sides of the fittings, and tighten them to the torque listed in this Chapter's Specifications.

7 Tighten the flare nuts with a flare nut wrench. Where the upper hose joins the master cylinder, use a new sealing washer and tighten the fitting as described in Section 7.

8 Flush the old brake fluid from the system, refill the system with the recommended fluid (see Chapter 1) and bleed the air from the system (see Section 8). Check the operation of the brakes carefully before riding the motorcycle.

7.5 This front caliper brake hose is secured to the strut by a retainer

Chapter 7 Brakes, wheels and tires

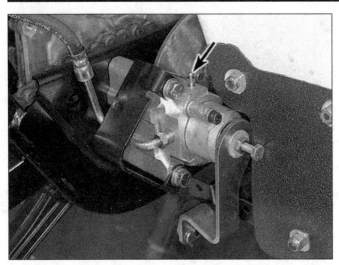

8.5a Auxiliary shaft caliper bleed valve (chain drive models)

8.5b Auxiliary shaft caliper bleed valve (shaft drive models)

8 Brake system bleeding

1 Bleeding the brake is simply the process of removing all the air bubbles from the brake fluid reservoir, the lines and the brake caliper. Bleeding is necessary whenever a brake system hydraulic connection is loosened, when a component or hose is replaced, or when the master cylinder or caliper is overhauled. Leaks in the system may also allow air to enter, but leaking brake fluid will reveal their presence and warn you of the need for repair.
2 To bleed the brake, you will need some new, clean brake fluid of the recommended type (see Chapter 1), a length of clear vinyl or plastic tubing, a small container partially filled with clean brake fluid, some rags and a wrench to fit the bleeder valves.
3 Cover the fuel tank and other painted components to prevent damage in the event that brake fluid is spilled.

Left-hand (main) brake system

Refer to illustrations 8.5a, 8.5b and 8.10

Note: *Polaris recommends on 1987 through 1995 models that the caliper pistons on the calipers not being bled be compressed into their bores with a C-clamp while the remaining caliper is bled. This requires that the calipers be dismounted (with the brake lines still connected) so the C-clamp can be installed. This step is not specified for 1996 and later models.*

4 Remove the reservoir cover and slowly pump the brake lever a few times, until no air bubbles can be seen floating up from the holes at the bottom of the reservoir. Doing this bleeds the air from the master cylinder end of the line. Reinstall the reservoir cover.
5 Attach one end of the clear vinyl or plastic tubing to the auxiliary shaft caliper bleeder valve and submerge the other end in the brake fluid in the container **(see illustrations)**.
6 Check the fluid level in the reservoir on the left handlebar. Do not allow the fluid level to drop below the lower mark during the bleeding process.
7 Carefully pump the brake lever three or four times and hold it while opening the caliper bleeder valve. **Note:** *This will be easier if you've got an assistant to help. If not, wrap a large rubber band tightly around the brake lever and handlebar after you've pumped the lever. The rubber band will hold the lever and pull it to the handlebar while you open the bleeder valve. When the valve is opened, brake fluid will flow out of the caliper into the clear tubing and the lever will move toward the handlebar.*
8 Retighten the bleeder valve, then release the brake lever gradually. Repeat the process until no air bubbles are visible in the brake fluid leaving the caliper and the lever is firm when applied. Remember to add fluid to the reservoir as the level drops. Use only new, clean brake fluid of the recommended type. Never reuse the fluid lost during bleeding.
9 Be sure to check the fluid level in the master cylinder reservoir frequently.
10 Repeat Steps 5 through 8 on the right front caliper **(see illustration)**.
11 Repeat Steps 5 through 8 on the left front caliper.
12 Replace the reservoir cover or cap, wipe up any spilled brake fluid and check the entire system for leaks. **Note:** *If bleeding is difficult, it may be necessary to let the brake fluid in the system stabilize for a few hours (it may be aerated). Repeat the bleeding procedure when the tiny bubbles in the system have settled out. If this doesn't work, Polaris recommends placing a C-clamp on the piston of each caliper, then pumping the piston at each caliper in and out several times by alternately squeezing the lever and tightening the C-clamp on the piston.*

Right-hand brake system (6X6)

Refer to illustration 8.15

13 The procedure is generally the same as for the left-hand brake system, but there are a couple of differences.
14 If the middle axle has two separate calipers, bleed the rear one first, then the front one **(see illustration 2.24)**.
15 Next, bleed the junction block on the frame **(see illustration)**. Early models didn't have a bleed valve. On these, you'll need to loosen the metal line's flare nut slightly to allow fluid and air to escape while bleeding.

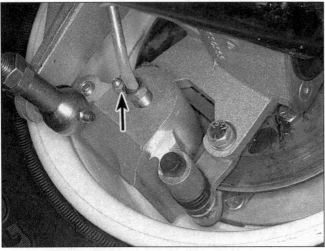

8.10 Front caliper bleed valve

8.15 The junction block on later 6X6 models has a bleed valve

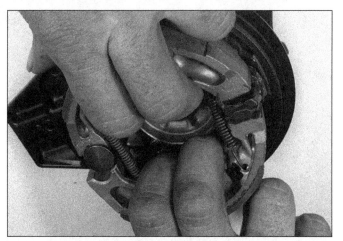

9.6 Spread the shoes and fold them into a V to release the spring tension

9 Brake drum and shoes - removal, inspection and installation

1 A drum brake is used on the front wheel of 1985 and 1986 three-wheel models.
Warning: *The dust created by the brake system may contain asbestos, which is harmful to your health. Never blow it out with compressed air and don't inhale any of it. An approved filtering mask should be worn when working on the brakes.*

Removal

1 Remove the wheel (see Section 00). Lift the brake panel out of the wheel.

Inspection

Refer to illustration 9.6

3 Check the brake drum for wear or damage. Measure the diameter at several points with a drum micrometer (or have this done by a Polaris dealer). If the measurements are uneven (indicating that the drum is out-of-round) or if there are scratches deep enough to snag a fingernail, replace the drum.
4 Check the linings for wear, damage and signs of contamination from road dirt or water. If the linings are visibly defective, replace them.
5 Measure the thickness of the lining material (just the lining material, not the metal backing) and compare with the value listed in the Chapter 1 Specifications. Replace the shoes if the material is worn to the minimum or less.
6 To remove the shoes, fold them toward each other to release the spring tension and lift them off the brake panel **(see illustration)**.
7 Check the ends of the shoes where they contact the brake cam and anchor pin. Replace the shoes if there's visible wear.
8 Check the brake cam, anchor pin and the inner seal on the brake panel for wear and damage and replace them if problems are found.

Installation

9 Apply high temperature brake grease to the brake cam, the anchor pin and the ends of the springs.
10 Hook the ends of the springs to the shoes. Position the shoes in a V on the brake panel, then fold them down into position **(see illustration 9.6)**. Make sure the ends of the shoes fit correctly on the cam and the anchor pin.
11 The remainder of installation is the reverse of the removal steps.

10 Mechanical rear caliper – removal, inspection and installation

1 The mechanical rear caliper is used on 1985 and 1986 models as well as 1987 2wd models. It's mounted on the transmission output shaft.
2 Remove the three mounting bolts. Lift the caliper off the brake disc, disconnect its cable and take it off.
3 Remove the ratchet adjuster from the caliper body. Push the outer brake pad out through the adjuster hole.
4 If the pads are worn to the groove around the circumference of each pad, replace them.
5 Clean the bore of the caliper with crocus cloth. If it's scored or severely corroded, replace the caliper.
6 Lubricate the caliper bore and the outer circumference of the moveable piston with a light coat of high-temperature brake grease.
7 Install the ratchet in the caliper and turn it clockwise until it's flush with the caliper.
8 Connect the cable to the ratchet. Assemble the ratchet arm and brake pads, then install them in the caliper as a unit.
9 Position the caliper on the transmission. Install the bolts and tighten them securely.

11 Front hubs (except 3-wheel models) – removal, inspection and installation

Removal

Refer to illustrations 11.3, 11.4 and 11.6

1 Securely block the rear wheels so the vehicle can't roll, jack up the front end and remove the wheels.
2 Remove the front brake caliper (Section 3). If you're working on an all-wheel drive model, drain the front hub fluid (see Chapter 1).
3 Remove the Allen bolts (if equipped) or pry off the hub cap **(see**

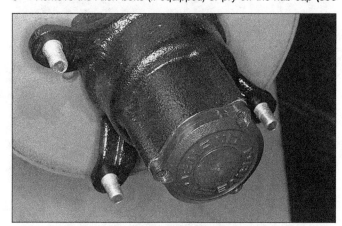

11.3 If the hub cap has Allen bolts, unscrew them

7-12 Chapter 7 Brakes, wheels and tires

11.4 Remove the cotter pin (if equipped) and unscrew the nut

11.6 Remove the outer bearing

illustration).

4   Remove the cotter pin and castellated nut or the self-locking nut and the washer **(see illustration)**. **Note:** *Some 1994 models, and all models from 1995 on, have left-hand threads on the left side of the vehicle. Turn the nut clockwise to remove it.*

5   If you're working on a 2wd model, pull the hub and bearings off the spindle.

6   If you're working on an all-wheel drive model, remove the outer bearing **(see illustration)**. Pull off the hub, then slide the inner bearing off the spindle.

Inspection
2wd models

7   Pry out the seal.

8   Spin the bearings with fingers without removing them from the hub. If they're rough, loose, or noisy, replace them. **Note:** *Driving the bearings out of the hub will ruin them, so don't so this unless you plan to install new bearings.*

9   Insert a soft metal drift into the hub from the outside. Pry it sideways to push the spacer out of the way, then tap the drift gently against the inner bearing, on opposite sides of the bearing, to drive it from the hub. Take the spacer out of the hub.

10   Insert the drift from the other side and remove the outer bearing in the same way.

11   Thoroughly clean all old grease from the spacer and the inside of the hub. Pack the new bearings with multi-purpose grease. Work the grease into the spaces between the bearing balls. Hold the outer race and rotate the bearing inner race as you pack it to distribute the grease.

12   Place the new outer bearing on the hub and drive it in with a 46 mm (1.18-inch) bearing driver. **Caution:** *Make sure the bearing driver presses only against the outer race of the bearing.*

13   Install the spacer in the hub, then drive in the remaining bearing with the same driver.

14   Tap a new seal into the inside of the hub until it is flush with the hub. The open end of the seal faces the bearing.

All-wheel drive models

Refer to illustrations 11.17a and 11.17b

15   Clean the bearings thoroughly and check them for wear and damage. Replace the bearings if any problems are found.

16   Check the Hilliard clutch for wear and damage, paying special attention to its garter spring. This is the device, consisting of six rollers and a cam, that rides inside the hub and engages the front wheel drive system. If there are any visible problems, have the Hilliard clutch inspected and repaired by a Polaris dealer or other qualified ATV shop. **Warning:** *The Hilliard clutch should not be disassembled by anyone without the proper tools and experience. A seemingly harmless mistake, such as overstretching the garter spring, can allow one of the front hubs to engage suddenly at high speed, causing the vehicle to go out of control.*

17   Check the seal, bearing and magnetic coil in the strut housing **(see illustrations)**. If the seal has been leaking, pry it out and tap in a

11.17a Check the seal and bearing in the strut housing

11.17b Make sure the magnetic coil is covered with sealant

TIRE CHANGING SEQUENCE

Deflate the tire and remove the valve core. Release the bead on the side opposite the tire valve with an ATV bead breaker, following the manufacturer's instructions. Make sure you have the correct blades for the tire size (using the wrong size blade may damage the wheel, the tire or the blade). Lubricate the bead with water before removal (don't use soap or any type of lubricant).

Turn the tire over and release the other bead.

If one side of the wheel has a smaller flange, remove and install the tire from that side. Use two tire levers to work the bead over the edge of the rim.

Before installing, ensure that tire is suitable for wheel. Take note of any sidewall markings such as direction of rotation arrows, then work the first bead over the rim flange.

Use tire levers to start the second bead over the rim flange.

Hold the bead while you work the last section of it over the rim flange. Install the valve core and inflate the tire, making sure not to overinflate it.

7-14   Chapter 7   Brakes, wheels and tires

12.2a Remove the trim cap . . .

12.2b . . . remove the cotter pin and unscrew the nut . . .

new one with a socket or bearing driver the same diameter as the seal. If the bearing or magnetic coil are worn or damaged, have them replaced by a Polaris dealer or other qualified ATV shop. Make sure the magnetic coil is completely covered by silicone sealant.

Installation
2wd models
18   Installation is the reverse of the removal steps, with the following additions:
   a) Grease the inside of the wheel bearing washer. This affects the torque reading. Tighten the wheel bearing nut to the torque listed in this Chapter's Specifications. Don't tighten it, loosen it and then retighten it to a lower torque, as is common practice with tapered roller bearings. The ball bearings used in 2wd hubs have no preload adjustment.
   b) Use a new cotter pin.

4wd models
19   Installation is the reverse of the removal steps, with the following additions:
   a) If the bearing uses a self-locking nut, measure the minimum torque while it is being tightened with an inch-pound torque wrench. If it isn't at least 75 inch-lbs, discard the nut and install a new one. To get the final torque, keep tightening until the torque reading exceeds to minimum reading by 100 inch-lbs. Turn the FRONT hubs several turns in each direction by jacking up the rear of the vehicle and turning the rear axle. Then loosen the nut and retighten it to the final torque (minimum torque plus 100 inch-lbs).
   b) If the bearing uses a castellated nut and cotter pin, tighten the nut to 100 inch-lbs. Turn the FRONT hubs several turns in each direction by jacking up the rear of the vehicle and turning the rear axle. Then loosen the nut and retighten it to 100 inch-lbs. Install a new cotter pin.
   c) Fill the hubs with the specified fluid (see Chapter 1).

12   Rear hubs – removal and installation

Refer to illustrations 12.2a, 12.2b and 12.3

1   Securely block the front wheel so the vehicle won't roll. Jack up the rear end and remove the rear wheel(s).
2   Remove the trim cap from the hub **(see illustration)**. Remove the cotter pin and undo the nut **(see illustration)**.
3   Pull the hub off the axle splines **(see illustration)**. If it won't come easily, tap it off with a mallet. If it's stuck, apply penetrating oil between the hub and splines and let it soak in.

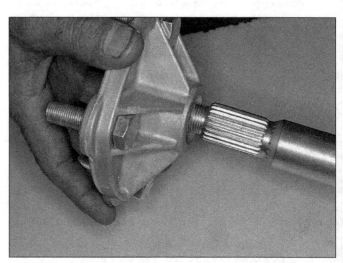

12.3 . . . and slide the hub off the splines

4   Installation is the reverse of the removal steps. Tighten the nut to the torque listed in this Chapter's Specifications and secure it with a new cotter pin.

13   Tires – general information

1   Tubeless tires are used as standard equipment on these vehicles. Unlike motorcycle tires, they run at very low air pressures and are completely unsuitable for use on pavement. Inflating ATV tires to excessive pressures will rupture them, making replacement of the tire necessary.
2   The force required to break the seal between the rim and bead of the tire is substantial, much more than required for motorcycle tires, and is beyond the capabilities of an individual working with normal tire irons or even a normal bead breaker. A special bead breaker is required for ATV tires; it produces a great deal of force and concentrates it in a relatively small area.
3   Also, repair of the punctured tire and replacement on the wheel rim requires special tools, skills and experience that the average do-it-yourselfer lacks.
4   For these reasons, if a puncture or flat occurs in an ATV tire, the wheel should be removed from the vehicle and taken to a dealer service department or a repair shop for repair or replacement of the tire. The illustrations on the following page can be used as a guide to tire replacement in an emergency, provided the necessary bead breaker is available.

14 Wheels – inspection, removal and installation

Inspection
1 Clean the wheels thoroughly to remove mud and dirt that may interfere with the inspection procedure or mask defects. Make a general check of the wheels and tires as described in Chapter 1.
2 The wheels should be visually inspected for cracks, dents, flat spots on the rim and other damage. Since tubeless tires are involved, look very closely for dents in the area where the tire bead contacts the rim. Dents in this area may prevent complete sealing of the tire against the rim, which leads to deflation of the tire over a period of time.
3 If damage is evident, the wheel will have to be replaced with a new one. Never attempt to repair a damaged wheel.

Removal
4 Securely block the wheel at the opposite end of the vehicle from the wheel being removed, so the vehicle can't roll.
5 Loosen the lug nuts on the wheel being removed. Jack up one end of the vehicle and support it securely on jackstands.
6 Remove the lug nuts and pull the wheel off.

Installation
Refer to illustration 14.7
7 Position the wheel on the studs. Make sure the directional arrow on the tire points in the forward rotating direction of the wheel **(see illustration)**.
8 Install the wheel nuts with their tapered sides toward the wheel. This is necessary to locate the wheel accurately on the hub.
9 Snug the wheel nuts evenly in a criss-cross pattern.
10 Remove the jackstands, lower the vehicle and tighten the wheel nuts, again in a criss-cross pattern.

14.7 The arrow on the tire must point in the forward rotating direction of the wheel

Notes

Chapter 8
Body and frame

Contents

	Section		Section
Dump bed (6X6 and 4X6) – removal and installation	4	General information	1
Frame - general information, inspection and repair	11	Headlight cover – removal and installation	7
Front cargo rack – removal and installation	9	Rear cargo rack – removal and installation	3
Front fenders - removal and installation	10	Rear fenders – removal and installation	5
Front skid plate – removal and installation	8	Seat - removal and installation	2
Fuel tank cover – removal and installation	6		

1 General information

This Chapter covers the procedures necessary to remove and install the body panels and other body parts. Since many service and repair operations on these vehicles require removal of the panels and/or other body parts, the procedures are grouped here and referred to from other Chapters.

In the case of damage to the panels or other body parts, it is usually necessary to remove the broken component and replace it with a new (or used) one. The material that the plastic body parts are composed of doesn't lend itself to conventional repair techniques. There are, however, some shops that specialize in "plastic welding", so it would be advantageous to check around first before throwing the damaged part away.

Four-wheel and six-wheel Polaris ATVs are manufactured with three basic body styles. Since 1996, they have been referred to as Gen II, Gen III and Gen IV, but the styles apply to earlier four-wheel and six-wheel models as well.

The Gen II body style includes Trail Boss, Magnum, 6X6, and Sportsman 4X4 models, as well as Xplorer models through 1995.

The Gen III body style includes Trail Blazer, Sport and Scrambler models.

The Gen IV body style includes 1997 Xpress 300 and 400, Xplorer 300, 400 and 500 and Sportsman 500 models.

Note: *When attempting to remove any body panel, first study the panel closely, noting any fasteners and associated fittings, to be sure of returning everything to its correct place on installation. In some cases, the aid of an assistant will be required when removing panels, to help avoid damaging the surface. Once the visible fasteners have been removed, try to lift off the panel as described but DO NOT FORCE the panel - if it will not release, check that all fasteners have been removed and try again. Where a panel engages another by means of tabs and slots, be careful not to break the tabs or to damage the bodywork. Remember that a few moments of patience at this stage will save you a lot of money in replacing broken panels!*

2 Seat - removal and installation

Refer to illustration 2.1

1 To unlock the seat, flip up the latch lever and lift the back end of the seat **(see illustration)**. Pull the seat to the rear.

2 Installation is the reverse of removal. Make sure that the posts at the rear of the seat are inserted into their holes **(see illustration 2.1)**.

2.1 The seat posts fit in the holes in the fenders (arrows)

Chapter 8 Body and frame

3.2a The rear cargo rack is secured by nuts and bolts at the rear (Trail Boss 250 shown) . . .

3.2b . . . and at the front (Trail Boss 250 shown)

3 Rear cargo rack – removal and installation

Refer to illustrations 3.2a and 3.2b

1   Remove the seat (Section 2).
2   The cargo rack is secured by four posts, two at the front and two at the rear. Remove the nuts and bolts and lift the posts out **(see illustrations)**.
3   Installation is the reverse of the removal steps.

4 Dump bed (6X6 and 4X6) – removal and installation

Refer to illustrations 4.1 and 4.2

1   Support the dump bed in the open position and remove its strut **(see illustration)**.
2   Remove the dump bed mounting nuts and take it off the frame **(see illustration)**.
3   Installation is the reverse of the removal steps.

4.1 Detach both ends of the strut (arrows) . . .

5 Rear fenders – removal and installation

Refer to illustrations 5.2, 5.3 and 5.4

1   Remove the rear cargo rack (Section 3).

4.2 . . . and remove the dump bed mounting nuts (arrows)

5.2 The mudflaps are secured to the fenders and footrests by nuts and bolts; don't lose the reinforcing strips

Chapter 8 Body and frame

5.3 The rear fender unit is secured by four bolts (shown) or six bolts, depending on model

5.4 Remove the screws at the sides of the rear fenders

2   Remove the nuts and reinforcing plates and detach the mud flaps from the fenders **(see illustration)**.
3   Remove the fender mounting bolts beneath the seat (six on Gen II models, four on Gen III and Gen IV) **(see illustration)**.
4   Remove the mounting screws from the sides of the fenders (if equipped) **(see illustration)**.
5   Make sure all fasteners have been removed and lift the fenders off.
6   Installation is the reverse of the removal steps.

6 Fuel tank cover – removal and installation

Refer to illustration 6.4

1   The fuel tank cover is used on Gen II models.
2   Remove the seat (Section 2).
3   Remove the ignition key bracket (see Chapter 5). Unscrew the fuel filler cap.
4   Remove the screws (two on each side of the cover) **(see illustration)**. Carefully disengage the cover from the tabs on the side panels and lift it off.
5   Installation is the reverse of the removal steps.

7 Headlight cover – removal and installation

Refer to illustration 7.1

1   If you're working on a Gen II model, remove the seat and fuel tank cover (Sections 2 and 6). Remove two Torx screws at the rear of the cover and one bolt on each side **(see illustration)**. Lift the cover off and disconnect the headlight connector.
2   If you're working on a Gen III model, remove the oil tank cap. Carefully unhook the retaining tabs at the rear and on each side, then lift up the cover and disconnect the headlight connector.
3   If you're working on a Gen IV model, remove five Phillips screws securing the upper half of the cover (two at the front and three at the rear). Lift off the top half of the cover. Disconnect the speedometer cable, choke cable, main key switch and the warning light electrical connector. Remove three more screws and lift off the bottom half of the cover.
4   Installation is the reverse of the removal steps, with the following addition: Adjust the headlight aim (see Chapter 5).

6.4 Remove the fuel tank cover screws and disengage the tabs (arrows)

7.1 The headlight cover is secured by bolts (left) and Torx screws (right)

8 Front skid plate – removal and installation

Refer to illustration 8.1

1 To remove the front skid plate, unscrew its mounting bolts (see illustration).
2 Installation is the reverse of the removal steps.

9 Front cargo rack – removal and installation

Refer to illustrations 9.2a and 9.2b

1 The front cargo rack is used on Gen II and Gen IV models.
2 To remove the rack, remove its mounting bolts and nuts (see illustrations).
3 Installation is the reverse of the removal steps.

10 Front fenders - removal and installation

Gen II models

1 Remove the seat (see Section 2) and the side panels.
2 Remove the fuel tank cover, headlight cover and front cargo rack (Sections 6, 7 and 9).
3 Remove the fuel pump mounting bracket and the two rivets beneath it (see Chapter 4).
4 Detach the footrests from the fender on both sides of the vehicle. Lift the fenders off.
5 Installation is the reverse of removal.

Gen III models

6 Remove the fuel filler cap and headlight cover.
7 Detach the footrests from the front fenders on both sides of the vehicle.
8 Remove the four fender mounting screws, two inside at the front and two at the rear.
9 Remove the faceplate for the main key switch and remove the switch.
10 Lift the fenders off.

Gen IV models

11 Remove the seat (see Section 2) and the side panels.
12 Remove the front cargo rack (Section 9), the front bumper and the front panel.
Remove three screws on each side securing the front fenders to the mudflaps.

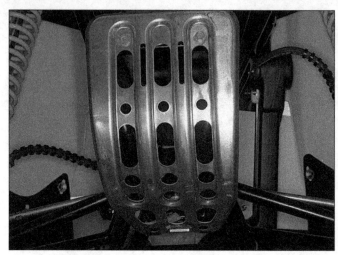

8.1 The front skid plate is bolted under the front of the machine

13 Remove one inner screw on each side securing the front fenders to the footrests.
14 Remove two screws from under the front panel. Lift the fenders off.
Installation is the reverse of the removal steps.

11 Frame - general information, inspection and repair

1 All models use a frame made of steel tubing.
2 The frame shouldn't require attention unless accident damage has occurred. In most cases, frame replacement is the only satisfactory remedy for such damage. A few frame specialists have the jigs and other equipment necessary for straightening the frame to the required standard of accuracy, but even then there is no simple way of assessing to what extent the frame may have been over-stressed.
3 After the machine has accumulated a lot of miles, the frame should be examined closely for signs of cracking or splitting at the welded joints. Corrosion can also cause weakness at these joints. Loose engine mount bolts can cause elongation of the bolt holes and can fracture the engine mounting points. Minor damage can often be repaired by welding, depending on the nature and extent of the damage.
Remember that a frame that is out of alignment will cause handling problems. If misalignment is suspected as the result of an accident, it will be necessary to strip the machine completely so the frame can be thoroughly checked.

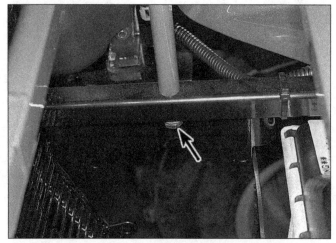

9.2a Here's one type of front cargo rack support post . . .

9.2b . . . and here's another type

Wiring diagrams

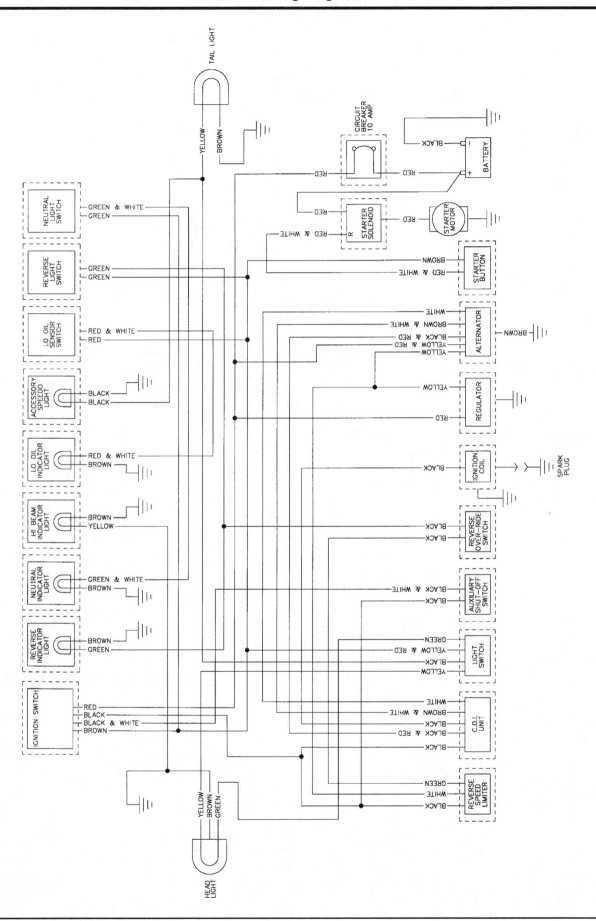

Wiring diagram - 1985 through 1988 models

9-2 Wiring diagrams

Terminal board - 1989 through 1995 (except Scrambler and Xplorer)

Wiring diagrams

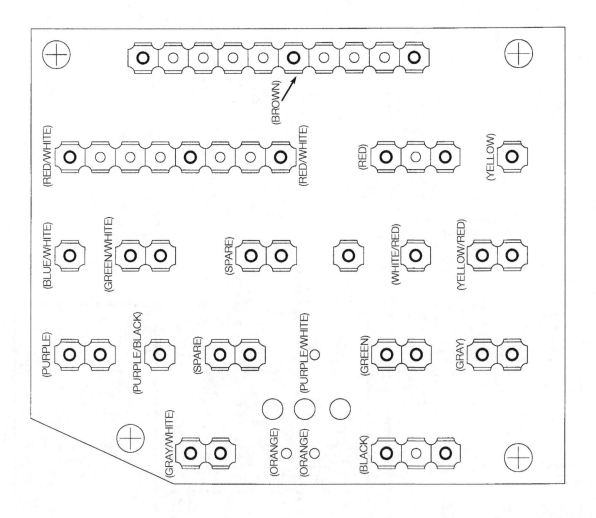

Terminal board - 1995 Scrambler and Xplorer; all 1996 and later models

Wiring diagrams

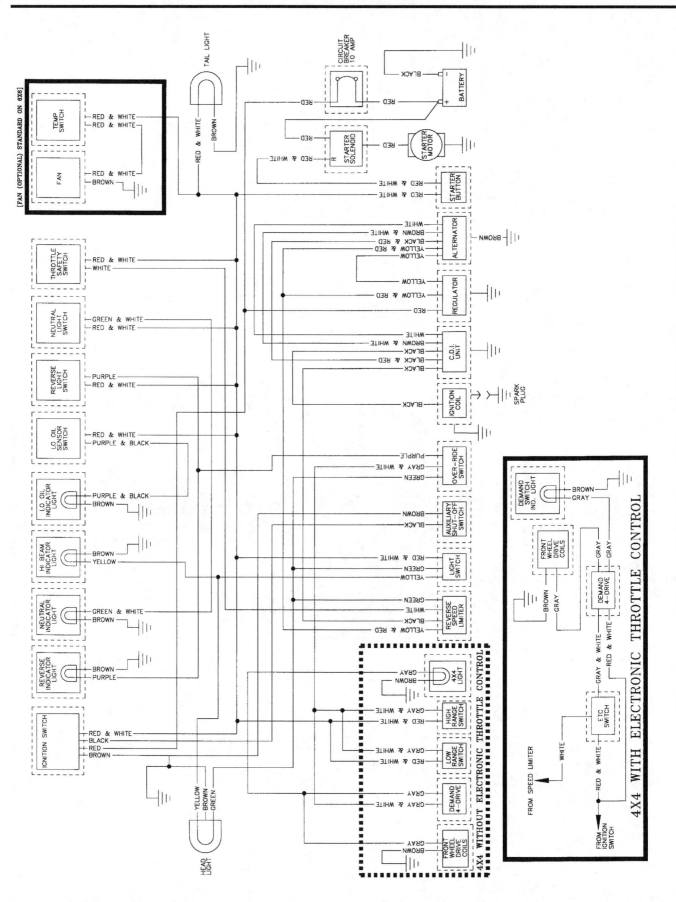

Wiring diagram - 1989 through 1995 Trail Boss; 1989 through 1993 Big Boss except 1993 350 6X6

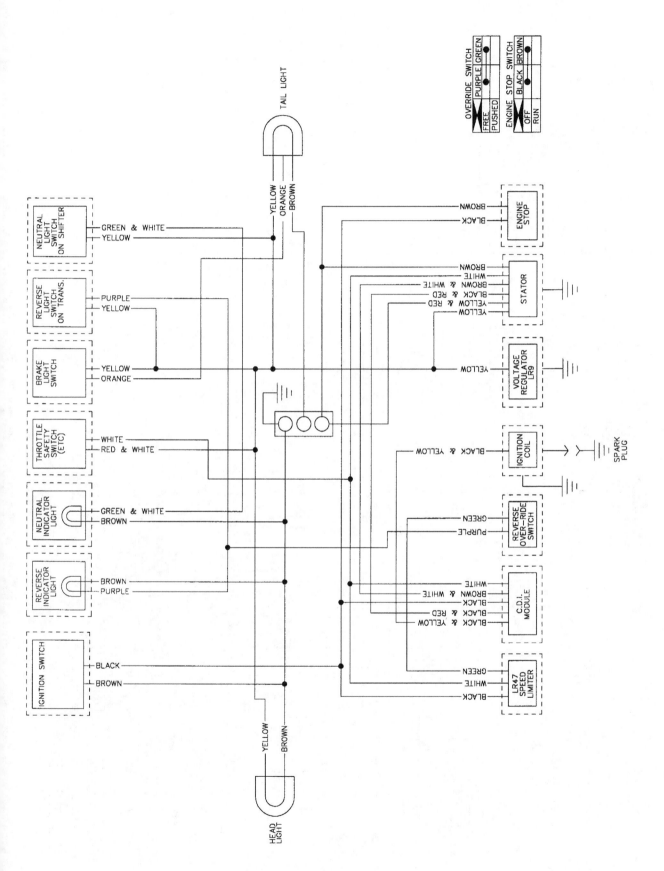

Wiring diagram - 1990 through 1995 Trail Blazer

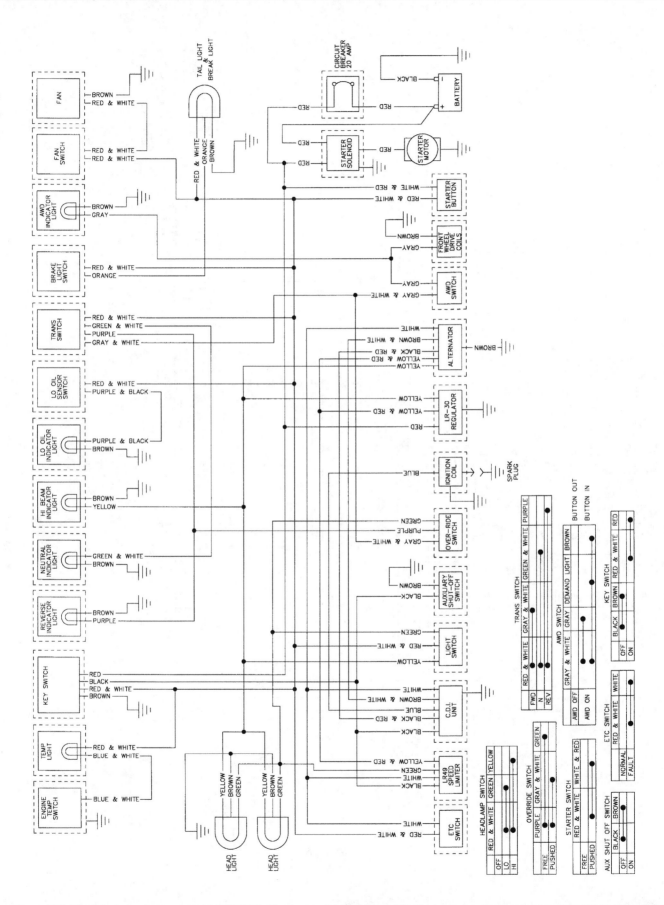

Wiring diagram – 1994 and 1995 400 Sport, 1995 Scrambler

Wiring diagrams

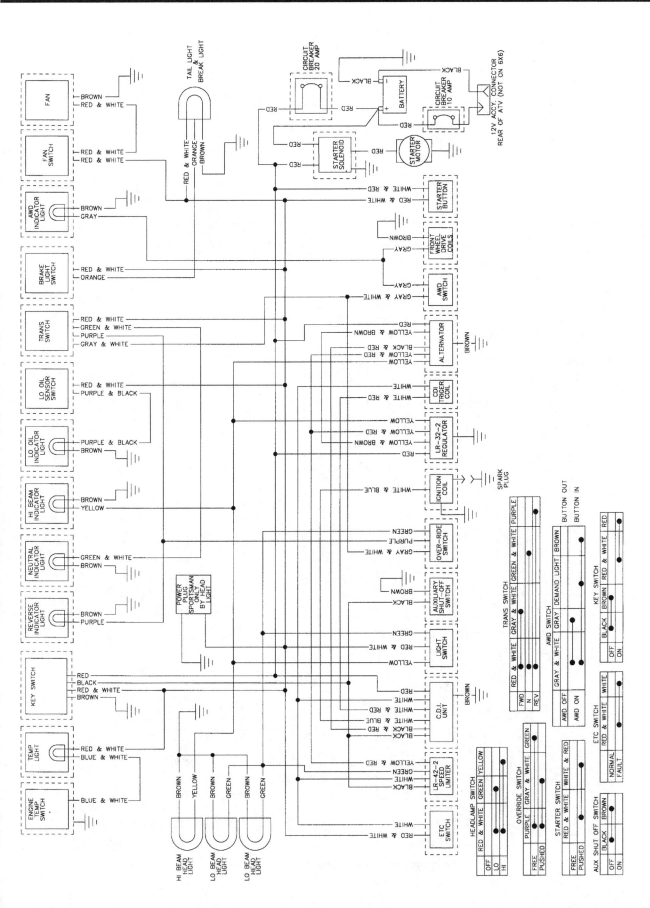

Wiring diagram – 1993 Trail Boss 350 6X6, all 1994 and 1995 models except 400 Sport

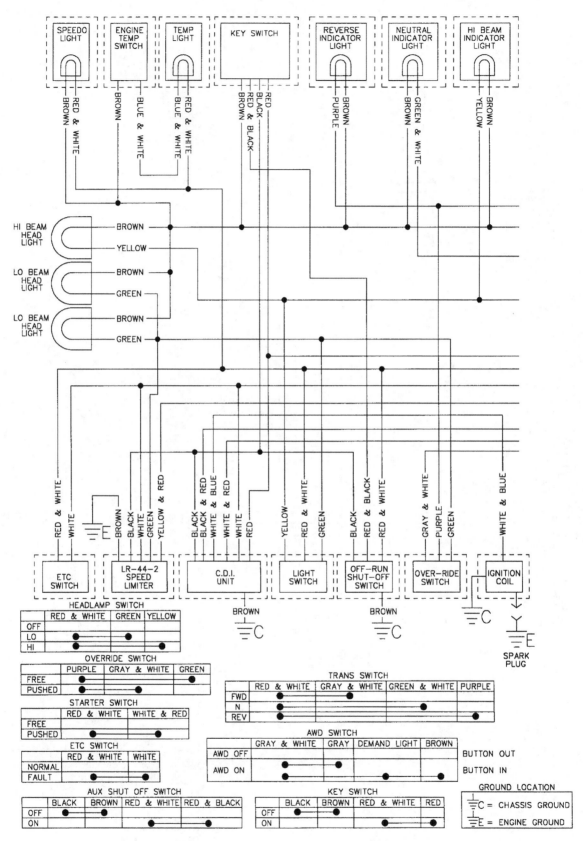

Wiring diagram - All 1996 and later models (1 of 2)

Wiring diagrams

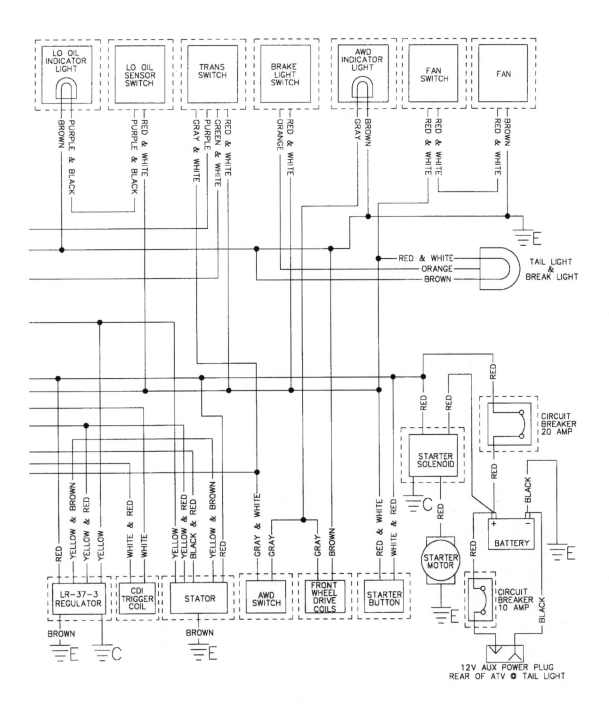

Wiring diagram - All 1996 and later models (2 of 2)

Notes

Conversion factors

Length (distance)
Inches (in)	X	25.4	= Millimetres (mm)	X 0.0394	= Inches (in)
Feet (ft)	X	0.305	= Metres (m)	X 3.281	= Feet (ft)
Miles	X	1.609	= Kilometres (km)	X 0.621	= Miles

Volume (capacity)
Cubic inches (cu in; in^3)	X	16.387	= Cubic centimetres (cc; cm^3)	X 0.061	= Cubic inches (cu in; in^3)
Imperial pints (Imp pt)	X	0.568	= Litres (l)	X 1.76	= Imperial pints (Imp pt)
Imperial quarts (Imp qt)	X	1.137	= Litres (l)	X 0.88	= Imperial quarts (Imp qt)
Imperial quarts (Imp qt)	X	1.201	= US quarts (US qt)	X 0.833	= Imperial quarts (Imp qt)
US quarts (US qt)	X	0.946	= Litres (l)	X 1.057	= US quarts (US qt)
Imperial gallons (Imp gal)	X	4.546	= Litres (l)	X 0.22	= Imperial gallons (Imp gal)
Imperial gallons (Imp gal)	X	1.201	= US gallons (US gal)	X 0.833	= Imperial gallons (Imp gal)
US gallons (US gal)	X	3.785	= Litres (l)	X 0.264	= US gallons (US gal)

Mass (weight)
Ounces (oz)	X	28.35	= Grams (g)	X 0.035	= Ounces (oz)
Pounds (lb)	X	0.454	= Kilograms (kg)	X 2.205	= Pounds (lb)

Force
Ounces-force (ozf; oz)	X	0.278	= Newtons (N)	X 3.6	= Ounces-force (ozf; oz)
Pounds-force (lbf; lb)	X	4.448	= Newtons (N)	X 0.225	= Pounds-force (lbf; lb)
Newtons (N)	X	0.1	= Kilograms-force (kgf; kg)	X 9.81	= Newtons (N)

Pressure
Pounds-force per square inch (psi; lbf/in^2; lb/in^2)	X	0.070	= Kilograms-force per square centimetre (kgf/cm^2; kg/cm^2)	X 14.223	= Pounds-force per square inch (psi; lbf/in^2; lb/in^2)
Pounds-force per square inch (psi; lbf/in^2; lb/in^2)	X	0.068	= Atmospheres (atm)	X 14.696	= Pounds-force per square inch (psi; lbf/in^2; lb/in^2)
Pounds-force per square inch (psi; lbf/in^2; lb/in^2)	X	0.069	= Bars	X 14.5	= Pounds-force per square inch (psi; lbf/in^2; lb/in^2)
Pounds-force per square inch (psi; lbf/in^2; lb/in^2)	X	6.895	= Kilopascals (kPa)	X 0.145	= Pounds-force per square inch (psi; lbf/in^2; lb/in^2)
Kilopascals (kPa)	X	0.01	= Kilograms-force per square centimetre (kgf/cm^2; kg/cm^2)	X 98.1	= Kilopascals (kPa)

Torque (moment of force)
Pounds-force inches (lbf in; lb in)	X	1.152	= Kilograms-force centimetre (kgf cm; kg cm)	X 0.868	= Pounds-force inches (lbf in; lb in)
Pounds-force inches (lbf in; lb in)	X	0.113	= Newton metres (Nm)	X 8.85	= Pounds-force inches (lbf in; lb in)
Pounds-force inches (lbf in; lb in)	X	0.083	= Pounds-force feet (lbf ft; lb ft)	X 12	= Pounds-force inches (lbf in; lb in)
Pounds-force feet (lbf ft; lb ft)	X	0.138	= Kilograms-force metres (kgf m; kg m)	X 7.233	= Pounds-force feet (lbf ft; lb ft)
Pounds-force feet (lbf ft; lb ft)	X	1.356	= Newton metres (Nm)	X 0.738	= Pounds-force feet (lbf ft; lb ft)
Newton metres (Nm)	X	0.102	= Kilograms-force metres (kgf m; kg m)	X 9.804	= Newton metres (Nm)

Vacuum
Inches mercury (in. Hg)	X	3.377	= Kilopascals (kPa)	X 0.2961	= Inches mercury
Inches mercury (in. Hg)	X	25.4	= Millimeters mercury (mm Hg)	X 0.0394	= Inches mercury

Power
Horsepower (hp)	X	745.7	= Watts (W)	X 0.0013	= Horsepower (hp)

Velocity (speed)
Miles per hour (miles/hr; mph)	X	1.609	= Kilometres per hour (km/hr; kph)	X 0.621	= Miles per hour (miles/hr; mph)

Fuel consumption*
Miles per gallon, Imperial (mpg)	X	0.354	= Kilometres per litre (km/l)	X 2.825	= Miles per gallon, Imperial (mpg)
Miles per gallon, US (mpg)	X	0.425	= Kilometres per litre (km/l)	X 2.352	= Miles per gallon, US (mpg)

Temperature
Degrees Fahrenheit = (°C x 1.8) + 32   Degrees Celsius (Degrees Centigrade; °C) = (°F - 32) x 0.56

*It is common practice to convert from miles per gallon (mpg) to litres/100 kilometres (l/100km), where mpg (Imperial) x l/100 km = 282 and mpg (US) x l/100 km = 235

Notes

Index

A

About this manual, 0-5
Acknowledgments, 0-2
Air cleaner
    drain tube, cleaning, 1-9
    pre-filter check and main element replacement, 1-8
Air cleaner housing, removal and installation, 4-15
Air ducts, removal and installation, 2C-2
Alternator, check and replacement, 5-14
ATV chemicals and lubricants, 0-30
Axle, rear, removal, inspection and installation, 6-20

B

Battery
    charging, 5-4
    electrolyte level/specific gravity, check, 1-10
    inspection and maintenance, 5-3
Body and frame, 8-1 through 8-4
    cargo rack, removal and installation
        front, 8-4
        rear, 8-2
    dump bed (6X6 and 4X6), removal and installation, 8-2
    fenders, removal and installation
        front, 8-4
        rear, 8-2
    frame, general information, inspection and repair, 8-4
    front skid plate, removal and installation, 8-4
    headlight cover, removal and installation, 8-3
    fuel tank cover, removal and installation, 8-3
    seat, removal and installation, 8-1
Brake fluid, change, 1-30
Brake lever and pedal freeplay, check and adjustment, 1-11
Brake switch, check and replacement, 5-6
Brake system, general check, 1-11
Brakes, wheels and tires, 7-1 through 7-16
    brake discs, inspection, removal and installation, 7-6
    brake drum and shoes, removal, inspection and installation, 7-11
    brake hoses and lines, inspection and replacement, 7-9
    brake pads, replacement, 7-2
    brake system bleeding, 7-10
    front brake master cylinder, removal, overhaul and installation, 7-7
    front hubs (except 3-wheel models), removal, inspection and installation, 7-11
    hydraulic brake calipers
        overhaul, 7-5
        removal and installation, 7-4
    mechanical rear caliper, removal, inspection and installation, 7-11
    rear hubs, removal and installation, 7-14
    tires, general information, 7-14
    wheels, inspection, removal and installation, 7-15
Buying parts, 0-8

C

Cam chain tensioner, removal and installation, four-stroke engines, 2B-7
Carburetors
    disassembly, cleaning and inspection, 4-9
    overhaul, general information, 4-7
    reassembly and float height check, 4-15
    removal and installation, 4-7
Cargo rack, removal and installation
    front, 8-4
    rear, 8-2

CDI magneto, check, removal and installation, 5-19
Charging system
   output test, 5-13
   testing, general information and precautions, 5-13
Choke cable, removal and installation, 4-18
Choke, check and freeplay adjustment, 1-14
Circuit breaker, check, 5-4
Clutches, inner cover and seal, removal and installation, 2C-5
Coolant hoses, replacement, 3-6
Coolant reservoir, removal and installation, 3-2
Cooling fan and thermostat switch, check and replacement, 3-2
Cooling system, 3-1 through 3-6
   coolant hoses, replacement, 3-6
   coolant reservoir, removal and installation, 3-2
   cooling fan and thermostat switch, check and replacement, 3-2
   draining, flushing and refilling, 1-28
   inspection, 1-28
   radiator cap, check, 3-2
   radiator, removal and installation, 3-4
   thermostat (4-stroke models), removal, check and installation, 3-3
   water pump, removal, inspection and installation, 3-4
Counterbalancer fluid, change, 1-27
Counterbalancer, removal, inspection and installation
   four-stroke engines, 2B-26
   two-stroke engines (350 and 400), 2A-12
Crankcase
   disassembly and reassembly
      four-stroke engines, 2B-22
      two-stroke engines, 2A-15
   inspection and servicing
      four-stroke engines, 2B-25
      two-stroke engines, 2A-16
   pressure and vacuum, check, two-stroke engines, 2A-3
Crankshaft and connecting rod, removal, inspection and installation, 2A-16
   four-stroke engines, 2B-26
   two-stroke engines, 2A-16
Cylinder compression, check, 1-22
Cylinder head and valves, disassembly, inspection and reassembly, four-stroke engines, 2B-12
Cylinder head covers, removal, inspection and installation, four-stroke engines, 2B-6
Cylinder head, removal, inspection and installation
   four-stroke engines, 2B-11
   two-stroke engines, 2A-4
Cylinder, removal, inspection and installation
   four-stroke engines, 2B-16
   two-stroke engines, 2A-7

D

Drive and driven clutches, disassembly, inspection and reassembly, 2C-7
Drive belt
   removal, inspection and installation, 2C-4
   width and play check, 2C-4
Drive chain and sprockets
   check, adjustment and lubrication, 1-24
   removal, cleaning, inspection and installation, 6-12
Driveaxle boots, inspection, 1-12
Driveaxles
   boot replacement and U-joint overhaul, 6-17
   removal and installation, 6-17
Dump bed (6X6 and 4X6), removal and installation, 8-2

E

Eccentrics, removal, inspection and installation, 6-14
Electrical and ignition system, 5-1 through 5-20
   alternator, check and replacement, 5-14
   battery
      charging, 5-4
      inspection and maintenance, 5-3
   brake switch, check and replacement, 5-6
   CDI magneto, check, removal and installation, 5-19
   CDI unit, check, removal and installation, 5-19
   charging system
      output test, 5-13
      testing, general information and precautions, 5-13
   circuit breaker, check, 5-4
   electrical troubleshooting, 5-3
   handlebar switches
      check, 5-7
      removal and installation, 5-8
   headlight aim, check and adjustment, 5-5
   headlight bulb, replacement, 5-5
   ignition coil, check, removal and installation, 5-18
   ignition main (key) switch, check and replacement, 5-7
   ignition system, check, 5-17
   ignition timing, general information and check, 5-19
   indicator bulbs, replacement, 5-7
   lighting system, check, 5-4
   neutral switch, check and replacement, 5-8
   regulator/rectifier, check and replacement, 5-16
   speedometer cable and drive, removal and installation, 5-17
   starter circuit, check and component replacement, 5-8
   starter motor
      disassembly, inspection and reassembly, 5-9
      removal and installation, 5-8

starter pinion, removal, inspection and
  installation, 5-13
taillight/brake light bulb, replacement, 5-6
wiring diagrams, 5-19
Engine oil/filter, change, 1-17
Engine: Four-stroke engines, 2B-1 through 2B-30
  cam chain tensioner, removal and installation, 2B-7
  counterbalancer, removal and installation, 2B-26
  crankcase components, inspection and
    servicing, 2B-25
  crankcase, disassembly and reassembly, 2B-22
  crankshaft and connecting rod, removal, inspection
    and installation, 2B-26
  cylinder head and valves, disassembly, inspection
    and reassembly, 2B-12
  cylinder head covers, removal, inspection and
    installation, 2B-6
  cylinder head, removal and installation, 2B-11
  cylinder, removal, inspection and installation, 2B-16
  disassembly and reassembly, general
    information, 2B-6
  external oil lines, check valve and tank, removal and
    installation, 2B-21
  major engine repair, general note, 2B-4
  oil pump, removal, inspection and installation, 2B-23
  operations possible with the engine in the
    frame, 2B-4
  operations requiring engine removal, 2B-4
  piston rings, installation, 2B-20
  piston, removal, inspection and installation, 2B-17
  recoil starter, removal, inspection and
    installation, 2B-28
  removal and installation, 2B-4
  rocker arms and camshaft, removal, inspection and
    installation, 2B-7
  valves/valve seats/valve guides, servicing, 2B-12
Engine: Two-stroke engines, 2A-1 through 2A-18
  counterbalancer (350 and 400 models), removal,
    inspection and installation, 2A-12
  crankcase
    disassembly and reassembly, 2A-15
    pressure and vacuum, check, 2A-3
  cylinder head, removal, inspection and
    installation, 2A-4
  cylinder, removal, inspection and installation, 2A-7
  disassembly and reassembly, general
    information, 2A-4
  major engine repair, general note, 2A-2
  oil pump
    check valve test, cable adjustment and pump
      test, 2A-8
    removal, inspection, installation and
      bleeding, 2A-11
  operations possible with the engine in the
    frame, 2A-2
  operations requiring engine removal, 2A-2

  piston and rings, removal, inspection and
    installation, 2A-8
  recoil starter, removal, inspection and
    installation, 2A-11
  reed valve, removal, inspection and installation, 2A-6
  removal and installation, 2A-3
Exhaust system
  inspection, 1-20
  removal and installation, 4-19
**External oil lines, check valve and tank, removal and
  installation, four-stroke engines, 2B-21**
EZ Shift shifter
  disassembly, inspection and reassembly, 2D-11
  removal and installation, 2D-11

F

Fasteners, check, 1-23
Fenders, removal and installation
  front, 8-4
  rear, 8-2
Fluid levels, check, 1-5
  brake fluid, 1-6
  counterbalancer oil (350 and 400 models), 1-6
  engine coolant, 1-6
  engine oil, 1-5
  front gearcase oil (shaft drive models), 1-7
  front hub fluid (4wd models), 1-7
  transmission oil, 1-7
Four-stroke engines, 2B-1 through 2B-30
  cam chain tensioner, removal and installation, 2B-7
  counterbalancer, removal and installation, 2B-26
  crankcase components, inspection and
    servicing, 2B-25
  crankcase, disassembly and reassembly, 2B-22
  crankshaft and connecting rod, removal, inspection
    and installation, 2B-26
  cylinder head and valves, disassembly, inspection
    and reassembly, 2B-12
  cylinder head covers, removal, inspection and
    installation, 2B-6
  cylinder head, removal and installation, 2B-11
  cylinder, removal, inspection and installation, 2B-16
  disassembly and reassembly, general
    information, 2B-6
  external oil lines, check valve and tank, removal and
    installation, 2B-21
  major engine repair, general note, 2B-4
  oil pump, removal, inspection and installation, 2B-23
  operations possible with the engine in the
    frame, 2B-4
  operations requiring engine removal, 2B-4
  piston rings, installation, 2B-20
  piston, removal, inspection and installation, 2B-17

recoil starter, removal, inspection and installation, 2B-28
removal and installation, 2B-4
rocker arms and camshaft, removal, inspection and installation, 2B-7
valves/valve seats/valve guides, servicing, 2B-12

Front forks (3-wheel models)
disassembly, inspection and reassembly, 6-6
removal and installation, 6-6

Front struts
ball-joint replacement, 6-5
removal and installation, 6-4
spring and cartridge replacement, 6-5

Fuel and exhaust systems, 4-1 through 4-22
air cleaner housing, removal and installation, 4-15
carburetors
  disassembly, cleaning and inspection, 4-9
  overhaul, general information, 4-7
  reassembly and float height check, 4-15
  removal and installation, 4-7
choke cable, removal and installation, 4-18
exhaust system, removal and installation, 4-19
idle fuel/air mixture adjustment, 4-6
pump
  disassembly, inspection and reassembly, 4-21
  removal and installation, 4-21
tank
  cleaning and repair, 4-6
  removal and installation, 4-4
throttle cable, removal, installation and adjustment, 4-17

Fuel system, check and filter cleaning, 1-19

G

Gearcase (front) and driveshaft (shaft drive models), removal and installation, 6-19
Gearcase oil (front) (shaft drive models), change, 1-19
General specifications, 0-8

H

Handlebar, removal, inspection and installation, 6-2
Handlebar switches
check, 5-7
removal and installation, 5-8
Headlight aim, check and adjustment, 5-5
Headlight bulb, replacement, 5-5
Headlight cover, removal and installation, 8-3
Hub fluid, front, change, 1-27
Hubs, front (except 3-wheel models), removal, inspection and installation, 7-11

Hydraulic brake calipers
overhaul, 7-5
removal and installation, 7-4

I

Identification numbers, 0-6
Idle fuel/air mixture adjustment, 4-6
Idle speed and throttle operation, check and adjustment, 1-13
Ignition
coil, check, removal and installation, 5-18
main (key) switch, check and replacement, 5-7
system, check, 5-17
timing, check, 1-20
timing, general information and check, 5-19
Indicator bulbs, replacement, 5-7
Introduction to the Polaris ATV, 0-5
Introduction to tune-up and routine maintenance, 1-5

L

Lighting system, check, 5-4
Lubrication, general, 1-15

M

Maintenance techniques, tools and working facilities, 0-22
Major engine repair, general note
four-stroke engines, 2B-4
two-stroke engines, 2A-2
Master cylinder, removal, overhaul and installation, 7-7
Mechanical rear caliper, removal, inspection and installation, 7-11

N

Neutral switch, check and replacement, 5-8

O

Oil pump
check valve test, cable adjustment and pump test, two-stroke engines, 2A-8
removal, inspection, installation and bleeding
  four-stroke engines, 2B-23
  two-stroke engines, 2A-11

Operations possible with the engine in the frame
    four-stroke engines, 2B-4
    two-stroke engines, 2A-2
Operations requiring engine removal
    four-stroke engines, 2B-4
    two-stroke engines, 2A-2
Outer cover and seal, removal, inspection and installation, 2C-2

P

Piston and rings, removal, inspection and installation
    four-stroke engines, 2B-17
    two-stroke engines, 2A-8
Piston rings, installation, four-stroke engines, 2B-20
Polaris Variable Transmission (PVT), 2C-1 through 2C-12
    air ducts, removal and installation, 2C-2
    clutches, inner cover and seal, removal and installation, 2C-5
    drive and driven clutches, disassembly, inspection and reassembly, 2C-7
    drive belt, removal, inspection and installation, 2C-4
    drive belt, width and play check, 2C-4
    outer cover and seal, removal, inspection and installation, 2C-2
    pulleys, alignment and offset check, 2C-5
Pulleys, alignment and offset check, 2C-5
PVT, check and clean, 1-30

R

Radiator cap, check, 3-2
Radiator, removal and installation, 3-4
Rear hubs, removal and installation, 7-14
Recoil housing, draining, 1-10
Recoil starter, removal, inspection and installation, 2A-11
    four-stroke engines, 2B-28
    two-stroke engines, 2A-11
Recommended start-up and break-in procedure, two-stroke engines, 2A-17
Reed valve, removal, inspection and installation, two-stroke engines, 2A-6
Regulator/rectifier, check and replacement, 5-16
Rocker arms and camshaft, removal, inspection and installation, four-stroke engines, 2B-7
Routine maintenance intervals, 1-4

S

Safety first!, 0-28
Seat, removal and installation, 8-1
Shift linkage, adjustment, 2D-9
Shock absorbers, rear, removal and installation, 6-8
Skid plate, front, removal and installation, 8-4
Spark plug, replacement, 1-21
Speedometer cable and drive, removal and installation, 5-17
Stabilizer struts (shaft drive models), removal, inspection and installation, rear, 6-10
Starter circuit, check and component replacement, 5-8
Starter motor
    disassembly, inspection and reassembly, 5-9
    removal and installation, 5-8
Starter pinion, removal, inspection and installation, 5-13
Steering system, inspection and toe adjustment, 1-24
Steering, suspension and final drive, 6-1 through 6-22
    drive chain and sprockets, removal, cleaning, inspection and installation, 6-12
    driveaxles
        boot replacement and U-joint overhaul, 6-17
        removal and installation, 6-17
    eccentrics, removal, inspection and installation, 6-14
    front forks (3-wheel models)
        disassembly, inspection and reassembly, 6-6
        removal and installation, 6-6
    front struts
        ball-joint replacement, 6-5
        removal and installation, 6-4
        spring and cartridge replacement, 6-5
    gearcase (front) and driveshaft (shaft drive models), removal and installation, 6-19
    handlebar, removal, inspection and installation, 6-2
    rear axle, removal, inspection and installation, 6-20
    rear shock absorbers, removal and installation, 6-8
    rear stabilizer struts (shaft drive models), removal, inspection and installation, 6-10
    steering shaft and tie-rods, removal, inspection and installation, 6-2
    suspension arms, removal, inspection and installation, 6-9
    swingarm
        bearings, check, 6-11
        removal and installation, 6-11
Suspension, check, 1-24
Swingarm
    bearings, check, 6-11
    removal and installation, 6-11

T

Taillight/brake light bulb, replacement, 5-6

Thermostat (4-stroke models), removal, check and installation, 3-3
Throttle cable, removal, installation and adjustment, 4-17
Tires, general information, 7-14
Tires/wheels, general check, 1-12
Tools, 0-23
Transmission, 2D-1 through 2D-14
   disassembly, inspection and reassembly, 2D-4
   EZ Shift shifter
      disassembly, inspection and reassembly, 2D-11
      removal and installation, 2D-11
   removal and installation, 2D-2
   shift linkage, adjustment, 2D-9
Transmission oil, change, 1-27
Troubleshooting, 0-30
Tune-up and routine maintenance, 1-1 through 1-30
Two-stroke engines, 2A-1 through 2A-18
   counterbalancer (350 and 400 models), removal, inspection and installation, 2A-12
   crankcase
      disassembly and reassembly, 2A-15
      inspection and servicing, 2A-16
   crankcase pressure and vacuum, check, 2A-3
   crankshaft and connecting rod, removal, inspection and installation, 2A-16
   cylinder head, removal, inspection and installation, 2A-4
   cylinder, removal, inspection and installation, 2A-7
   disassembly and reassembly, general information, 2A-4
   major engine repair, general note, 2A-2
   oil pump
      check valve test, cable adjustment and pump test, 2A-8
      removal, inspection, installation and bleeding, 2A-11
   operations possible with the engine in the frame, 2A-2
   operations requiring engine removal, 2A-2
   piston and rings, removal, inspection and installation, 2A-8
   recommended start-up and break-in procedure, 2A-17
   reed valve, removal, inspection and installation, 2A-6
   removal and installation, 2A-3

V

Valve clearances, check and adjustment, 1-22
Valves/valve seats/valve guides, servicing, four-stroke engines, 2B-12

W

Water pump, removal, inspection and installation, 3-4
Wheels, inspection, removal and installation, 7-15
Wiring diagrams, 5-19
Working facilities, 0-27